A Student's Guide to Atomic Physics

This concise and accessible book provides a detailed introduction to the fundamental principles of atomic physics at an undergraduate level. Concepts are explained in an intuitive way, and the book assumes only a basic knowledge of quantum mechanics and electromagnetism. With a compact format specifically designed for students, the first part of the book covers the key principles of the subject, including quantum theory of the hydrogen atom, radiative transitions, the shell model of multi-electron atoms, spin–orbit coupling, and the effects of external fields. The second part provides an introduction to four key applications of atomic physics: lasers, cold atoms, solid-state spectroscopy, and astrophysics. This highly pedagogical text includes worked examples and end-of-chapter problems to allow students to test their knowledge, as well as numerous diagrams of key concepts, making it perfect for undergraduate students looking for a succinct primer on the concepts and applications of atomic physics.

MARK FOX is Professor of Physics at the University of Sheffield. He is also a fellow of the Optical Society of America and the Institute of Physics. His research focuses on optics and photonics, and he specializes in solid-state atoms and quantum dots. He has authored two highly successful books: *Optical Properties of Solids* (2nd edition, 2010) and *Quantum Optics: An Introduction* (2005).

Group→ ↓Period	1	2	3		4	5	6	7	8	9	10	11	12	13	14	15	16	17	18
1	1 H																		2 He
2	3 Li	4 Be												5 B	6 C	7 N	8 O	9 F	10 Ne
3	11 Na	12 Mg												13 Al	14 Si	15 P	16 S	17 Cl	18 Ar
4	19 K	20 Ca	21 Sc		22 Ti	23 V	24 Cr	25 Mn	26 Fe	27 Co	28 Ni	29 Cu	30 Zn	31 Ga	32 Ge	33 As	34 Se	35 Br	36 Kr
5	37 Rb	38 Sr	39 Y		40 Zr	41 Nb	42 Mo	43 Tc	44 Ru	45 Rh	46 Pd	47 Ag	48 Cd	49 In	50 Sn	51 Sb	52 Te	53 I	54 Xe
6	55 Cs	56 Ba	57 La	*	72 Hf	73 Ta	74 W	75 Re	76 Os	77 Ir	78 Pt	79 Au	80 Hg	81 Tl	82 Pb	83 Bi	84 Po	85 At	86 Rn
7	87 Fr	88 Ra	89 Ac	**	104 Rf	105 Db	106 Sg	107 Bh	108 Hs	109 Mt	110 Ds	111 Rg	112 Cn	113 Nh	114 Fl	115 Mc	116 Lv	117 Ts	118 Og
				*	58 Ce	59 Pr	60 Nd	61 Pm	62 Sm	63 Eu	64 Gd	65 Tb	66 Dy	67 Ho	68 Er	69 Tm	70 Yb	71 Lu	
				**	90 Th	91 Pa	92 U	93 Np	94 Pu	95 Am	96 Cm	97 Bk	98 Cf	99 Es	100 Fm	101 Md	102 No	103 Lr	

Other books in the Student's Guide series

A Student's Guide to Waves, Daniel Fleisch, Laura Kinnaman
A Student's Guide to Entropy, Don S. Lemons
A Student's Guide to Dimensional Analysis, Don S. Lemons
A Student's Guide to Numerical Methods, Ian H. Hutchinson
A Student's Guide to Langrangians and Hamiltonians, Patrick Hamill
A Student's Guide to the Mathematics of Astronomy, Daniel Fleisch, Julia Kregonow
A Student's Guide to Vectors and Tensors, Daniel Fleisch
A Student's Guide to Maxwell's Equations, Daniel Fleisch
A Student's Guide to Fourier Transforms, J. F. James
A Student's Guide to Data and Error Analysis, Herman J. C. Berendsen

A Student's Guide to Atomic Physics

MARK FOX
University of Sheffield

CAMBRIDGE
UNIVERSITY PRESS

CAMBRIDGE
UNIVERSITY PRESS

University Printing House, Cambridge CB2 8BS, United Kingdom

One Liberty Plaza, 20th Floor, New York, NY 10006, USA

477 Williamstown Road, Port Melbourne, VIC 3207, Australia

314-321, 3rd Floor, Plot 3, Splendor Forum, Jasola District Centre, New Delhi - 110025, India

79 Anson Road, #06-04/06, Singapore 079906

Cambridge University Press is part of the University of Cambridge.

It furthers the University's mission by disseminating knowledge in the pursuit of education, learning and research at the highest international levels of excellence.

www.cambridge.org
Information on this title: www.cambridge.org/9781107188730
DOI: 10.1017/9781316981337

First published 2018

A catalogue record for this publication is available from the British Library

Library of Congress Cataloging in Publication data
Names: Fox, Mark (Anthony Mark), author.
Title: A student's guide to atomic physics / Mark Fox (University of Sheffield).
Description: Cambridge, United Kingdom ; New York, NY : Cambridge University Press, 2018. | Includes bibliographical references and index.
Identifiers: LCCN 2017051568| ISBN 9781107188730 (hbk.) | ISBN 1107188733 (hbk.) | ISBN 9781108446310 (pbk.) | ISBN 1108446310 (pbk.)
Subjects: LCSH: Nuclear physics. | Atomic theory.
Classification: LCC QC173 .F675 2018 | DDC 539.7–dc23 LC record available at https://lccn.loc.gov/2017051568

ISBN 978-1-107-18873-0 Hardback
ISBN 978-1-108-44631-0 Paperback

Additional resources for this publication at www.cambridge.org/9781107188730.

Contents

Preface

Undergraduate students come across the concepts of atomic physics at various stages during their degree programs. For example, the Bohr model is a central part of introductory courses on quantum physics, while the hydrogen atom is a key element in a first course on quantum mechanics. After that, the more advanced topics could either be a component of a second, broad quantum physics module, or a stand-alone unit. This book is designed for the latter approach, without necessarily excluding its usefulness for the former, where it might be used, for example, in conjunction with a text on nuclear physics.

The book evolved from a detailed set of lecture notes prepared for a third-year module at the University of Sheffield. The notes were prepared to respond to the lack of a short text at the right level. The subject material was either scattered across various chapters of large quantum physics texts, or was included in introductory sections of more advanced texts. Neither case was particularly suited to the needs of the students.

The range of topics included within the book aims to cover the core curriculum on atomic physics set out by the Institute of Physics, and might be useful either to second- or third-year students within the United Kingdom, depending on how a particular university subdivides the syllabus. For readers outside the United Kingdom, the text is pitched at intermediate-level students. It assumes basic familiarity with the techniques of quantum mechanics, but does not have the depth required for masters-level courses.

The course notes have been freely available on the Internet for several years, and I was approached by several publishers who thought they could form the basis for a textbook. Having already written two textbooks, I was well aware of the extra effort required to turn a set of lecture notes into a book and resisted the approaches I received. However, I then discovered the Cambridge Student's Guide series, and realized that it is the right place for the material.

Its inclusion within the series makes it clear that the book does not claim to be an authoritative reference work, but rather an intermediate-level text aimed at explaining the basic concepts to undergraduate students.

The text is divided into two parts:

- Part I: Fundamental Principles (Chapters 1–8)
- Part II: Applications of Atomic Physics (Chapters 9–12)

The first part should be useful for undergraduate students at most universities, as it covers the core concepts of university-level atomic physics. The second part will find varied use, depending on how a particular university organizes its course. Chapter 9 covers most of the basic ideas required for the laser-physics component of Institute of Physics (IOP) curriculum. Chapter 10 gives a brief introduction to the techniques of laser cooling that underpin a large sector of modern atomic physics research. Chapter 11 reflects the author's own background in semiconductor physics and solid-state lasers. The final chapter arose from the suggestions of the manuscript reviewers, and its writing involved a fascinating learning experience for the author.

Texts within the Cambridge Student's Guide series are deliberately kept short. For this reason, some nonessential material that was in the first draft of the manuscript has been moved to an online supplement. The sections where additional notes are available online are identified by the ➋ symbol in the margin. Another key feature of the series is the inclusion of worked examples and exercises. Solutions to the exercises are available from the online resources.

I am very grateful to numerous people who have helped in various ways to bring the book to fruition. First, I would like to thank the generations of students at the University of Sheffield who have taken the course and provided feedback on the notes. I am also grateful to my colleagues at the University of Sheffield, on whom I have bounced ideas and with whom I have clarified concepts. Among these, I would like to single out Professor Paul Crowther, who provided invaluable help with Chapter 12. My knowledge of astrophysics was very limited before I wrote the chapter, and his critical reading of the manuscript has both greatly improved it and also ironed out deficiencies in my understanding. I would also like to thank people around the world who provided feedback on the Internet version of the notes, especially Dr. André Xuereb, from the University of Malta, for his comments on the 2013 version.

Second, I would like to thank the people who taught me atomic physics at the University of Oxford, especially my tutor, Professor Roger Cashmore, and my lecturers, Dr. Alan Corney and Dr. Kem Woodgate. I regard this book as an introduction to their excellent texts, which are both still in print and included in the References. The structure of Part I broadly follows a set of lecture notes by Professors Paul Ewart and Derek Stacey at the University of Oxford, although the final ordering of material departs a little from their plan. Professor Stacey also provided comments on Part I of the manuscript, which have helped to iron out some potentially confusing statements.

Next, I would like to thank Dr. Nicholas Gibbons at Cambridge University Press for introducing me to the *Student's Guide* series and supporting the project. I am especially grateful to him for finding a very helpful set of reviewers at the syndicate approval stage. These anonymous reviewers provided numerous helpful suggestions. In particular, the final chapter is included on their suggestion, while much of Chapter 1 is a response to one of the reviewers. This reviewer pointed out that my original notes took several basic concepts for granted, and this prompted me to rewrite the first three sections to provide fundamental definitions.

Finally, I would like to thank Dr. John Pantazis, from Amptek, Inc., for providing the data in Figure 4.6(a), and Róisín Munnelly at Cambridge Unversity Press for her role as Content Manager. Her patience in seeing the project through to completion is much appreciated.

Symbols

The list gives the main symbols used in the text, excluding some that are used infrequently and are defined *in situ*. In some cases, it is necessary to use the same symbol to represent different quantities. Whenever this occurs, it should be obvious from the context which meaning is intended.

a_0	Bohr radius
a_{H}	Bohr radius of hydrogen
A	area
A_{ij}	Einstein A coefficient
$\boldsymbol{B}$	magnetic field (flux density)
B_{ij}	Einstein B coefficient
d	distance
e	magnitude of electron charge
E	energy
$\boldsymbol{\mathcal{E}}$	electric field
$\boldsymbol{F}$	force, total angular momentum
$g(E)$	density of states at energy E
$g(\nu)$	spectral line-shape function
g	degeneracy
g_J	Landé g-factor
g_{N}	nuclear g-factor
g_{s}	electron spin g-factor
h	Planck's constant
$\hbar$	$h/2\pi$
$\hat{H}$	Hamiltonian
H'	perturbation
i	electrical current

I	moment of inertia, optical intensity, nuclear spin
$\boldsymbol{I}$	nuclear angular momentum
I_z	z component of nuclear angular momentum
$\boldsymbol{j}$	angular momentum (single electron)
J	exchange constant
$\boldsymbol{J}$	total angular momentum
$\boldsymbol{l}$	orbital angular momentum (single electron)
l_z	z component of orbital angular momentum (single electron)
$\boldsymbol{L}$	orbital angular momentum
m	mass, magnetic quantum number
m^*	effective mass
m_{H}	mass of hydrogen atom
M_{ij}	matrix element
n	refractive index
N	number of atoms per unit volume
$\boldsymbol{p}$	electric dipole moment, linear momentum
P	power, pressure
q	charge
r	radius
$\boldsymbol{r}$	position vector
R	reflectivity
$\mathbb{R}$	pumping rate per unit volume
R_{H}	Rydberg energy of hydrogen
$\boldsymbol{s}$	spin angular momentum (single electron)
s_z	z component of spin angular momentum (single electron)
$\boldsymbol{S}$	spin angular momentum
t	time
T	temperature
u	initial velocity
$u(\nu)$	spectral energy density at frequency ν
$\boldsymbol{v}$	velocity
V	voltage, potential energy, volume
W_{ij}	transition rate
x	position coordinate
$\hat{\mathbf{x}}$	unit vector along the x-axis
y	position coordinate
$\hat{\mathbf{y}}$	unit vector along the y-axis
Y_{l,m_l}	spherical harmonic function
z	position coordinate, Doppler redshift

$\hat{\mathbf{z}}$	unit vector along the z-axis
Z	atomic number
α	fine-structure constant, absorption coefficient, polarizability
γ	gyromagnetic ratio, gain coefficient
$\mathbf{\Gamma}$	torque
δ	frequency detuning
$\delta_{k,k'}$	Kronecker delta function
$\delta(x)$	Dirac delta function
ϵ_{r}	relative permittivity
θ	polar angle
λ	wavelength
λ_{deB}	de Broglie wavelength
$\boldsymbol{\mu}$	magnetic dipole moment
ν	frequency
$\overline{\nu}$	wave number
τ	lifetime
τ_{c}	collision time
ϕ	azimuthal angle
ψ, Ψ	wave function
ω	angular frequency

Quantum Numbers

In atomic physics, lower- and uppercase letters refer to individual electrons or whole atoms respectively.

F	hyperfine total angular momentum
I	nuclear spin
j, J	total electron angular momentum
l, L	orbital angular momentum
m	magnetic
M_F	z component of hyperfine angular momentum
M_I	z component of nuclear spin
m_j, M_J	z component of electron total angular momentum
m_l, M_L	z component of orbital angular momentum
m_s, M_S	z component of spin angular momentum
n	principal
s, S	spin

Part I

Fundamental Principles

1
Preliminary Concepts

Atomic physics is the subject that studies the inner workings of the atom. It remains one of the most important testing grounds for quantum theory and is therefore a very active area of research, both for its contribution to fundamental physics and to technology. Furthermore, many other branches of science rely heavily on atomic physics, especially astrophysics, laser physics, solid-state physics, quantum information science, and chemistry. So much so, that Richard Feynman once wrote (1964):

> If, in some cataclysm, all scientific knowledge were to be destroyed, and only one sentence passed on to the next generation of creatures, what statement would contain the most information in the fewest words? I believe it is the *atomic hypothesis* (or atomic fact, or whatever you wish to call it) that *all things are made of atoms – little particles that move around in perpetual motion, attracting each other when they are a little distance apart, but repelling upon being squeezed into one another*. In that one sentence you will see an *enormous* amount of information about the world, if just a little imagination and thinking are applied.

The task of atomic physics is to understand the structure of atoms, and hence to explain experimental observations such as the wavelengths of spectral lines. For all elements apart from hydrogen, we have to deal with a complicated many-body problem consisting of a nucleus and more than one electron. Atomic physics proceeds by a series of approximations that make this problem tractable. Before we set about this task, it is first necessary to cover a number of important basic concepts and definitions.

1.1 Quantized Energy States in Atoms

The first basic concept we need is that of **bound states**. Atoms are held together by the attractive force between the positively charged nucleus and

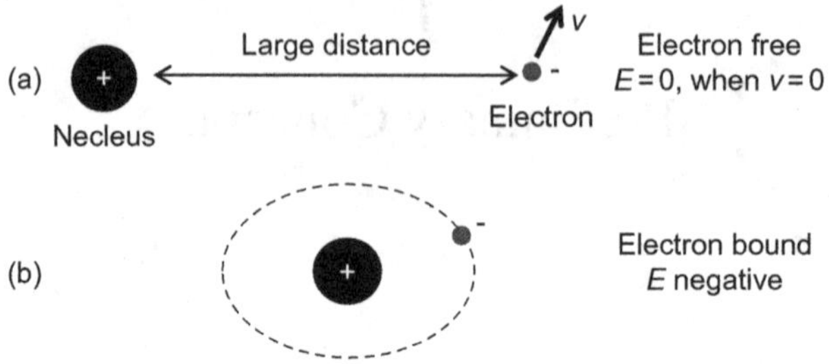

Figure 1.1 (a) Unbound state with the electron far from the nucleus. The electron moves freely with velocity (**v**) independent of the presence of the nucleus. (b) Bound electron state with negative energy.

the negatively charged electrons: the electrons are *bound* to the atom, rather than being *free* to move though space. In the limit where the electron is very far away from the nucleus, the attractive force is negligible; the electron is free to move with velocity ($\boldsymbol{v}$) without any influence from the nucleus, as illustrated schematically in Figure 1.1(a). It is natural to define the energy (E) of this free (or *unbound*) state as being zero when $\boldsymbol{v} = 0$. When the electron moves closer to the nucleus, it begins to experience an attractive force, leading to the formation of a stable bound state as illustrated in Figure 1.1(b). The energy of the bound state is lower than that of the free electron since it requires energy to pull the electron away from the nucleus. The amount of energy required is called the **binding energy** of the electron. With our definition of $E = 0$ corresponding to the unbound state, the absolute energy (E) of the bound state must be negative, with the binding energy equal to $-E = |E|$.

The early understanding of the atom was built around the solar system analogy, that the planets orbit around the sun under the influence of the attractive gravitational force. While it will not be appropriate to push this analogy too far on account of the need to use quantum mechanics rather than Newtonian mechanics to describe the motion, it does provide a useful starting point. In the same way that the planets arrange themselves into orbits at varying radii from the sun, the electrons in an atom are arranged in a series of quantized states around the nucleus. The planets nearest the sun are very strongly bound and have small radii with fast periods. The outer planets, by contrast, are less strongly bound, and have large radii and long periods. Similarly, the electrons are arranged into orbital **shells** around the nucleus. The electrons nearest the nucleus are very strongly bound, while those further away are more weakly bound. The arrangement of the electrons within these quantized shells around the nucleus is the basis of the **shell model** of the atom discussed in Chapter 4.

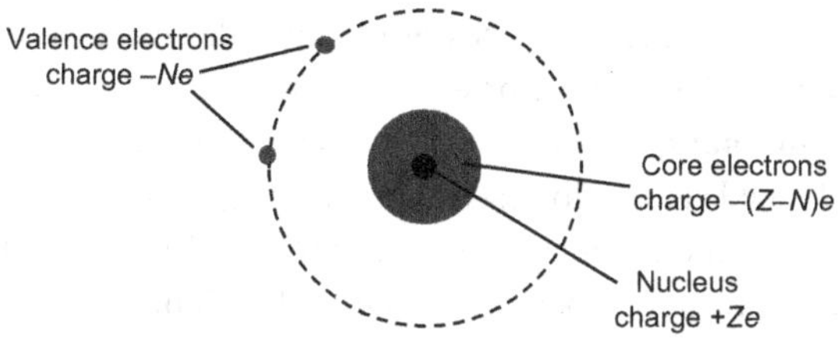

Figure 1.2 Arrangement of electrons into core and valence shells within a neutral atom of atomic number Z with N valence electrons.

Elements are identified by their atomic number Z, which defines the number of protons in the nucleus. Since the charge of the proton is $+e$, where e is the magnitude of the electron charge, the charge of the nucleus is equal to $+Ze$. Free atoms are normally found in a neutral electrical state, which means that they have Z bound electrons (charged atoms are discussed in Section 1.2). The electrons in the outermost shell are called **valence electrons**. It is these valence electrons that take part in chemical bonding, with their number N determining the chemical valency of the atom. The remaining $(Z - N)$ electrons are in inner shells, and are called **core electrons**, as illustrated in Figure 1.2. These core electrons are very strongly bound and can only be accessed by using high-energy (e.g., X-ray) photons, as discussed in Section 4.4.3. The optical spectra of the atom are determined by the valence electrons, which are, therefore, the main focus of atomic physics.

The energies of bound states in atoms are frequently quoted in **electron volt** (eV) units. One electron volt is the energy acquired by an electron when it is accelerated by a voltage of 1 volt. Thus 1 eV = e J, where $-e \approx -1.6 \times 10^{-19}$ C is the charge of the electron. This is a convenient unit, because the binding energies of the valence electrons in atoms are typically a few eV. The core electrons, however, have much larger binding energies, typically in the keV range for atoms with large Z.

1.2 Ionization States and Spectroscopic Notation

In the previous section, we considered the case of a neutral atom in which Z electrons are bound to a nucleus containing Z protons. Charged atoms also exist in which the number of electrons is different to Z. Such charged atoms are called **ions**. In atomic physics, we deal almost exclusively with positively charged ions, in which the number of bound electrons is less than Z.

In chemistry, however, it is also necessary to consider negative ions, in which the atom binds more than Z electrons.

The **ionization energy** of an atom (also sometimes called the **ionization potential**) is defined as the lowest energy required to remove an electron. The electrons are bound to the atom in shells with different quantized binding energies, and the ionization energy is equal to the binding energy of the least strongly bound electron. In practice, this will be one of the valence electrons.

Hydrogen is the first element and has $Z = 1$. Since it only binds one electron, it only has one ionization energy. All other atoms have more than one bound electron, and therefore have more than one ionization energy. An atom with atomic number Z has Z **ionization states**, and hence Z ionization energies. The nth ionization energy is defined as the energy required to remove the nth electron from the atom, according to the following sequence:

$$\begin{array}{lcll} \mathrm{A} & \rightarrow & \mathrm{A}^{+} + \mathrm{e}^{-} & \text{first ionization energy} \\ \mathrm{A}^{+} & \rightarrow & \mathrm{A}^{2+} + \mathrm{e}^{-} & \text{second ionization energy} \\ & \vdots & & \vdots \\ \mathrm{A}^{(Z-1)+} & \rightarrow & \mathrm{A}^{Z+} + \mathrm{e}^{-} & Z\text{th ionization energy,} \end{array}$$

where A^{n+} represents an atom, A, that has lost n electrons from the neutral state, with A^{Z+} corresponding to an isolated nucleus. Each ionization state has a unique spectrum, which allows the atom to be identified from analysis of its spectral lines.

In normal laboratory conditions at temperature T (with $T \sim 300\,\mathrm{K}$), the thermal energy $k_{\mathrm{B}}T$ is significantly smaller than the first ionization energy of the atom. This means that atoms are normally in the neutral state. In order to study ions, we either have to raise the temperature significantly (e.g., in a flame), or we have to deliberately strip off the electrons (e.g., in a collision with another charged particle in a discharge tube). In astrophysics, however, we study the spectra of atoms in stars, where the temperature is always very high and highly ionized states are routinely found.

Astronomers have been studying the spectra of atoms and ions for a long time, using the characteristic spectral lines of the elements to determine the composition of stars. In order to categorize the spectral lines, **spectroscopic notation** was introduced to identify the different ionization states of the atoms. In this notation, the nth ionization state of atom A is written A (n+1), where (n+1) is written in capital Roman numerals. Thus, A I is the neutral state of the atom, A II is the first ionization state A^{+}, and so on. Spectroscopic notation is widely used in astrophysics and also in important databases of atomic physics (see Section 1.5). Table 1.1 shows how the notation is applied to the element sodium (chemical symbol Na), which has an atomic number of 11.

Table 1.1 *Ionization states of the element sodium (chemical symbol Na), which has an atomic number of 11.*

Atom/ion	Spectroscopic notation	Number of electrons
Na	Na I	11
Na^+	Na II	10
Na^{2+}	Na III	9
⋮	⋮	⋮
Na^{11+}	Na XII	0

1.3 Ground States and Excited States

A neutral atom with atomic number Z has Z electrons bound to the nucleus. As mentioned in Section 1.1 and discussed in detail in Chapter 4, the quantized electron states are arranged in shells around the nucleus. The Pauli exclusion principle, which will be discussed in Chapters 4 and 6, dictates that each shell can only hold a specific number of electrons. The electrons therefore fill up the shells in sequence of increasing energy, moving to a higher energy shell once the lower energy shell is full. Eventually, all the electrons have been bound. The final state of the atom with its electrons filling up the lowest available energy shells in accordance with the Pauli exclusion principle is called the **ground state** of the atom.

The ground state of a typical atom is shown schematically in Figure 1.3. As before, we assume that there are N valence electrons, and therefore $(Z - N)$ core electrons. The diagram is drawn for the specific case of the neutral magnesium atom, where $Z = 12$ and $N = 2$. Each horizontal line indicates a quantized energy state, and the vertical axis is energy. The zero of energy is defined as the point at which the electron is free and all the quantized bound states have negative energy, as discussed in Section 1.1. The shading for the free states indicates that the energy is not quantized: the electron is free to move with arbitrary kinetic energy, and so can have any positive energy. The free states are therefore said to form a **continuum**: there is a continuous spectrum of energies that are possible, with no breaks due to quantization.

It is important to note that the energy axis in Figure 1.3 is not linear. The core shell states have very large negative energies, and should really be way off the bottom of the page. Since the core electrons play no part in the optical spectra, they are usually omitted from atomic energy-level diagrams; this will be the policy adopted from here onward, unless we are specifically considering the core electrons (as we do in Section 4.4.3).

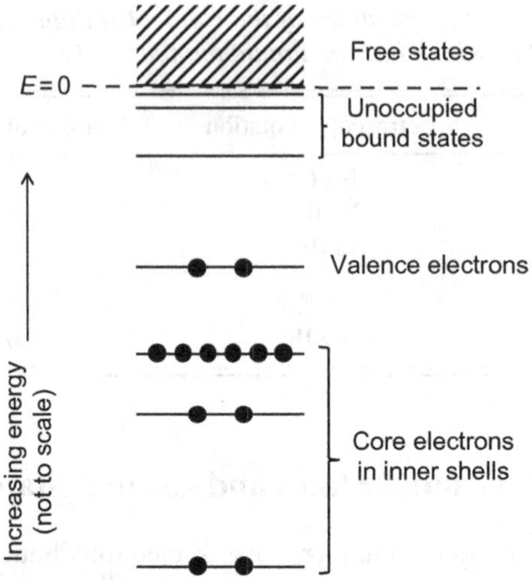

Figure 1.3 Arrangement of the electrons in the ground state of an atom. The electrons fill up the atomic shells in order of increasing energy until all the electrons have been accounted. The shading for the free states indicates that the energy is not quantized: it forms a continuum. The diagram is drawn for the case of the neutral Mg atom ($Z = 12$), which has 12 electrons. Note that the energy scale is not linear. The core shells are very strongly bound, and their large negative energies would be way off the page on a linear scale. These core electron states are usually omitted from atomic energy-level diagrams.

There are an infinite number of quantized bound states in an atom, but only a small number (the ones with lowest energy) are occupied in the ground state configuration of the atom. All of the other states lie at higher energy. The **excited states** of the atom are obtained by promoting valence electrons to these unoccupied states at higher energy. If there is more than one valence electron, then the excited states are obtained by promoting just *one* of the valence electrons to a higher energy state, as shown in Figure 1.4. Despite the large number of these excited states, we usually only need to consider the first few to explain the most important features of the optical spectra. The large number of other excited states at higher energies are increasingly weakly bound, and eventually merge into the continuum of free states available to unbound electrons. This means that the infinitieth excited state corresponds to the ionization limit, which provides a method to define the energy of the ground state electron configuration. This energy is identified in Figure 1.4, and can be determined experimentally by measuring the first ionization energy of the atom.

The energy gap between the ground state of an atom and its first excited state is typically much larger than the thermal energy $k_B T$ at room temperature.

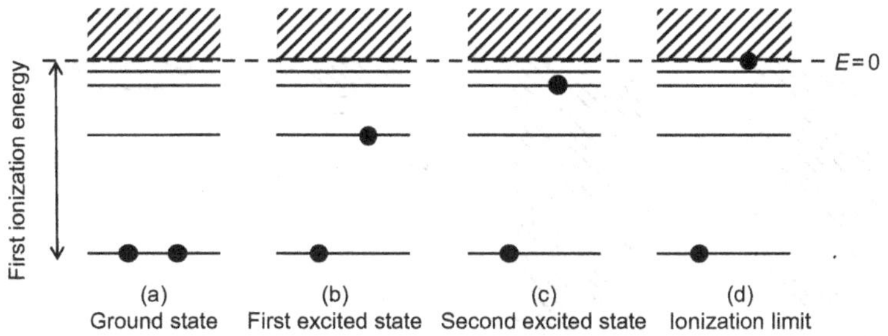

Figure 1.4 Ground and excited states of an atom with two valence electrons. (a) Ground state. (b) First excited state. (c) Second excited state. (d) Ionization limit, equivalent to the infinitieth excited state. Note that the ground state is the same as in Figure 1.3, except that the core electrons are no longer shown.

This means that the atom will normally be in its ground state. In order to promote the atom to its excited states, energy must be imparted to it. This is typically done by placing the atom in a discharge tube, and applying voltage to cause collisions with electrons flowing down the tube. The atom can also be promoted to a specific excited state by absorption of a photon (see section 1.4).

For atoms that have two or more valence electrons, it is reasonable to ask why we only consider excited states in which only one electron is promoted to higher energy. For example, in Figure 1.4, the second excited state has one electron in the lowest level and the other in the third, rather than both electrons in the second level. We only consider these states because it costs more energy to promote both electrons than to completely remove the first electron: the ionized state has a lower energy than the unionized one with two electrons in higher levels. It is therefore easier to ionize the atom than to excite both electrons simultaneously.

The state of the atom after one electron has been removed corresponds to the singly charged ion A^+. The method of defining a ground state and excited states starts again for this ion, with the ground state of the ion corresponding to the ionization state of the neutral atom. For example, the ionization limit of the neutral helium atom ($Z = 2$) corresponds to the ground state of the He^+ ion. (See discussion of Figure 6.2 in Chapter 6.) If the atom has more than two electrons, this process keeps repeating itself, with the ground state of the ion A^{n+} corresponding to the ionization limit of the ion $A^{(n-1)+}$. Each ionization state has its own characteristic sequence of energy levels, which can be determined by analysis of the optical spectra, as discussed in section 1.4.

The correspondence between the ionization limit of one ionization state and the ground state of the next one is shown in Figure 1.5. It is apparent from this

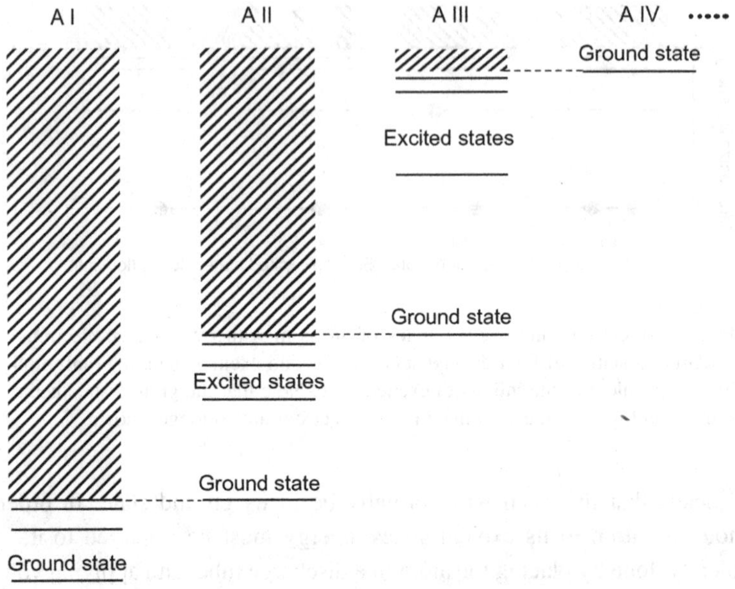

Figure 1.5 Correspondence between the ionization limit of an atom or ion and the ground state of the next ion in the sequence. Spectroscopic notation is used for the different ionization states: A I indicates the neutral atom, A II the singly charged ion, A III the doubly charged ion, and A IV the triply charged ion.

diagram that the definition of $E = 0$ is a *relative* one: $E = 0$ for one ionization state corresponds to a negative energy for the next one. (This distinction does not apply, of course, to hydrogen, as it only has one electron.) In absolute terms, the true zero of energy should be defined as the state with all Z electrons stripped from the nucleus. For a multi-electron atom, this would mean that the ground state of the neutral atom, together with its excited states, all have large negative energies in absolute terms. However, since the energies of the core electrons remain constant while the valence electrons are excited, it makes sense to subtract them and define the zero of energy for each ionization state as the energy to remove the first valence electron.

1.4 Atomic Spectroscopy

We can gain a great deal of knowledge about atoms from studying the way they interact with light, and in particular from measuring atomic spectra. The extreme precision with which optical spectral lines can be measured

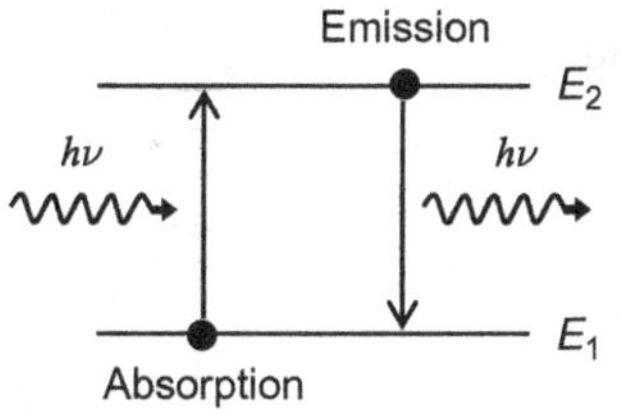

Figure 1.6 Absorption and emission transitions between two quantized energy states.

makes atomic physics the most precise branch of physics. For example, the frequencies of the spectral lines of hydrogen have been measured with extremely high accuracy, permitting the testing of small but important quantum phenomena that are normally unobservable.

The basis for atomic spectroscopy is the measurement of the energy of the photon absorbed or emitted when an electron jumps between two quantized bound states, as shown in Figure 1.6. These are called optical **transitions**. The frequency (ν) of the photon (and hence its wavelength, λ) is determined by the difference in energy of the two levels according to:

$$h\nu = \frac{hc}{\lambda} = E_2 - E_1 , \tag{1.1}$$

where E_1 and E_2 are the energies of the lower and upper levels respectively, h is Planck's constant, and c is the velocity of light. If the electron is initially in the lower level, it can only be promoted to the higher level by absorbing energy from a radiation field incident on the atom. The radiation must contain photons with frequency given by Eq. (1.1), and conservation of energy requires that one of these photons is removed from the beam as the electron makes its jump upward. This is the process of **absorption**. By contrast, if the electron is initially in the upper level, then it can spontaneously drop to the lower level by emitting a photon with frequency given by Eq. (1.1) without the need of an external radiation field. The process is therefore called **spontaneous emission**, or, simply, just **emission**. (The related process of *stimulated* emission will be discussed in Chapter 9.)

The bound states of atoms have quantized energies, and so the absorption and emission frequencies that are observed from a particular atom are discrete. The **absorption spectrum** can be measured by illuminating the atoms with a continuous range of frequencies, and analyzing the intensity that gets transmitted. Dips in the transmitted intensity will be observed at the frequencies that satisfy Eq. (1.1), as shown schematically in Figure 1.7(a). The factors

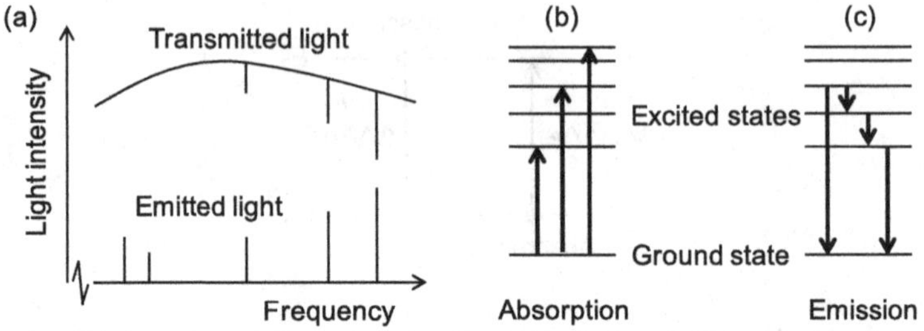

Figure 1.7 (a) Absorption and emission-line spectra. (b) Absorption transitions, starting from the ground state. (c) Emission transitions starting from one of the excited states accessed from the ground state. Different decay routes are possible, leading to additional frequencies in the emission spectrum.

that determine the width of these dips will be discussed in section 3.7. At this stage, all we need to know is that the width is usually very much smaller than the center frequency (e.g., width $\sim 10^9$ Hz, as opposed to a center frequency of $\sim 10^{14}$ Hz). The absorption dips usually just look like vertical downward lines unless a very high-resolution spectrometer is used, and they are typically called **absorption lines**. Similarly, the **emission spectrum** consists of narrow peaks that occur at the frequencies that obey Eq. (1.1), as indicated in Figure 1.7(a). These peaks are called **emission lines**. Atomic spectra are generally called **line spectra** to contrast them with absorption or emission bands, where a continuous range of frequencies is absorbed or emitted, as in the spectra of solids or molecules. (See Chapter 11 for a discussion of the spectra of solids.)

In section 1.3, we have seen that an atom is normally found in its ground state. In the absorption spectrum, we can therefore normally only observe transitions that start from the ground state, as shown in Figure 1.7(b). The frequencies of the absorption lines are given by Eq. (1.1) with E_1 equal to the energy of the ground state, and E_2 the energy of one of the excited states. In emission, by contrast, we start from an excited state. Let us consider the case where we start from one of the excited states that can be reached by absorption from the ground state, as shown in Figure 1.7(c). The electron might just drop back directly to the ground state, emitting a photon with the same frequency as the absorption line. However, the electron can also decay via intermediate states, emitting photons with frequencies in which both of E_2 and E_1 in Eq. (1.1) are the energies of excited states. The net result is that the emission spectrum has more lines than the absorption spectrum, as shown schematically in Figure 1.7(a).

Spectroscopists measure the wavelength of the photon emitted in an optical transition, and use that measurement to deduce energy differences.

The absolute energies of the quantized bound states are determined by fixing the energy of one of the levels by additional methods, and then determining the energies of the others relative to it. As discussed in Section 1.3, the energy of the ground state relative to the ionization limit is the natural reference point for the atom. The usual strategy is to determine the energy of the ground state (e.g., by measuring the ionization energy), and then to use it as a reference for the excited states to deduce their energies from the appropriate spectral lines. There will, of course, be many lines in the spectrum, and the individual transitions have to be identified by a process of logical deduction. For example, in Figure 1.7(a) it is obvious that the three lines with highest frequency in the emission spectrum terminate on the ground state. This is confirmed by the fact that they also appear in the absorption spectrum. The states involved in the other lines are worked out by trial and error until a self-consistent assignment is reached.

The larger number of lines in the emission spectrum makes it more interesting to investigate. Moreover, it is usually easier to measure emission than absorption in the laboratory, as all that is needed is a **discharge tube**. In such a device, a vacuum tube with electrodes at both ends is filled with a gas of the atoms under study, as shown in Figure 1.8. The negative electrode (the cathode) is heated to eject electrons, which then flow as a current to the positive electrode (the anode) when an external voltage V is applied. The atoms are excited by collisions with the electrons and emit photons as they relax to the ground state, either directly or in a cascade. The maximum energy that can be imparted to the atom is equal to eV, and this determines the states that can be accessed. If eV is larger than the ionization energy, ions will be present in the tube, and their characteristic spectra will also be observed.

The fact that each atom has a unique set of quantized energy levels, both in its neutral and ionized states, means that every element has a unique set

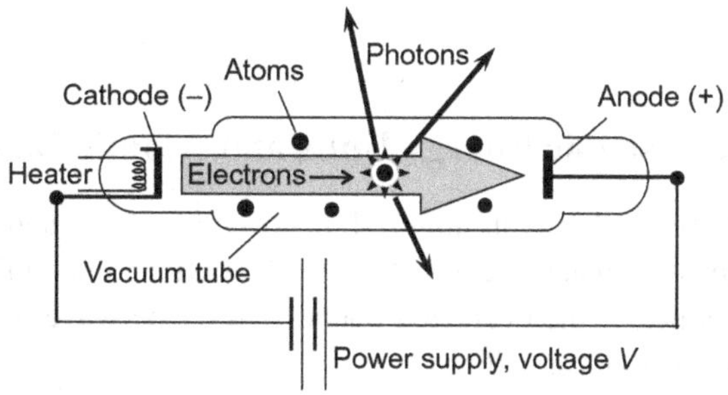

Figure 1.8 Electrical discharge tube for observing atomic emission spectra.

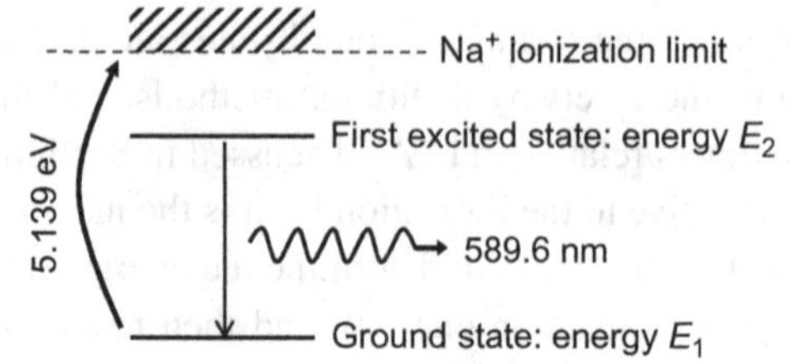

Figure 1.9 Energy levels and transition considered in Example 1.1.

of spectral lines, thereby providing a method to identify elements from their characteristic spectra. This technique is used extensively in astrophysics to learn about the composition of stars and other astronomical sources, as will be discussed in Chapter 12. (See especially section 12.3.)

Example 1.1 The first ionization energy of sodium is 5.139 eV, and the transition from the ground state to the first excited state occurs at 589.6 nm. What is the energy of the first excited state?

Solution: The information given in the example is summarized in Figure 1.9. We know that an energy of 5.139 eV is required to remove an electron from the ground state, and so the energy (E_1) of the ground state relative to the Na^+ ionization limit is −5.139 eV. The energy of the photon emitted in the transition from the first excited state to the ground state can be worked out from Eq. (1.1) as:

$$h\nu = \frac{hc}{589.6 \times 10^{-9}} = 3.368^{-19}\,\mathrm{J} = 2.103\,\mathrm{eV}\,.$$

The first excited state therefore lies 2.103 eV above the ground state, and its energy E_2 is thus:

$$E_2 = E_1 + h\nu = (-5.139 + 2.103)\,\mathrm{eV} = -3.036\,\mathrm{eV}\,.$$

1.5 Spectroscopic Energy Units and Atomic Databases

The close connection between atomic line spectra and the underlying level structure of the atom makes it convenient to use wave-number units (cm^{-1}) to specify the energies of the quantized bound states. The **wave number** ($\overline{\nu}$) is related to the energy (E) as follows:

$$\overline{\nu} = \frac{E}{hc}\,. \tag{1.2}$$

The standard international (SI) unit for wave number is m^{-1}. However, atomic spectroscopists usually use cm^{-1}, in which case it is necessary to specify c in cm/s in Eq. (1.2). Note that $1\,\mathrm{cm}^{-1} = 100\,\mathrm{m}^{-1}$; the cm^{-1} is a *larger* unit by a factor of 100. The conversion factor to the other convenient unit for atomic levels, namely the electron Volt, is:

$$1\,\mathrm{eV} = (e/hc)\,\mathrm{cm}^{-1} = 8065.54\,\mathrm{cm}^{-1}\,. \tag{1.3}$$

Note, again, that it is necessary to use c in cm/s here (i.e., $c = 2.998 \times 10^{10}$ cm/s) to get the conversion to cm^{-1} correct.

Wave-number units are particularly convenient for atomic spectroscopy. This is because they dispense with the need to introduce fundamental constants in our calculation of the wavelength. Thus the wavelength of the radiation emitted in a transition between two levels is simply given by (cf., Eq. [1.1]):

$$\frac{1}{\lambda} = \frac{E_2}{hc} - \frac{E_1}{hc} = \overline{\nu}_2 - \overline{\nu}_1\,, \tag{1.4}$$

where $\overline{\nu}_2$ and $\overline{\nu}_1$ are the energies of upper and lower levels, respectively, in cm^{-1} units, and λ is measured in cm.

The convenience of wave-number units means that most atomic databases use them to specify atomic energies. Moreover, these databases also usually use the ground-state level as the reference point, rather than the ionization limit. This point is clarified in Figure 1.10, where the two different definitions of energies are compared. On the left, we have the convention that has been followed so far, following section 1.1, where $E = 0$ is defined as the ionization limit, and all the bound-state energies E_n are negative. On the right, we have the alternative system used by spectroscopists, where $E = 0$ corresponds to the ground-state level. In this convention, the excited-state energies are positive, and specified in wave-number units relative to the ground state. The ionization limit in cm^{-1} is then $-E_1/hc = +|E_1|/hc$.

The National Institute of Standards and Technology (NIST) in the United States maintains a particularly important on-line resource of atomic data. An extremely detailed database is provided for the use of professional research scientists (see Kramida et al. [2016]), together with a simpler online *Handbook* (see Sansonetti et al. [2005].) Both databases use the system on the right of Figure 1.10, with the default unit being cm^{-1}. The *Handbook* includes data for the neutral atom and singly charged ion, while the professional database includes all the known ionization states. The ionization states are identified using the spectroscopic notation introduced in Section 1.2. (See, for example, Table 1.1.) Note that these databases usually give the *air* wavelength of transitions, while the wavelength worked out from the energy differences gives

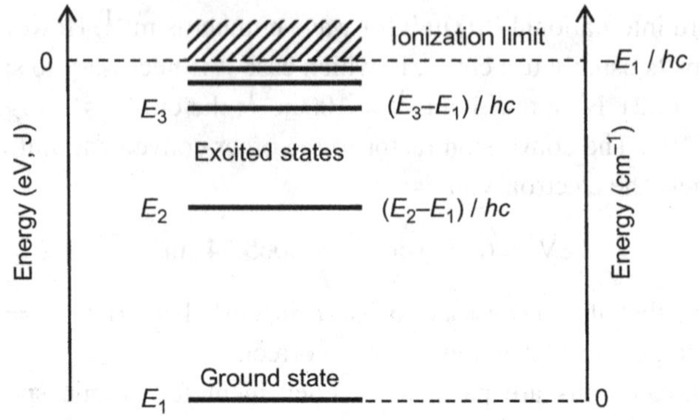

Figure 1.10 Different conventions for specifying atomic energies. On the left, we define $E = 0$ by the ionization limit, so that all the bound-state energies E_n are negative. On the right, we define $E = 0$ by the ground state, so that all the excited state energies E_n for $n > 1$ are positive. The convention on the right is the one frequently used in atomic databases, with the excited-state energies specified in cm^{-1}.

the *vacuum* wavelength. At optical frequencies, these two differ on average by a factor of 1.000277, due to the refractive index, n, of the air:

$$\lambda^{\text{vacuum}} = n(\lambda)\,\lambda^{\text{air}}\,. \tag{1.5}$$

Example 1.2 Hydrogen has two excited states with energies of 82,259 cm^{-1} and 97,492 cm^{-1}. What is the wavelength of the photon emitted in a transition between them?

Solution: The energy difference is:

$$\overline{\nu}_2 - \overline{\nu}_1 = (97{,}292 - 82{,}259)\,\text{cm}^{-1} = 15{,}233\,\text{cm}^{-1}\,.$$

The wavelength is then given by Eq. (1.4) as:

$$\lambda = \frac{1}{\overline{\nu}_2 - \overline{\nu}_1} = \frac{1}{15{,}233\,\text{cm}^{-1}} = 6.5647 \times 10^{-5}\,\text{cm} = 656.47\,\text{nm}\,.$$

Example 1.3 What is the energy of the first excited state considered in Example 1.1 in wave number units?

Solution: The energy can be worked out by applying Eq. (1.4) to Figure 1.9. With wave number units, we define the ground state as 0 cm^{-1}. The energy of the excited state is then simply given by:

$$E_2(\text{cm}^{-1}) = E_1(\text{cm}^{-1}) + \frac{1}{\lambda\,(\text{cm})} = 0 + \frac{1}{589.6 \times 10^{-7}} = 1.696 \times 10^4\,\text{cm}^{-1}\,.$$

1.6 Energy Scales in Atoms

In atomic physics it is traditional to order the interactions that occur inside the atom into a three-level hierarchy according to the scheme summarized in Table 1.2. The effect of this hierarchy on the observed atomic spectra is illustrated schematically in Figure 1.11.

Gross Structure

The first level of the hierarchy is called the **gross structure**, and covers the largest interactions within the atom, namely:

- the kinetic energy of the electrons in their orbits around the nucleus;
- the attractive electrostatic potential between the positive nucleus and the negative electrons; and
- the repulsive electrostatic interaction between the different electrons in a multi-electron atom.

Table 1.2 *Rough energy scales for the different interactions that occur within atoms. The numerical values apply to the valence electrons.*

	Energy scale		Contributing effects
	eV	cm^{-1}	
Gross structure	1 – 10	$10^4 - 10^5$	Electron–nuclear attraction Electron–electron repulsion Electron kinetic energy
Fine structure	0.001 – 0.01	10 – 100	Spin-orbit interactions and other relativistic corrections
Hyperfine structure	$10^{-6} - 10^{-5}$	0.01 – 0.1	Nuclear interactions

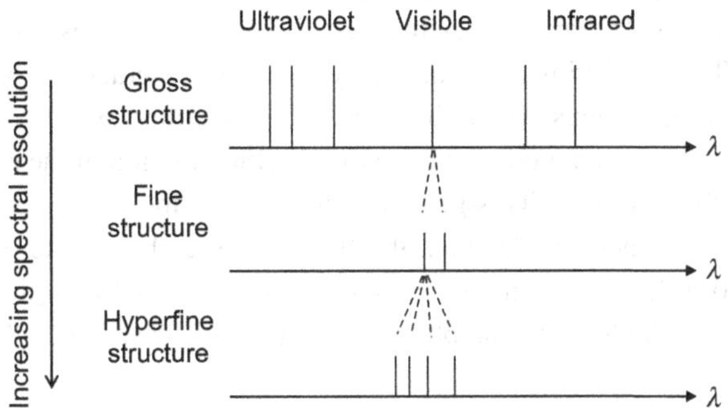

Figure 1.11 Hierarchy of spectral lines observed with increasing spectral resolution.

The size of these interactions gives rise to energies in the 1–10 eV range and upwards. They thus determine whether the photon that is emitted in a transition is in the infrared, visible, ultraviolet, or X-ray spectral regions, and more specifically, whether it is violet, blue, green, yellow, orange, or red for the case of a visible transition.

Fine Structure

Close inspection of atomic spectral lines reveals that some of them come as multiplets. For example, the strong yellow line of sodium is actually a doublet: there are two lines with wavelengths of 589.0 nm and 589.6 nm. This tells us that there are smaller interactions going on inside the atom in addition to the gross-structure effects. The gross-structure interactions determine that the emission line is yellow, but fine-structure effects cause the splitting into the doublet. In the case of the sodium yellow line, the fine-structure energy splitting is 2.1×10^{-3} eV or $17\,\text{cm}^{-1}$, which is smaller than the average transition energy (2.104 eV) by a factor of $\sim 10^{-3}$.

The main cause of fine structure is interactions between the spin of the electron and its orbital motion, as will be explained in Chapter 7. The spin-orbit interaction energy can be deduced by measuring the fine structure in the spectra, and in this way we can learn about the way the spin and the orbital motion of the atom couple together. In more advanced theories of the atom (e.g., the Dirac theory), it becomes apparent that the spin-orbit interaction is actually a relativistic effect.

Hyperfine Structure

Even closer inspection of the spectral lines with a very high resolution spectrometer reveals that the fine-structure lines are themselves split into more multiplets. These splittings are caused by hyperfine interactions between the electrons and the nucleus, as will be discussed in Section 7.8. The nuclear spin can interact with the magnetic field due to the orbital motion of the electron just as in spin-orbit coupling. This gives rise to shifts in the atomic energies that are about 2000 times smaller than the fine-structure shifts. The well-known 21 cm line of radio astronomy is caused by transitions between the hyperfine levels of atomic hydrogen. The photon energy in this case is 6×10^{-6} eV, or $0.05\ \text{cm}^{-1}$.

Exercises

1.1 Write down the ionization states of the following atoms that are iso-electronic (i.e., containing the same number of electrons) to neutral carbon (C I): oxygen, sodium, argon, iron.

1.2 How many valence and core electrons does Ca II have?

1.3 The upper and lower levels of a certain atomic transition have energies of 41197 and 21911 cm^{-1} respectively. Calculate the wavelength of the transition and its energy in eV units.

1.4 The upper and lower levels of one of the red lines of neon have energies of 149657 and 134041 cm^{-1} respectively. Calculate the wavelength of the transition between these levels, and its energy in eV units.

1.5 The first ionization energy of calcium is 6.113 eV, and the neutral atom has an excited state at 23,652 cm^{-1} relative to its ground state. What is the energy of the excited state relative to the Ca^+ ionization limit?

1.6 The first and second ionization energies of He are 198,311 and 438,909 cm^{-1} respectively. What is the energy in eV of the He ground state relative to He III?

1.7 The spectrum of the Mg^+ ion has a doublet with wavelengths of 279.553 and 280.271 nm. What is the fine-structure energy splitting in wave number and eV units?

1.8 The first and second ionization energies of neon are 21.6 and 41.0 eV, and the first accessible excited states of Ne I and Ne II are at 1.34×10^5 and 2.24×10^5 cm^{-1} respectively. Consider a neon discharge tube.

(a) What is the minimum voltage required to observe any emission lines?

(b) What would be the minimum voltage required to observe the full spectrum of Ne I?

(c) What is the minimum voltage required to observe any emission lines from Ne II?

(d) What would be the minimum voltage required to observe the full spectrum of Ne II?

2
Hydrogen

The quantum theory of hydrogen is the starting point for the whole subject of atomic physics. Bohr's derivation of the quantized energies was one of the triumphs of early quantum theory, and makes a useful introduction to the notion of quantized energies and angular momenta. We, therefore, give a brief review of the Bohr model before moving to the main subject of the chapter, namely: the solution of the Schrödinger equation for the electron-nucleus system.

2.1 The Bohr Model of Hydrogen

The Bohr model is part of the "old" quantum theory of the atom (i.e., pre-quantum mechanics). It includes the quantization of energy and angular momentum, but uses classical mechanics to describe the motion of the electron. With the advent of quantum mechanics, we realize that this is an inconsistent approach, and therefore should not be pushed too far. Nevertheless, the Bohr model does give the correct quantized energy levels of hydrogen, and also gives a useful parameter (the Bohr radius) for quantifying the size of atoms. Hence, it remains a useful starting point to understand the basic structure of atoms.

It is well known from classical physics that planetary orbits are characterized by their energy and angular momentum. We shall see that these are also key quantities in the quantum theory of the hydrogen atom. In 1911, Rutherford discovered the nucleus, which led to the idea of atoms consisting of electrons in classical orbits where the central forces are provided by the Coulomb attraction to the positive nucleus, as shown in Figure 2.1. The problem with this idea is that the electron in the orbit is constantly accelerating. Accelerating charges emit radiation called ***bremsstrahlung***, and so the electrons should be radiating all the time, losing energy. This would cause the electron to spiral into the

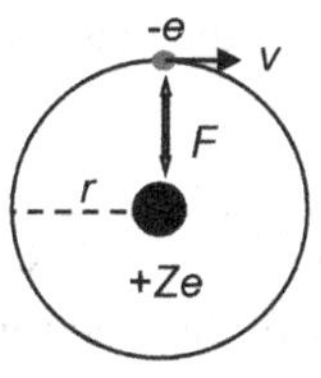

Figure 2.1 The Bohr model of the atom considers the electrons to be in orbit around the nucleus. The central force is provided by the Coulomb attraction. The angular momentum of the electron is quantized in integer units of $\hbar$.

nucleus, like an old satellite crashing to Earth. In 1913, Bohr resolved this issue by postulating that:

- The angular momentum L of the electron is quantized in units of $\hbar$ ($\hbar = h/2\pi$):

$$L = n\hbar\,, \tag{2.1}$$

  where n is an integer.
- The atomic orbits are stable, and light is only emitted or absorbed when the electron jumps from one orbit to another.

When Bohr made these hypotheses in 1913, his only scientific justification was their success in predicting the energy spectrum of hydrogen. With hindsight, we realize that the first assumption is equivalent to stating that the circumference of the orbit must correspond to a fixed number of de Broglie wavelengths:

$$2\pi r = \text{integer} \times \lambda_{\text{deB}} = n \times \frac{h}{p} = n \times \frac{h}{mv}\,, \tag{2.2}$$

which can be rearranged to give:

$$L \equiv mvr = n \times \frac{h}{2\pi}\,. \tag{2.3}$$

The second assumption is a consequence of the fact that the Schrödinger equation leads to time-independent solutions (i.e., *eigenstates*).

The derivation of the quantized energy levels proceeds as follows: Consider an electron orbiting a nucleus of mass m_{N} and charge $+Ze$. The central force is provided by the Coulomb force:

$$F = \frac{mv^2}{r} = \frac{Ze^2}{4\pi\epsilon_0 r^2}\,. \tag{2.4}$$

As with all two-body orbit systems, the mass m that enters here is the **reduced mass** (see Appendix A):

$$\frac{1}{m} = \frac{1}{m_\mathrm{e}} + \frac{1}{m_\mathrm{N}}, \tag{2.5}$$

where m_e and m_N are the masses of the electron and the nucleus, respectively. On rearranging Eq. (2.4) to obtain mv^2r and dividing by Eq. (2.3), we find:

$$v = \frac{Ze^2}{2\epsilon_0 nh}, \tag{2.6}$$

which implies, from Eq. (2.3), that:

$$r = \frac{n^2h^2 4\pi\epsilon_0}{mZe^2}. \tag{2.7}$$

The energy is then worked out from the sum of the kinetic and potential terms:

$$\begin{aligned} E_n &= \frac{1}{2}mv^2 - \frac{Ze^2}{4\pi\epsilon_0 r} \\ &= -\frac{mZ^2e^4}{8\epsilon_0^2h^2n^2}. \end{aligned} \tag{2.8}$$

Note that this is an example of the virial theorem of classical mechanics, where the kinetic and potential energies differ by sign and by a factor of two. The quantized energy can be written in the form:

$$E_n = -\frac{R'}{n^2}, \tag{2.9}$$

where R' is given by:

$$R' = \left(\frac{m}{m_\mathrm{e}}Z^2\right)R_\infty hc, \tag{2.10}$$

and $R_\infty hc$ is the **Rydberg energy**:

$$R_\infty hc = \frac{m_\mathrm{e}e^4}{8\epsilon_0^2h^2}. \tag{2.11}$$

The Rydberg energy is a fundamental constant and has a value of 2.17987×10^{-18} J, or 13.606 eV. The equivalent fundamental constant in wave-number units is called the **Rydberg constant**:

$$R_\infty = \frac{m_\mathrm{e}e^4}{8\epsilon_0^2ch^3}, \tag{2.12}$$

which has a value of 109,737 cm^{-1}. The subscript ∞ comes from considering an atom with an infinitely heavy nucleus (i.e., $m_\mathrm{N} \to \infty$), so that the reduced mass is identical with the electron mass.

R' is the effective Rydberg energy for the system in question. In the hydrogen atom, we have an electron orbiting around a proton of mass m_p. The reduced mass is therefore given by:

$$m = m_e \times \frac{m_p}{m_e + m_p} = 0.99946\, m_e \,, \tag{2.13}$$

and the effective Rydberg energy for hydrogen is:

$$R_H = \frac{me^4}{8\epsilon_0^2 h^2} = \left(\frac{m}{m_e}\right) R_\infty hc = 0.99946\, R_\infty hc \,. \tag{2.14}$$

Atomic spectroscopy is very precise, and 0.05% factors such as this are easily measurable. Furthermore, in other systems such as positronium (see Section 2.4), the reduced mass effect can be much larger.

The final result for hydrogen with $Z = 1$ is that the energy levels are given by:

$$E_n = -\frac{R_H}{n^2} \,, \tag{2.15}$$

where R_H is given by Eq. (2.14) and has a value of 13.60 eV. This tells us that the gross energy of the atomic states in hydrogen is of order 1 – 10 eV, or $10^4 - 10^5\,\text{cm}^{-1}$ in wave-number units. When high precision is not required, it is convenient just to use the symbol R_H for the Rydberg energy, and this is the policy frequently adopted throughout the book. When doing so, a maximum of only three significant figures should be used (i.e., $R_H = 13.6\,\text{eV}$), as R_H differs from the true Rydberg energy by $\approx$ 0.05% (see Eq. (2.14)). Note that the energies in Eq. (2.15) are all negative, as appropriate for bound states (see Section 1.1.)

The quantized velocity and radius given in Eqs. (2.6) and (2.7) can be rewritten in the forms:

$$v_n = \alpha \frac{Z}{n} c \tag{2.16}$$

and

$$r_n = \frac{n^2}{Z} \frac{m_e}{m} a_0 \,. \tag{2.17}$$

The two fundamental constants that appear here are the **Bohr radius**, a_0:

$$a_0 = \frac{h^2 \epsilon_0}{\pi m_e e^2} \,, \tag{2.18}$$

and the **fine-structure constant**, α:

$$\alpha = \frac{e^2}{2\epsilon_0 hc} \,, \tag{2.19}$$

Table 2.1 *Fundamental constants that arise from the Bohr model of the atom.*

Quantity	Symbol	Formula	Numerical Value
Rydberg energy	$R_\infty hc$	$m_e e^4/8\epsilon_0^2 h^2$	2.17987×10^{-18} J 13.6057 eV
Rydberg constant	R_∞	$m_e e^4/8\epsilon_0^2 h^3 c$	109,737 cm^{-1}
Bohr radius	a_0	$\epsilon_0 h^2/\pi e^2 m_e$	5.29177×10^{-11} m
Fine-structure constant	α	$e^2/2\epsilon_0 hc$	1/137.04

which are related to each other according to:

$$a_0 = \frac{\hbar}{m_e c}\frac{1}{\alpha}. \tag{2.20}$$

The Rydberg energy can be conveniently written in terms of a_0 as:

$$R_\infty hc = \frac{\hbar^2}{2m_e}\frac{1}{a_0^2}. \tag{2.21}$$

Table 2.1 summarizes the fundamental constants that arise from the Bohr model.

The radius of the hydrogen ground state can be worked out by putting $n = 1$ and $Z = 1$ into Eq. (2.17) to obtain $a_H = (m_e/m)a_0$. This is called the Bohr radius of hydrogen, and it is very close to a_0, differing only by the factor of $m_e/m = 1.0005$. Therefore, except when very high precision is required, it is common not to distinguish a_H from a_0.

The energies of the photons emitted in transition between the quantized levels of hydrogen can be deduced from Eq. (2.15):

$$h\nu = R_H\left(\frac{1}{n_1^2} - \frac{1}{n_2^2}\right), \tag{2.22}$$

where n_1 and n_2 are the quantum numbers of the two states involved, with n_1 being the lower one. Since $\nu = c/\lambda$, this can also be written in the form:

$$\frac{1}{\lambda} = \frac{m}{m_e}R_\infty\left(\frac{1}{n_1^2} - \frac{1}{n_2^2}\right). \tag{2.23}$$

In absorption, we start from the ground state, so we put $n_1 = 1$. In emission, we can have any combination where $n_1 < n_2$. Some of the series of spectral lines have been given special names. The emission lines with $n_1 = 1$ are called the **Lyman series**, those with $n_1 = 2$ are called the **Balmer series**, etc. (see Table 12.2). The Lyman and Balmer lines occur in the ultraviolet and visible spectral regions respectively.

A simple calculation can easily show that the Bohr model is not consistent with quantum mechanics. The linear momentum of the electron is given by:

$$p = mv = \left(\frac{\alpha Z}{n}\right) mc = \frac{n\hbar}{r_n}\,. \tag{2.24}$$

However, we know from the Heisenberg uncertainty principle that the precise value of the momentum must be uncertain. If we say that the uncertainty in the position of the electron is about equal to the radius of the orbit r_n, we find:

$$\Delta p \sim \frac{\hbar}{\Delta x} \approx \frac{\hbar}{r_n}\,. \tag{2.25}$$

On comparing Eqs. (2.24) and (2.25) we see that:

$$\Delta p \approx \frac{|p|}{n}\,. \tag{2.26}$$

This shows us that the magnitude of p is undefined except when n is large.

Example 2.1 Find the wavelength of the $n = 4 \rightarrow 2$ transition of hydrogen to four significant figures.

Solution: The transition wavelength can be worked out using Eq. (2.23) with $n_1 = 2$ and $n_2 = 4$:

$$\frac{1}{\lambda} = \frac{m}{m_e} R_\infty \left(\frac{1}{2^2} - \frac{1}{4^2}\right) = 0.99946 \times 109,737\,\text{cm}^{-1} \times \frac{3}{16} = 2.056 \times 10^4 \text{cm}^{-1}.$$

Hence, $\lambda = 4.863 \times 10^{-5}$ cm = 486.3 nm.

Example 2.2 Deuterium is an isotope of hydrogen with an extra neutron in the nucleus. Calculate the shift in the energy of the $n = 4 \rightarrow 2$ transition relative to hydrogen in wave number units.

Solution: The transition energy is shifted relative to hydrogen due to the different reduced mass. In the case of hydrogen, we have $m_\text{N} = m_\text{p}$, while for deuterium we have $m_\text{N} = 2m_\text{p}$. The difference in the reduced masses can be worked out using Eq. (2.5):

$$m_\text{D} - m_\text{H} = m_\text{e}\left(1 + \frac{m_\text{e}}{2m_\text{p}}\right)^{-1} - m_\text{e}\left(1 + \frac{m_\text{e}}{m_\text{p}}\right)^{-1} = \frac{m_\text{e}^2}{2m_\text{p}} \text{ (to first order)}\,.$$

The transition energy shift is then given from Eq. (2.23) as:

$$\Delta\overline{\nu} = \left(\frac{m_\text{D}}{m_\text{e}} - \frac{m_\text{H}}{m_\text{e}}\right) R_\infty \left(\frac{1}{2^2} - \frac{1}{4^2}\right) = \frac{m_\text{e}}{2m_\text{p}} \times 109,737\,\text{cm}^{-1} \times \frac{3}{16} = 5.6\,\text{cm}^{-1}.$$

2.2 The Quantum Mechanics of the Hydrogen Atom

In classical physics we are able to calculate the precise trajectory of orbits, but this is not possible in quantum mechanics. The best we can do is to find the wave functions, which give us the probability amplitudes from which all the measurable properties of the system can be calculated. The Bohr model presented in the previous section is a mixture of classical and quantum models, and is only self-consistent in the semi-classical limit at large n. A fully consistent solution needs to use quantum mechanics throughout. Our task, therefore, is to solve the Schrödinger equation for the hydrogen atom.

2.2.1 The Schrödinger Equation

The time-independent Schrödinger equation for hydrogen is given by:

$$\left(-\frac{\hbar^2}{2m}\nabla^2 - \frac{Ze^2}{4\pi\epsilon_0 r}\right)\Psi(r,\theta,\phi) = E\,\Psi(r,\theta,\phi)\,, \tag{2.27}$$

where the spherical polar coordinates (r,θ,ϕ) refer to the position of the electron relative to the nucleus. Spherical polar coordinates are used here because the spherical symmetry of the atom facilitates the solution of the Schrödinger equation by the separation of variables method. Since we are considering the motion of the electron relative to a stationary nucleus, the mass that appears in the Schrödinger equation is the *reduced* mass defined previously in Eq. (2.5) and discussed in more detail in Appendix A. As we have already seen in Eq. (2.13), the reduced mass of hydrogen has a value of about $0.9995m_e$, which is very close to m_e.

Written out explicitly in spherical polar coordinates, the Schrödinger equation becomes:

$$-\frac{\hbar^2}{2m}\left[\frac{1}{r^2}\frac{\partial}{\partial r}\left(r^2\frac{\partial\Psi}{\partial r}\right)+\frac{1}{r^2\sin\theta}\frac{\partial}{\partial\theta}\left(\sin\theta\frac{\partial\Psi}{\partial\theta}\right)+\frac{1}{r^2\sin^2\theta}\frac{\partial^2\Psi}{\partial\phi^2}\right]-\frac{Ze^2}{4\pi\epsilon_0 r}\Psi = E\,\Psi. \tag{2.28}$$

Our task is to find the wave functions, $\Psi(r,\theta,\phi)$, that satisfy this equation, and ultimately find the allowed quantized energies, E.

2.2.2 Separation of Variables

The solution of the Schrödinger equation proceeds by the method of separation of variables. This works because the Coulomb potential is an example of a **central field** in which the force only lies along the radial direction. This allows us to separate the motion into the radial and angular parts:

$$\Psi(r,\theta,\phi) = R(r)\,F(\theta,\phi)\,. \tag{2.29}$$

We can rewrite the Schrödinger equation in the following form:

$$-\frac{\hbar^2}{2m}\frac{1}{r^2}\frac{\partial}{\partial r}\left(r^2\frac{\partial \Psi}{\partial r}\right) + \frac{\hat{\boldsymbol{L}}^2}{2mr^2}\Psi - \frac{Ze^2}{4\pi\epsilon_0 r}\Psi = E\,\Psi\,, \tag{2.30}$$

where the "hat" symbol on $\hat{\boldsymbol{L}}^2$ indicates that we are representing an operator and not just a number. The explicit form of the $\hat{\boldsymbol{L}}^2$ operator is:

$$\hat{\boldsymbol{L}}^2 = -\hbar^2\left[\frac{1}{\sin\theta}\frac{\partial}{\partial\theta}\left(\sin\theta\frac{\partial}{\partial\theta}\right) + \frac{1}{\sin^2\theta}\frac{\partial^2}{\partial\phi^2}\right]. \tag{2.31}$$

This operator is derived from the **angular momentum operator**, $\hat{\boldsymbol{L}}$, which will be considered in detail in Chapter 5. At this stage, we just consider a few basic points relating to the solution of the hydrogen atom.

On substituting Eq. (2.29) into Eq. (2.30), and noting that $\hat{\boldsymbol{L}}^2$ only acts on θ and ϕ, we find:

$$-\frac{\hbar^2}{2m}\frac{1}{r^2}\frac{\mathrm{d}}{\mathrm{d}r}\left(r^2\frac{\mathrm{d}R}{\mathrm{d}r}\right)F + R\frac{\hat{\boldsymbol{L}}^2 F}{2mr^2} - \frac{Ze^2}{4\pi\epsilon_0 r}RF = E\,RF\,. \tag{2.32}$$

Multiply by r^2/RF and rearrange to obtain:

$$-\frac{\hbar^2}{2m}\frac{1}{R}\frac{\mathrm{d}}{\mathrm{d}r}\left(r^2\frac{\mathrm{d}R}{\mathrm{d}r}\right) - \frac{Ze^2 r}{4\pi\epsilon_0} - Er^2 = -\frac{1}{F}\frac{\hat{\boldsymbol{L}}^2 F}{2m}\,. \tag{2.33}$$

The left-hand side is a function of r only, while the right-hand side is only a function of the angular coordinates θ and ϕ. The only way this can be true is if both sides are equal to a constant. Let's call this constant $-\hbar^2 l(l+1)/2m$, where l is an arbitrary number that could be complex at this stage. This gives us, after a bit of rearrangement:

$$-\frac{\hbar^2}{2m}\frac{1}{r^2}\frac{\mathrm{d}}{\mathrm{d}r}\left(r^2\frac{\mathrm{d}R(r)}{\mathrm{d}r}\right) + \frac{\hbar^2 l(l+1)}{2mr^2}R(r) - \frac{Ze^2}{4\pi\epsilon_0 r}R(r) = ER(r)\,, \tag{2.34}$$

and

$$\hat{\boldsymbol{L}}^2 F(\theta,\phi) = \hbar^2 l(l+1)F(\theta,\phi)\,. \tag{2.35}$$

The task thus breaks down to solving two separate equations: one that describes the angular part of the wave function and the other dealing with the radial part.

2.2.3 The Angular Solution and the Spherical Harmonics

It is apparent from Eq. (2.35) that the angular function $F(\theta,\phi)$ is an eigenfunction of the $\hat{\boldsymbol{L}}^2$ operator. These eigenfunctions are known as the **spherical harmonic** functions. The spherical harmonics satisfy the equation:

$$\hat{\boldsymbol{L}}^2 Y(\theta,\phi) \equiv -\hbar^2 \left[\frac{1}{\sin\theta}\frac{\partial}{\partial\theta}\left(\sin\theta\frac{\partial}{\partial\theta}\right) + \frac{1}{\sin^2\theta}\frac{\partial^2}{\partial\phi^2} \right] Y(\theta,\phi) = L^2 Y(\theta,\phi)\,, \tag{2.36}$$

where $L^2 = \hbar^2 l(l+1)$ is the eigenvalue of $\hat{\boldsymbol{L}}^2$.

The solution of Eq. (2.36) is considered in Appendix B. In summary, the solution begins by doing a second separation of variables, with $Y(\theta,\phi)$ written as a product of separate functions of θ and ϕ, i.e. $Y(\theta,\phi) = f(\theta)g(\phi)$. We then derive separate equations in θ and ϕ, with the ϕ equation being:

$$\frac{\mathrm{d}^2 g}{\mathrm{d}\phi^2} = -m^2\, g\,, \tag{2.37}$$

where m^2 is a new separation constant that has to be introduced. This equation is easily solved to obtain:

$$g(\phi) = \text{constant} \times \mathrm{e}^{im\phi}\,. \tag{2.38}$$

The wave function must have a single value for each value of ϕ, which implies $g(\phi+2\pi) = g(\phi)$, and hence that m must be an integer. We therefore conclude that the angular wave functions are of the form:

$$Y(\theta,\phi) = f(\theta)\,\mathrm{e}^{im\phi}\,, \tag{2.39}$$

where m is any positive or negative integer, including 0. This makes it apparent that the wave functions are also eigenfunctions of the operator that describes the z-component of the angular momentum, namely $\hat{L}_z$ (see Eq. [5.8]):

$$\hat{L}_z = -\mathrm{i}\hbar\frac{\partial}{\partial\phi}\,. \tag{2.40}$$

We can see this by operating on $Y(\theta,\phi)$ with $\hat{L}_z$:

$$\begin{aligned}
\hat{L}_z Y(\theta,\phi) &= -i\hbar\frac{\partial}{\partial\phi}\left(f(\theta)\mathrm{e}^{im\phi}\right)\,, \\
&= -\mathrm{i}\hbar f(\theta)\,\frac{\mathrm{d}}{\mathrm{d}\phi}\mathrm{e}^{im\phi}\,, \\
&= -\mathrm{i}\hbar f(\theta)\cdot im\,\mathrm{e}^{im\phi}\,, \\
&= m\hbar\, Y(\theta,\phi)\,.
\end{aligned} \tag{2.41}$$

This shows that the wave functions are eigenvalues of $\hat{L}_z$ with eigenvalue $m\hbar$. The symbol m is called the **magnetic quantum number**, for reasons that will become apparent in Chapter 8 when we consider the effect of external magnetic fields. Note that the same symbol m is used to represent both the mass and the magnetic quantum number. Its meaning should be clear from the context, and, if necessary, we can add a subscript to the quantum number to distinguish it: m_l.

The solution for the θ part of the wave function is, unfortunately, not so simple. The final result is that the spherical harmonic functions are of the form:

$$Y_{lm}(\theta, \phi) = \text{normalization constant} \times P_l^m(\cos\theta)\, \mathrm{e}^{\mathrm{i}m\phi}\,, \tag{2.42}$$

where $P_l^m(\cos\theta)$ is a special function called an associated Legendre polynominal: e.g., $P_0^0(\cos\theta) = \text{constant}$, $P_1^0(\cos\theta) = \cos\theta$, $P_1^{\pm 1}(\cos\theta) = \sin\theta$, etc. The demonstration that the first few of these are indeed solutions of Eq. (2.36) is considered in Exercise 2.2

The indices l and m on the spherical harmonics are the separation constants introduced to solve the equations, and solutions only exist when they are integers, with $l \geq 0$ and $-l \leq m \leq +l$. (The integer quantum number l that appears here is called the **angular momentum quantum number**.) The value of l is usually designated by a letter in atomic physics, with $s, p, d, f, \ldots$ denoting $l = 0, 1, 2, 3, \ldots$, respectively. See, for example, Tables 2.3 or 4.2.

The first few spherical harmonic functions are listed in Table 2.2. Representative polar plots of the wave functions are shown in Figure 2.2. The spherical harmonics are **orthonormal** to each other, satisfying:

$$\int_{\theta=0}^{\pi}\int_{\phi=0}^{2\pi} Y_{lm}^*(\theta,\phi) Y_{l'm'}(\theta,\phi)\ \sin\theta\, \mathrm{d}\theta\, \mathrm{d}\phi = \delta_{l,l'}\, \delta_{m,m'}\,. \tag{2.43}$$

Table 2.2 *Spherical harmonic functions.*

l	m	$Y_{lm}(\theta,\phi)$
0	0	$\sqrt{\frac{1}{4\pi}}$
1	0	$\sqrt{\frac{3}{4\pi}}\cos\theta$
1	± 1	$\mp\sqrt{\frac{3}{8\pi}}\sin\theta\,\mathrm{e}^{\pm\mathrm{i}\phi}$
2	0	$\sqrt{\frac{5}{16\pi}}(3\cos^2\theta - 1)$
2	± 1	$\mp\sqrt{\frac{15}{8\pi}}\sin\theta\cos\theta\,\mathrm{e}^{\pm\mathrm{i}\phi}$
2	± 2	$\sqrt{\frac{15}{32\pi}}\sin^2\theta\,\mathrm{e}^{\pm 2\mathrm{i}\phi}$

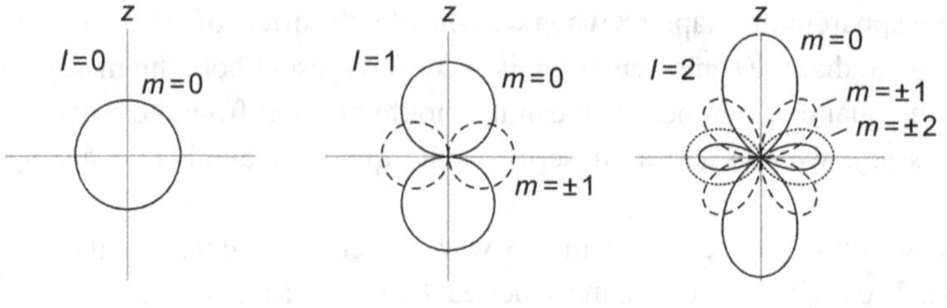

Figure 2.2 Polar plots of the spherical harmonics with $l \leq 2$. The plots are to be imagined with spherical symmetry about the z axis. In these polar plots, the value of the function for a given angle is plotted as a function of the distance from the origin.

The symbol $\delta_{k,k'}$ is called the **Kronecker delta function**. It has a value of 1, if $k = k'$ and 0 if $k \neq k'$. The $\sin\theta$ factor in Eq. (2.43) comes from the volume increment in spherical polar coordinates (see Eq. [2.55].)

On putting all this together, we see that the spherical harmonics (and hence the wave functions of the hydrogen atom) are eigenfunctions of both the $\hat{\boldsymbol{L}}^2$ and $\hat{L}_z$ operators:

$$\hat{\boldsymbol{L}}^2 Y_{lm}(\theta,\phi) = l(l+1)\hbar^2 Y_{lm}(\theta,\phi)\,, \tag{2.44}$$

and

$$\hat{L}_z Y_{lm}(\theta,\phi) = m\hbar\, Y_{lm}(\theta,\phi)\,. \tag{2.45}$$

In quantum mechanics, the allowed values of measurable quantities such as L^2 and L_z are found by solving eigenvalue equations. We can therefore interpret Eqs. (2.44) and (2.45) as stating that the quantized states of the hydrogen atom have quantized angular momenta with $L^2 = l(l+1)\hbar^2$ and a z-component of $m\hbar$. We can recall that L was quantized in integer units of $\hbar$ in the Bohr model (see Eq. [2.3]). The full quantum treatment shows that the Bohr value is only valid in the classical limit where n is large and l approaches its maximum value of $(n-1)$, so that $L = \sqrt{l(l+1)}\hbar \sim \sqrt{(n-1)n}\hbar \sim n\hbar$.

The quantum-mechanical angular momentum states can be represented pictorially in the **vector model** shown in Figure 2.3. The angular momentum is represented as a vector of length, $\sqrt{l(l+1)}\hbar$, angled in such a way that its component along the z-axis is equal to $m\hbar$: we cannot specify the exact direction of $\boldsymbol{L}$, only $|\boldsymbol{L}|^2$ and L_z. As will be discussed in Section 5.2.1, the x- and y-components of the angular momentum are not known, because they do not commute with $\hat{L}_z$.

The quantization of the magnitude of the angular momentum $|\boldsymbol{L}|^2$ with well-defined eigenvalues reflects the fact that the angular momentum of a classical

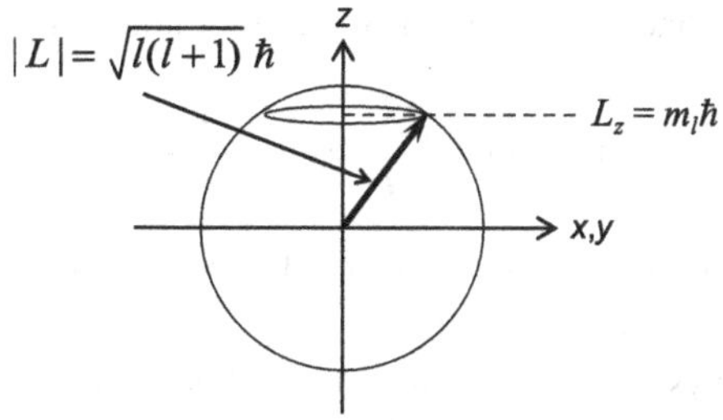

Figure 2.3 Vector model of the angular momentum in an atom. The angular momentum is represented by a vector of length, $\sqrt{l(l+1)}\hbar$, with z-component equal to $m_l\hbar$.

particle interacting with a central field (i.e., one with a radial force parallel to $\boldsymbol{r}$) is a **constant of the motion**. This follows because the torque on the particle is zero, and so $\boldsymbol{L}$ must be a conserved quantity. (See discussion in Section 5.2.1.)

2.2.4 The Radial Wave Functions

We now return to solving the radial equation, with the additional constraint that the separation constant l in Eq. (2.34) is the angular momentum quantum number l, and can only have positive integer values or be zero. On substituting $R(r) = P(r)/r$ into Eq. (2.34), we find:

$$\left[-\frac{\hbar^2}{2m}\frac{\mathrm{d}^2}{\mathrm{d}r^2} + \frac{\hbar^2 l(l+1)}{2mr^2} - \frac{Ze^2}{4\pi\epsilon_0 r}\right]P(r) = EP(r)\,. \tag{2.46}$$

This now makes physical sense. It is a Schrödinger equation of the form:

$$\hat{H}P(r) = EP(r)\,, \tag{2.47}$$

where the energy operator $\hat{H}$ (i.e., the Hamiltonian) is given by:

$$\hat{H} = -\frac{\hbar^2}{2m}\frac{\mathrm{d}^2}{\mathrm{d}r^2} + V_{\text{effective}}(r)\,. \tag{2.48}$$

The first term in Eq. (2.48) is the **radial kinetic energy** given by:

$$\text{K.E.}_{\text{radial}} = \frac{p_r^2}{2m} = -\frac{\hbar^2}{2m}\frac{\mathrm{d}^2}{\mathrm{d}r^2}\,.$$

The second term is the **effective potential energy**:

$$V_{\text{effective}}(r) = \frac{\hbar^2 l(l+1)}{2mr^2} - \frac{Ze^2}{4\pi\epsilon_0 r}\,, \tag{2.49}$$

which has two components. The first is the orbital kinetic energy:

$$\text{K.E.}_{\text{orbital}} = \frac{L^2}{2I} = \frac{\hbar^2 l(l+1)}{2mr^2},$$

where $I \equiv mr^2$ is the moment of inertia. The second is the usual potential energy due to the Coulomb interaction.

This analysis shows that the orbital motion adds quantized kinetic energy to the radial motion. For $l > 0$, the orbital kinetic energy will always be larger than the Coulomb energy at small r, and so the effective potential will be positive near $r = 0$. This has the effect of keeping the electron away from the nucleus, and explains why states with $l > 0$ have nodes at the origin. (See below.)

The radial wave function $R(r)$ that we require can be found by solving Eq. (2.34), with l constrained by the angular equation to be an integer ≥ 0. The solution is given in Section B.2 in Appendix B. The mathematics is somewhat complicated, and here we just quote the main results. Solutions are only found if we introduce an integer quantum number n, which must be $> l$, and therefore positive. The functional form of $R(r)$ depends on both n and l, with:

$$R_{nl}(r) = C_{nl} \cdot (\text{polynomial in } r \text{ depending on } n \text{ and } l) \cdot \mathrm{e}^{-Zr/na}, \tag{2.50}$$

where $a \equiv a_{\mathrm{H}} = (m_{\mathrm{e}}/m)\, a_0 = 5.29 \times 10^{-11}$ m is the $n = 1$ Bohr radius of hydrogen given in Eq. (2.17). C_{nl} is a normalization constant, worked out from the normalization condition:

$$\int_{r=0}^{\infty} R_{nl}^2\, r^2\, \mathrm{d}r = 1. \tag{2.51}$$

The factor of r^2 that appears here is discussed in connection with Eq. (2.56) below. The polynomial functions in Eq. (2.50) are of order $(n-1)$, with $(n-1)$ nodes. If $l = 0$ and $n > 1$, all the nodes occur at finite r, but if $l > 0$, one of the nodes is at $r = 0$. A list of the first few radial functions is given in Table 2.3, and representative wave functions are plotted in Figure 2.4. Note that the radial wave functions are all real.

2.2.5 The Full-Wave Function and Energy

The full-wave function for hydrogen is obtained by substituting the results of Sections 2.2.3 and 2.2.4 into Eq. (2.29):

$$\Psi_{nlm}(r,\theta,\phi) = R_{nl}(r)\, Y_{lm}(\theta,\phi), \tag{2.52}$$

Table 2.3 *Radial wave functions of the hydrogen atom; $a = (m_e/m)\,a_0$, where a_0 is the Bohr radius (5.29×10^{-11} m). The wave functions are normalized according to Eq. (2.51).*

Spectroscopic name	n	l	$R_{nl}(r)$
1s	1	0	$(Z/a)^{3/2} 2 \exp(-Zr/a)$
2s	2	0	$(Z/2a)^{3/2} \left(1 - \frac{Zr}{2a}\right) \exp(-Zr/2a)$
2p	2	1	$(Z/2a)^{3/2} \frac{2}{\sqrt{3}} \left(\frac{Zr}{2a}\right) \exp(-Zr/2a)$
3s	3	0	$(Z/3a)^{3/2} 2 \left[1 - 2\frac{Zr}{3a} + \frac{2}{3}\left(\frac{Zr}{3a}\right)^2\right] \exp(-Zr/3a)$
3p	3	1	$(Z/3a)^{3/2} (4\sqrt{2}/3) \left(\frac{Zr}{3a}\right) \left(1 - \frac{1}{2}\frac{Zr}{3a}\right) \exp(-Zr/3a)$
3d	3	2	$(Z/3a)^{3/2} (2\sqrt{2}/3\sqrt{5}) \left(\frac{Zr}{3a}\right)^2 \exp(-Zr/3a)$

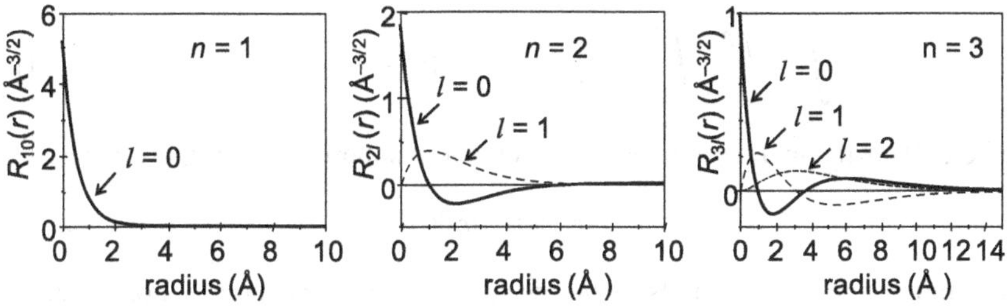

Figure 2.4 The radial wave functions $R_{nl}(r)$ for the hydrogen atom with $Z = 1$. Note that the axes for the three graphs are not the same.

where $R_{nl}(r)$ is one of the radial functions given in Eq. (2.50), and $Y_{lm}(\theta, \phi)$ is a spherical harmonic function discussed in Section 2.2.3. The quantum numbers obey the following rules:

- n can have any integer value ≥ 1.
- l can have positive integer values from zero up to $(n - 1)$.
- m can have integer values from $-l$ to $+l$.

These rules drop out of the mathematical solutions. Functions that do not obey these rules will not satisfy the Schrödinger equation for the hydrogen atom.

The energy of the system is found to be (see Eq. (B.35) in Appendix B.2):

$$E_n = -\frac{mZ^2e^4}{8\epsilon_0^2 h^2}\frac{1}{n^2}, \tag{2.53}$$

which is the same as the Bohr formula given in Eq. (2.8). The energy only depends on the principal quantum number n, which means that all the l states for a given value of n are **degenerate** (i.e., have the same energy), even though the radial wave functions depend on both n and l. This degeneracy with respect to l is called "accidental," and it is a consequence of the fact that the electrostatic energy has a precise $1/r$ dependence in hydrogen. In more complex atoms, the electrostatic energy will depart from a pure $1/r$ dependence due to the shielding effect of inner electrons, and the gross energy will depend on l as well as n – even before we start thinking of higher-order fine-structure effects. We shall see an example of how this works when we consider alkali atoms in Section 4.5. Note, also, that the energy does not depend on the magnetic quantum number m at all. Hence, the m states for each value of l are degenerate in the gross structure of all atoms in the absence of external fields.

The wave functions are normalized so that:

$$\int_{r=0}^{\infty}\int_{\theta=0}^{\pi}\int_{\phi=0}^{2\pi} \Psi^*_{n,l,m}\Psi_{n',l',m'}\,\mathrm{d}V = \delta_{n,n'}\delta_{l,l'}\delta_{m,m'}\,, \tag{2.54}$$

where $\mathrm{d}V$ is the incremental volume element in spherical polar coordinates:

$$\mathrm{d}V = r^2\sin\theta\,\mathrm{d}r\,\mathrm{d}\theta\,\mathrm{d}\phi\,. \tag{2.55}$$

The radial and angular parts of the wave function are separately normalized, as given by Eqs. (2.43) and (2.51). The **radial probability density** function $P_{nl}(r)$ is the probability that the electron is found between r and $r+\mathrm{d}r$:

$$\begin{aligned} P_{nl}(r)\,\mathrm{d}r &= \int_{\theta=0}^{\pi}\int_{\phi=0}^{2\pi} \Psi^*\Psi\; r^2\sin\theta\,\mathrm{d}r\,\mathrm{d}\theta\,\mathrm{d}\phi \\ &= R_{nl}(r)^2 r^2\,\mathrm{d}r\int_{\theta=0}^{\pi}\int_{\phi=0}^{2\pi} Y^*_{lm}(\theta,\phi)Y_{lm}(\theta,\phi)\;\sin\theta\,\mathrm{d}\theta\,\mathrm{d}\phi \\ &= R_{nl}(r)^2\,r^2\,\mathrm{d}r\,, \end{aligned} \tag{2.56}$$

where we used Eq. (2.43) in the third line. The factor of r^2 that appears here is just related to the surface area of the radial shell of radius r (i.e., $4\pi r^2$). Some representative radial probability functions are sketched in Figure 2.5. It is easy to show that the single-peaked ground-state 1s wave function, with $n=1$ and $l=0$, peaks at the Bohr radius. (See Exercise 2.4.)

Expectation values of measurable quantities with quantum mechanical operator $\hat{A}$ are calculated as follows:

$$\langle\hat{A}\rangle = \iiint \Psi^*\hat{A}\Psi\,\mathrm{d}V\,. \tag{2.57}$$

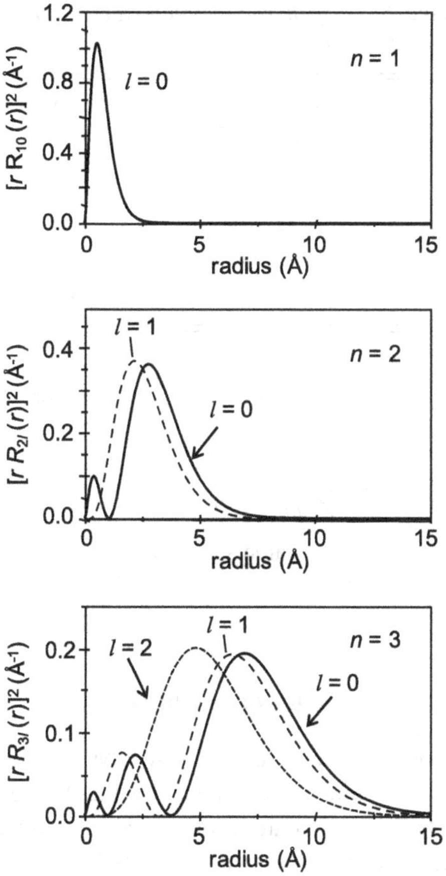

Figure 2.5 Radial probability functions for the first three n states of the hydrogen atom with $Z = 1$. Note that the radial probability is equal to $r^2 R_{nl}(r)^2$, not just to $R_{nl}(r)^2$. Note, also, that the horizontal axes are the same for all three graphs, but not the vertical axes.

The expectation value of the radius is therefore given by:

$$
\begin{aligned}
\langle r \rangle &= \iiint \Psi^* r \Psi \, \mathrm{d}V \,, \\
&= \int_{r=0}^{\infty} R_{nl}^* r R_{nl} r^2 \mathrm{d}r \int_{\theta=0}^{\pi} \int_{\phi=0}^{2\pi} Y_{lm}^*(\theta, \phi) Y_{lm}(\theta, \phi) \, \sin\theta \, \mathrm{d}\theta \, \mathrm{d}\phi \,, \\
&= \int_{r=0}^{\infty} R_{nl}^2 r^3 \mathrm{d}r \,,
\end{aligned}
\tag{2.58}
$$

where we again used Eq. (2.43). For the 1s ground state, we find $\langle r \rangle = 3a/2$. (See Exercise 2.5.) Hence the Bohr radius corresponds to the peak of the radial

Table 2.4 *Degeneracy of the l states of the hydrogen atom.*

l	Spectroscopic name	Degeneracy
0	s	2
1	p	6
2	d	10
3	f	14
$\vdots$		$\vdots$
l		$2(2l+1)$

probability density, but only two-thirds of the expectation value. The general result for $\langle r \rangle$ is:

$$\langle r \rangle = \frac{n^2}{Z} \frac{m_e}{m} a_0 \left(\frac{3}{2} - \frac{l(l+1)}{2n^2} \right) . \tag{2.59}$$

This only approaches the Bohr value given in Eq. (2.17) for the states with $l = n - 1$ at large n.

2.3 Degeneracy and Spin

The fact that electrons are **spin-1/2 particles** will be very important at various stages of this book, for example, in understanding the Pauli exclusion principle that underpins the shell model of atoms discussed in Chapter 4. At this stage, we just note that each electron has two spin states specified by the quantum number m_s, where $m_s = \pm 1/2$. The electron spin does not appear in the Schrödinger equation given in Eq. (2.28), which means that each quantum state defined by the quantum numbers (n, l, m_l) has a degeneracy of two due to the two allowed spin states. (Note that we have added a subscript l to the magnetic quantum number m to distinguish it clearly from m_s.)

We noted above that the m_l states of the hydrogen atom are degenerate (i.e., have the same energy) in the absence of external fields. Since each l state has $(2l + 1)$ m_l levels, the full degeneracy of each l state including the spin degeneracy is $2 \times (2l+1) = 2(2l+1)$, as listed in Table 2.4. In hydrogen, the l states are also degenerate. The degeneracy of the energy levels in hydrogen is therefore obtained by summing up the total number of (l, m_l, m_s) states that are possible for a given value of n:

$$\text{degeneracy} = 2 \times \sum_{l=0}^{n-1} (2l+1) = 2n^2 . \tag{2.60}$$

2.4 Hydrogen-Like Atoms

The theory of the hydrogen atom can be applied to any system that consists of a single negative particle orbiting around a single positive one. There is a great variety of these **hydrogenic atoms**, and they can be treated by the theory developed here – but with the appropriate reduced mass included and the appropriate value of Z. In some cases (notably in solid-state physics), it might also be necessary to replace ϵ_0 by $\epsilon_r\epsilon_0$ to account for the relative dielectric constant of the medium.

Here are some examples of hydrogenic atoms:

- **Antihydrogen.** This consists of a positron bound to an antiproton. It should be exactly the same as hydrogen. Experiments are under way at CERN to make antihydrogen and measure the energy levels with very high precision. The discovery of a small difference in the spectra of hydrogen and antihydrogen might help to answer the question why there is no antimatter in our known universe. At present, it is known that the energy levels are identical to about one part in 10^{10}.
- **Ionized atoms.** Ionized atoms with $Z > 1$ in which all of the electrons have been stripped off apart from the last one (i.e., $A^{(Z-1)+}$). The simplest example is He^+, where $Z = 2$. We then have Li^{2+} ($Z = 3$), Be^{3+} ($Z = 4$), ..., etc. These would be written He II, Li III, Be IV, etc., in the spectroscopic notation explained in Section 1.2.
- **Positronium.** This consists of an electron bound to a positron. Since the positive particle has mass m_e, the reduced mass is $0.5m_e$. In solid-state physics, an exciton consists of an electron bound to a hole. The reduced mass is worked out from the effective masses of the electrons and holes, and the dielectric constant of the medium has to be included. (See Section 11.3.2.)
- **Impurity levels in semiconductors.** These are modeled as electrons or holes bound to a positive or negative impurity atom. The impurity is bound to the crystal and therefore can be treated as having infinite mass. The effective mass must be used, and the dielectric constant of the medium. See Section 11.3.1.
- **Muonium.** This consists of an electron bound to a μ^+. The nucleus has mass $207m_e$, and hence $m = 0.995m_e$.
- **Muonic hydrogen.** This is a μ^- bound to a proton. The reduced mass is $186m_e$.

Another interesting application of hydrogen theory is to the study of **Rydberg atoms**, which are atoms in very highly excited states called Rydberg states, e.g., with $n \sim 100$. In the case of a neutral Rydberg atom with atomic

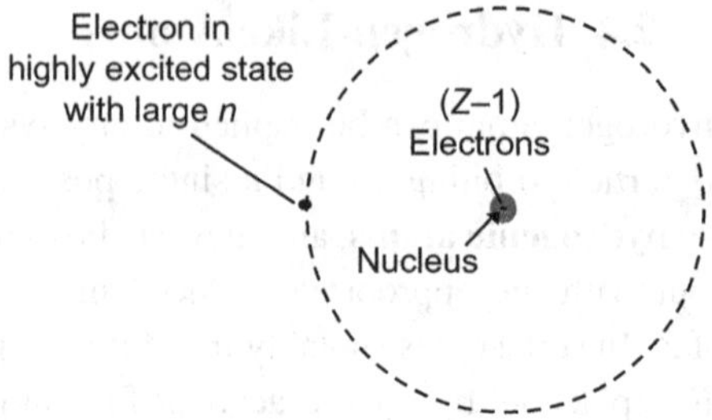

Figure 2.6 Rydberg atom. One of the electrons of a multi-electron atom is in a highly excited state far from the nucleus. The remaining $(Z-1)$ electrons are in tightly bound states close to the nucleus.

number Z, there are $(Z-1)$ electrons in tightly bound states close to the nucleus and one electron in a very large radius state far from the nucleus, as shown in Figure 2.6. The single outer electron has very low probability of overlapping the other electron wave functions, and so the central charge cloud close to the nucleus behaves as a net charge of $+e$, just as in hydrogen. The energies of the Rydberg states can thus be modeled as hydrogenic. With such large quantum numbers, the transition energies are in the microwave or radio-wave spectral regions, and these are important in radio-frequency astronomy. (See Section 12.4.3.) Since the radii are large and the binding energies are small, the behavior of Rydberg atoms is close to the semi-classical limit. Precision atomic spectroscopy can then test the convergence of classical and quantum theories in the limit of large n.

Example 2.3 Calculate the frequency of the $n = 100 \rightarrow 99$ transition in hydrogen.

Solution: Since $\nu = c/\lambda$, the frequency can be worked out from Eq. (2.23):

$$\begin{aligned}\nu &= \frac{m}{m_{\rm e}} c R_\infty \left(\frac{1}{99^2} - \frac{1}{100^2}\right) \\ &= 0.99946 * 2.998 \times 10^{10} * 109{,}737 * 2.03 \times 10^{-6} \\ &= 6.67 \times 10^9\,\text{Hz} \equiv 6.67\,\text{GHz}\,.\end{aligned}$$

Exercises

2.1 Substitute $\Psi(r,\theta,\phi) = C\exp(-r/a)$ into the hydrogen Schrödinger equation with $Z = 1$ to show that it is a solution if $a = (m_{\rm e}/m)\,a_0$. (C is a constant.) What is the energy of this state? Find the value of C that normalizes the wave function.

2.2 Find values of l for which the following functions are solutions of the angular equation (Eq. [2.36]) with $L^2 = \hbar^2 l(l+1)$: (a) $Y(\theta, \phi) = C$; (b) $Y(\theta, \phi) = C \cos\theta$; (c) $Y(\theta, \phi) = C \sin\theta e^{-i\phi}$. ($C$ is a constant.) In each case, state the value of L^2 and L_z.

2.3 Find values of β and l for which the function $R(r) = Cr^2 \exp(-\beta r)$ is a solution of the radial equation (Eq. [2.34]), where C is a constant. Given that the radial wave functions vary as $\exp(-Zr/na)$, where $a = (m_e/m)\, a_0$, deduce the value of n for this state, and verify that the energy agrees with the Bohr formula.

2.4 Show that the peak of the radial probability density of the 1s wave function occurs at the Bohr radius.

2.5 Substitute the 1s radial wave function into Eq. (2.58) to show that the expectation value of the radius is equal to $3a/2$ for the hydrogen ground state.

2.6 Find the probability that an electron in the 1s state of hydrogen has $r \leq a$, where a is the Bohr radius.

2.7 Find the expectation value of the potential energy of the electron in the ground state of hydrogen. Deduce the expectation value of the kinetic energy.

2.8 The radius of the ^{40}Ar nucleus is approximately 4.3×10^{-15} m. Estimate the probability that the electron in the ground state of the $^{40}Ar^{17+}$ ion (i.e., Ar XVIII) lies within the nucleus. (Argon has $Z = 18$.)

2.9 Write down the quantum numbers of the degenerate states with energy $-R_H/16$ in hydrogen. Verify that the total number of these states satisfies Eq. (2.60).

2.10 Find the wavelength of the $n = 5 \rightarrow 2$ transition in (a) positronium, (b) He^+, (c) muonium, (d) muonic hydrogen.

2.11 Find the values of $\langle r \rangle$ for the 2s and 2p orbitals of positronium, stating your answer in nm units.

2.12 Find the frequency of the $n = 120 \rightarrow 118$ transition in hydrogen. What would be the frequency shift of the equivalent transition in ^{4}He?

2.13 Find a formula for the Zth ionization energy of an element in terms of R_H.

3
Radiative Transitions

We can learn a great deal about atoms by analyzing the photons emitted or absorbed in transitions between quantized energy levels. It is therefore important to understand the processes that govern the radiative transition rates. This will lead to the concept of selection rules that determine whether a particular transition is allowed or not, and also to a discussion of the physical mechanisms that affect the shape of the spectral lines that are observed in atomic spectra.

3.1 Classical Theories of Radiating Dipoles

The main focus of this chapter will be on the quantum theory of radiative transitions, but it is helpful first to give a brief review of the classical theory based on electromagnetism. These classical models were developed in the late nineteenth century following the electromagnetic theories of James Clerk Maxwell and the experimental work of Heinrich Hertz on radio waves. At that time, the electron and the nucleus had not yet been discovered, but we can now benefit from hindsight to understand more clearly how the classical theory works. This allows us to model the atom as a heavy nucleus with electrons attached to it by springs with different spring constants, as shown in Figure 3.1(a). The spring represents the binding force between the nucleus and the electrons, and the values of the spring constants determine the resonant frequencies of each of the electrons. Every atom therefore has several different natural frequencies.

The nucleus is heavy, and so it does not move very easily at high frequencies. However, the electrons can readily vibrate about their mean position, as illustrated in Figure 3.1(b). The vibrations of the electron create a fluctuating **electric dipole**. In general, electric dipoles consist of two opposite charges of $\pm q$ separated by a distance d. The dipole moment $\boldsymbol{p}$ is defined by:

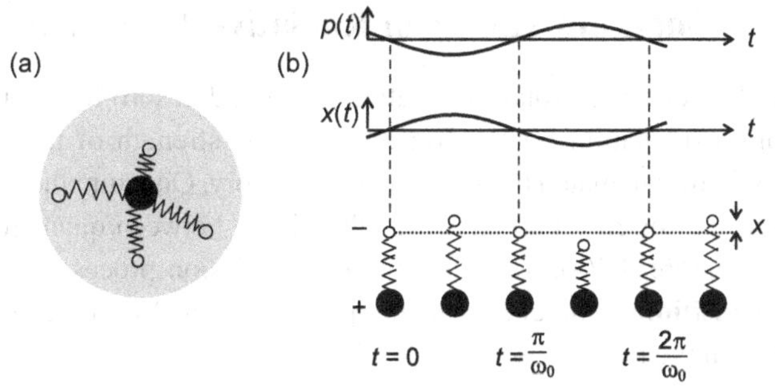

Figure 3.1 (a) Classical atoms can be modeled as electrons bound to a heavy nucleus by springs with characteristic force constants. (b) The vibrations of an electron at its natural resonant frequency, ω_0, creates an oscillating electric dipole.

$$\boldsymbol{p} = q\boldsymbol{d} \,, \tag{3.1}$$

where $\boldsymbol{d}$ is a vector of length d pointing from $-q$ to $+q$. In the case of atomic dipoles, the positive charge may be considered as being stationary, and so the time dependence of $\boldsymbol{p}$ is just determined by the movement of the electron:

$$p(t) = -ex(t) \,, \tag{3.2}$$

where $x(t)$ is the time dependence of the electron displacement.

It is well known that oscillating electric dipoles emit electromagnetic radiation at the oscillation frequency. We can then understand that an electron excited into vibration can act like an aerial and emit electromagnetic waves at its natural frequency ω_0. This is the classical explanation of why atoms emit characteristic colors when excited electrically in a discharge tube. Furthermore, it is easy to see that an incoming electromagnetic wave at frequency ω_0 can drive the natural vibrations of the atom through the oscillating force exerted on the electron by the electric field of the wave. This transfers energy from the wave to the atom, which causes absorption at the resonant frequency. Hence, the atom is also expected to absorb strongly at its natural frequencies.

The classical theories have to assume that each electron has several natural frequencies of varying strengths in order to explain the observed spectra. This is particularly obvious in the case of hydrogen, which has only one electron but many spectral lines. There was no classical explanation for the origin of the atomic dipoles, and it is therefore not surprising that we run into contradictions such as this when we try to patch up the model by applying our knowledge of electrons and nuclei gained by hindsight.

3.2 Quantum Theory of Radiative Transitions

The classical theory can explain why atoms emit and absorb light, but it does not offer any explanation for the frequency or the strength of the radiation. These can only be calculated by using quantum theory. Quantum theory tells us that atoms absorb or emit photons when they jump between quantized states, as shown in Figure 3.2(a). The absorption or emission processes are called **radiative transitions**. The energy of the photon is equal to the difference in energy between the two levels:

$$h\nu = E_2 - E_1 \,. \tag{3.3}$$

Our task here is to calculate the rate at which these transitions occur.

The **transition rate** W_{12} can be calculated from the initial and final wave functions of the states involved by using **Fermi's golden rule**:

$$W_{12} = \frac{2\pi}{\hbar}|M_{12}|^2 g(h\nu)\,, \tag{3.4}$$

where M_{12} is the **matrix element** for the transition and $g(h\nu)$ is the **density of states**. The matrix element is written $\langle 2|H'|1\rangle$ in the shorthand **Dirac notation**, and it is equal to the overlap integral:

$$M_{12} \equiv \langle 2|H'|1\rangle = \int \psi_2^*(\boldsymbol{r})\, H'(\boldsymbol{r})\, \psi_1(\boldsymbol{r})\, \mathrm{d}^3\boldsymbol{r}\,, \tag{3.5}$$

where H' is the perturbation that causes the transition. For the case of an optical transition, H' represents the interaction between the atom and the light wave. There are a number of mechanisms that cause atoms to absorb or emit light, and the strongest of these is the electric-dipole (E1) interaction. We therefore discuss E1 transitions first, leaving the discussion of other higher-order effects to Section 3.5.

The density of states factor is defined so that $g(h\nu)\,\mathrm{d}E$ is the number of *final* states per unit volume that fall within the energy range E to $E + \mathrm{d}E$, where

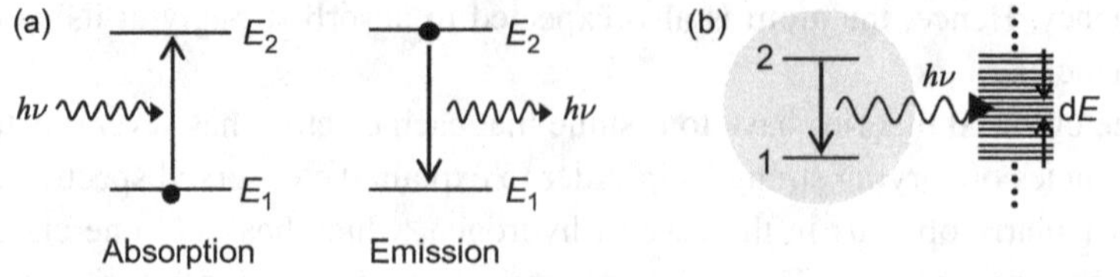

Figure 3.2 (a) Absorption and emission transitions in an atom. (b) Emission into a continuum of photon modes during a radiative transition between discrete atomic states.

$E = h\nu$. In the standard case of transitions between quantized levels in an atom, the initial and final electron states are discrete. In this case, the density of states factor that enters the golden rule is the density of *photon* states. In free space, the photons can have any frequency and there is a continuum of states available, as illustrated in Figure 3.2(b). The atom can therefore always emit a photon and it is the matrix element that determines whether the transition probability is zero or not. Hence we concentrate on the matrix element from now on, although we can make two useful general points relating to the density of states factor.

(i) It is well known, from the theory of back-body radiation, for example, that the density of photon states in free space is given by:

$$g(h\nu) = \frac{8\pi\nu^2}{hc^3}\,. \tag{3.6}$$

This shows that $g(h\nu) \propto \nu^2$, and we therefore expect that the transition rate should, in general, increase with the frequency. Thus X-ray transitions are expected to be much faster than transitions at optical frequencies, which is, in fact, what is normally observed.

(ii) In solid-state physics, we might need to consider transitions between electronic bands rather than between discrete states. We then have to consider the density of electron states, as well as the density of photon states, when we calculate the transition rate. We will come back to this point when we consider solid-state systems in Chapter 11.

3.3 Electric Dipole (E1) Transitions

Electric-dipole transitions are the quantum-mechanical equivalent of the classical dipole oscillator discussed in Section 3.1. We assume that the atom is irradiated with light, and makes a jump from level 1 to 2 by absorbing a photon. The interaction energy between an electric dipole $\boldsymbol{p}$ and an external electric field $\boldsymbol{\mathcal{E}}$ is given by:

$$E = -\boldsymbol{p} \cdot \boldsymbol{\mathcal{E}}\,. \tag{3.7}$$

We presume that the nucleus is heavy, and so we only need to consider the effect on the electron. Hence we put $\boldsymbol{p} = -e\boldsymbol{r}$, where $\boldsymbol{r}$ is the position vector of the electron (*cf*, Eq. [3.2]), to rewrite the electric dipole perturbation as:

$$H' = +e\boldsymbol{r} \cdot \boldsymbol{\mathcal{E}}\,, \tag{3.8}$$

and $\mathcal{E}$ is the electric field of the light wave. This can be simplified to:

$$H' = e(x\mathcal{E}_x + y\mathcal{E}_y + z\mathcal{E}_z)\,, \tag{3.9}$$

where $\mathcal{E}_x$ is the component of the field amplitude along the x-axis, etc. Now, atoms are small compared to the wavelength of light, and so the amplitude of the electric field will not vary significantly over the dimensions of an atom. We can therefore take $\mathcal{E}_x$, $\mathcal{E}_y$, and $\mathcal{E}_z$ in Eq. (3.9) to be constants in the calculation, and just evaluate the following integrals:

$$\begin{aligned} M_{12} &\propto \int \psi_1^*\, x\, \psi_2\, \mathrm{d}^3\boldsymbol{r} \qquad x\text{–polarized light}\,, \\ M_{12} &\propto \int \psi_1^*\, y\, \psi_2\, \mathrm{d}^3\boldsymbol{r} \qquad y\text{–polarized light}\,, \\ M_{12} &\propto \int \psi_1^*\, z\, \psi_2\, \mathrm{d}^3\boldsymbol{r} \qquad z\text{–polarized light}\,. \end{aligned} \tag{3.10}$$

Integrals of this type are called **dipole moments**. The dipole moment is a key parameter that determines the transition rate for the electric-dipole process.

At this stage it is helpful to give a hand-waving explanation for why electric dipole transitions lead to the emission of light. To do this we need to to consider the time-dependence of the wave functions. This naturally drops out of the *time-dependent* Schrödinger equation:

$$\hat{H}(\boldsymbol{r})\, \Psi(\boldsymbol{r},t) = \mathrm{i}\hbar \frac{\partial}{\partial t} \Psi(\boldsymbol{r},t)\,, \tag{3.11}$$

where $\hat{H}(\boldsymbol{r})$ is the Hamiltonian of the system. The solutions of Eq. (3.11) are of the form:

$$\Psi(\boldsymbol{r},t) = \psi(\boldsymbol{r})\, \mathrm{e}^{-\mathrm{i}Et/\hbar}\,, \tag{3.12}$$

where $\psi(\boldsymbol{r})$ satisfies the *time-independent* Schrödinger equation:

$$\hat{H}(\boldsymbol{r})\, \psi(\boldsymbol{r}) = E\, \psi(\boldsymbol{r})\,. \tag{3.13}$$

During a transition between two quantum states of energies E_1 and E_2, the electron will be in a superposition state with a wave function given by:

$$\begin{aligned} \Psi(\boldsymbol{r},t) &= c_1 \Psi_1(\boldsymbol{r},t) + c_2 \Psi_2(\boldsymbol{r},t) \\ &= c_1 \psi_1(\boldsymbol{r})\, \mathrm{e}^{-\mathrm{i}E_1 t/\hbar} + c_2 \psi_2(\boldsymbol{r})\, \mathrm{e}^{-\mathrm{i}E_2 t/\hbar}\,, \end{aligned} \tag{3.14}$$

where c_1 and c_2 are the amplitude coefficients. The expectation value of the position of the electron is given by:

$$\langle x \rangle = \int \Psi^*\, x\, \Psi\, \mathrm{d}^3\boldsymbol{r}\,. \tag{3.15}$$

With Ψ given by Eq. 3.14 we obtain:

$$\langle x \rangle = c_1^* c_1 \int \psi_1^* \, x \, \psi_1 \, \mathrm{d}^3\boldsymbol{r} + c_2^* c_2 \int \psi_2^* \, x \, \psi_2 \, \mathrm{d}^3\boldsymbol{r} \tag{3.16}$$
$$+ c_1^* c_2 \mathrm{e}^{-\mathrm{i}(E_2-E_1)t/\hbar} \int \psi_1^* \, x \, \psi_2 \, \mathrm{d}^3\boldsymbol{r} \; + \; c_2^* c_1 \mathrm{e}^{-\mathrm{i}(E_1-E_2)t/\hbar} \int \psi_2^* \, x \, \psi_1 \, \mathrm{d}^3\boldsymbol{r} \, .$$

This shows that if the dipole moment defined in Eq. (3.10) is nonzero, then the electron wave-packet oscillates in space at angular frequency $(E_2 - E_1)/\hbar$. The oscillation of the electron wave-packet creates an oscillating electric dipole, which then radiates light at angular frequency $(E_2 - E_1)/\hbar$, as required.

3.4 Selection Rules for E1 Transitions

Electric-dipole transitions can only occur if the **selection rules** summarized in Table 3.1 are satisfied. Transitions that obey these E1 selection rules are called **allowed** transitions. If the selection rules are not satisfied, the matrix element (i.e., the dipole moment) is zero, and we then see from Eq. (3.4) that the transition rate is zero. The origins of these rules are discussed within this section.

Table 3.1 *Electric-dipole (E1) selection rules for the quantum numbers of the states involved in the transition. More specific rules for the polarizations are given in Table 8.2.*

Quantum number	Selection rule	Polarization
Parity	Changes	
l	$\Delta l = \pm 1$	
m	$\Delta m = 0, \pm 1$	Unpolarized light
	$\Delta m = 0$	Linear polarization $\parallel z$
	$\Delta m = \pm 1$	Linear polarization in (x, y) plane
	$\Delta m = +1$	σ^+ circular polarization
	$\Delta m = -1$	σ^- circular polarization
s	$\Delta s = 0$	
m_s	$\Delta m_s = 0$	

Parity

The parity of a function refers to the sign change under inversion about the origin. Thus if $f(-\boldsymbol{r}) = f(\boldsymbol{r})$, we have even parity, whereas if $f(-\boldsymbol{r}) = -f(\boldsymbol{r})$, we have odd parity. Now atoms are spherically symmetric, which implies that:

$$|\psi(-\boldsymbol{r})|^2 = |\psi(+\boldsymbol{r})|^2 \,. \tag{3.17}$$

Hence we must have that:

$$\psi(-\boldsymbol{r}) = \pm\psi(+\boldsymbol{r}) \,. \tag{3.18}$$

In other words, the wave functions have either even or odd parity. The dipole moment of the transition is given by Eq. (3.10). Since x, y, and z are odd functions, the product $\psi_1^*\psi_2$ must be an odd function if M_{12} is to be nonzero. Hence ψ_1 and ψ_2 must have different parities.

The Orbital Quantum Number l

The parity of the spherical harmonic functions is equal to $(-1)^l$. (Some examples demonstrating this are considered in Exercise 3.1.) Hence the parity selection rule implies that Δl must be an odd number. Detailed evaluation of the overlap integrals tightens this rule to $\Delta l = \pm 1$. This can be seen as a consequence of the fact that the angular momentum of a photon is $\pm\hbar$, with the sign depending on whether we have a left or right circularly polarized photon. Conservation of angular momentum therefore requires that the angular momentum of the atom must change by one unit.

The Magnetic Quantum Number m

The dipole moment for the transition can be written out explicitly:

$$M_{12} \propto \int_{r=0}^{\infty}\int_{\theta=0}^{\pi}\int_{\phi=0}^{2\pi} \Psi^*_{n',l',m'}\, \boldsymbol{r}\, \Psi_{n,l,m}\, r^2 \sin\theta \,\mathrm{d}r\,\mathrm{d}\theta\,\mathrm{d}\phi \,. \tag{3.19}$$

We consider here just the ϕ part of this integral:

$$M_{12} \propto \int_0^{2\pi} \mathrm{e}^{-\mathrm{i}m'\phi}\, \boldsymbol{r}\, \mathrm{e}^{\mathrm{i}m\phi}\,\mathrm{d}\phi \,, \tag{3.20}$$

where we have made use of the fact that (see Eqs. [2.52] and [2.42]):

$$\Psi_{n,l,m}(r,\theta,\phi) \propto \mathrm{e}^{\mathrm{i}m\phi} \,. \tag{3.21}$$

Consider z-polarized light. Since $z = r\cos\theta$ and has no dependence on ϕ, the dipole matrix element in Eq. (3.10) simplifies to:

$$M_{12} \propto \int_0^{2\pi} \mathrm{e}^{-\mathrm{i}m'\phi}\, z\, \mathrm{e}^{\mathrm{i}m\phi}\,\mathrm{d}\phi \propto \int_0^{2\pi} \mathrm{e}^{-\mathrm{i}m'\phi}\cdot 1 \cdot \mathrm{e}^{\mathrm{i}m\phi}\,\mathrm{d}\phi \,. \tag{3.22}$$

Hence, we must have that $m' = m$ if M_{12} is to be nonzero. If the light is polarized in the x direction, we have integrals like:

$$M_{12} \propto \int_0^{2\pi} \mathrm{e}^{-\mathrm{i}m'\phi}\, x\, \mathrm{e}^{\mathrm{i}m\phi}\, \mathrm{d}\phi \propto \int_0^{2\pi} \mathrm{e}^{-\mathrm{i}m'\phi} \cdot (\mathrm{e}^{+\mathrm{i}\phi} + \mathrm{e}^{-\mathrm{i}\phi}) \cdot \mathrm{e}^{\mathrm{i}m\phi}\, \mathrm{d}\phi\,. \quad (3.23)$$

This is because $x = r\sin\theta\cos\phi = r\sin\theta\,(\mathrm{e}^{+\mathrm{i}\phi} + \mathrm{e}^{-\mathrm{i}\phi})/2$. The integral is nonzero for $m' - m = \pm 1$, which implies $\Delta m = \pm 1$. A similar rule applies for y-polarized light, as $y = r\sin\theta\sin\phi = r\sin\theta\,(\mathrm{e}^{+\mathrm{i}\phi} - \mathrm{e}^{-\mathrm{i}\phi})/2\mathrm{i}$. The rule can be tightened by saying that $\Delta m = +1$ for σ^+ circularly polarized light propagating in the z direction, and $\Delta m = -1$ for σ^- circularly polarized light. (See Exercise 3.2.)

In the absence of an applied magnetic field (or some other perturbation that defines the z direction), the internal axes of the atom can be defined arbitrarily. The atom will therefore emit all possible polarizations, leading to the observation of $\Delta m = 0, \pm 1$ transitions. On the other hand, when the z-axis is defined by an external magnetic field, the direction of the electric field of the light relative to z is physically significant, and the Δm transitions have different polarizations. This point is developed in detail in Chapter 8, and is summarized in Table 8.2.

Spin

The photon does not interact with the electron spin. Therefore, the spin state of the atom does not change during the transition. This implies that the spin quantum numbers s and m_s are unchanged.

Example 3.1 What E1-allowed transitions are possible in emission for an electron in the following states of hydrogen: (a) 3s, (b) 3p, and (c) 3d?

Solution: The quantized levels of hydrogen with $n = 1$, 2, and 3 are illustrated in Figure 3.3. Here, the levels are displaced vertically according to their energy, and horizontally according to their l-value. (Such a diagram is called a **Grotrian diagram**.) The states are labeled by their n-value and the spectroscopic letter that indicates the l-value. (See Table 2.4.) As discussed in Section 2.2.5, l can take values from 0 to $(n-1)$. The levels with the same n but different l are degenerate in hydrogen, but this will not be the case for other atoms.

During emission, the electron moves downward in energy. On applying the selection rule on l, namely $\Delta l = \pm 1$, we realize that the electron can make transitions to any state of lower energy that differs in l by ± 1.

(a) A 3s electron has $l = 0$. It can therefore only make transitions to p states with $l = 1$. The only one available at lower energy is 2p. Hence there is only one possible transition: 3s $\rightarrow$ 2p.

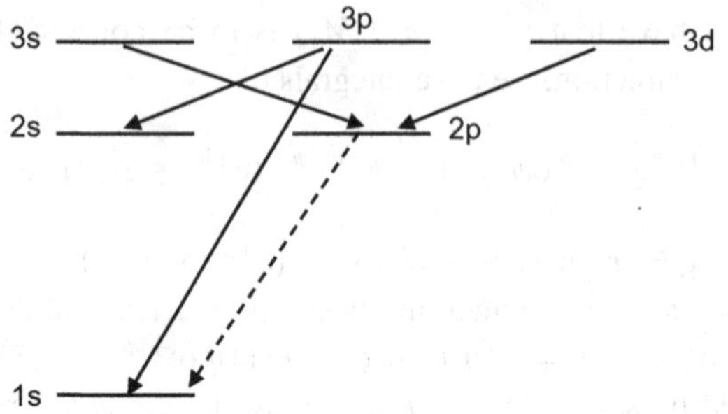

Figure 3.3 Allowed E1 transitions for the hydrogen $n = 3$ levels. The E1–allowed decay from the 2p level is shown by the dashed arrow.

(b) A 3p electron has $l = 1$. It can therefore make transitions to s states with $l = 0$, or to d states with $l = 2$. There are two s states available at lower energy, and no d states. Hence there are two possible transitions: 3p → 2s and 3p → 1s.

(c) A 3d electron has $l = 2$, and can therefore make transitions to p states with $l = 1$ or f states with $l = 3$. There is one p state available at lower energy, and no f states. Hence there is only one possible transition: 3d → 2p.

These four transitions are shown as arrows in Figure 3.3. In two cases, the electron ends up in the 2p state, which can decay to the 1s ground state by the allowed E1 transitions shown by the dashed arrow. In the case where the electron ends up in the 2s level, no further allowed E1 transitions are possible. In normal laboratory conditions, the atom could easily exchange angular momentum with another atom in an elastic collision, and then decay to the ground state from the 2p level. If collisions are unlikely, as they might be in some astrophysical environments, the atom in the 2s state decays by other, lower probability processes, as will be discussed in Section 12.4.

3.5 Higher-Order Transitions

An electron in an excited state has a natural tendency to drop to the lowest energy state. If direct decay is not possible, then it might be possible to decay by intermediate steps. For example, the 3s → 1s decay cannot occur by an E1 transition, but an atom in the 3s level can easily de-excite by two allowed E1 transitions: namely 3s → 2p, followed by 2p → 1s. This is not possible for states like the 2s level of hydrogen, as discussed in Example 3.1. Excited states that cannot relax to a lower level by an allowed E1 transition are said to be **metastable**.

An electron in a metastable state might be able to de-excite by making a **forbidden** transition. The use of the word "forbidden" is somewhat misleading here. It really means "electric-dipole forbidden." The transitions are perfectly possible; they just occur at a slower rate.

The next two strongest interactions between the photon and the atom, after the electric-dipole interaction, give rise to **magnetic dipole** (M1) and **electric quadrupole** (E2) transitions. These have different selection rules to E1 processes, and therefore may be allowed for metastable states when E1 transitions are forbidden. An example is the parity selection rule. Parity is unchanged in M1 and E2 transitions, and so transitions can occur between states of the same parity – unlike E1 processes. (The detailed selection rules for M1 and E2 transitions will be discussed in Section 12.2.2.) M1 and E2 transitions are second-order processes and have much smaller probabilities than E1 transitions. This means that they occur on a slower timescale.

In extreme cases it may happen that all types of radiative transitions involving the emission of a single photon are forbidden. In these cases, the atom must de-excite by transferring its energy to other atoms in collisional processes or by multi-photon emission.

3.6 Radiative Lifetimes

An atom in an excited state has a spontaneous tendency to de-excite by a radiative transition involving the emission of a photon. This follows from statistical physics: atoms with excess energy tend to want to get rid of it. This process is called **spontaneous emission**.

Consider a transition from an upper level labeled 2 to a lower level labeled 1, as, for example, in Figure 3.2(a). Let us suppose that there are N_2 atoms in level 2 at time t. We use quantum mechanics to calculate the transition rate from level 2 to 1, and then write down a rate equation for N_2 as follows:

$$\frac{\mathrm{d}N_2}{\mathrm{d}t} = -AN_2 \,. \tag{3.24}$$

This merely says that the total number of atoms making transitions is proportional to the number of atoms in the excited state and to the quantum mechanical probability. The parameter A that appears in Eq. (3.24) is determined by the transition probability and gives the spontaneous decay rate. It is called the **Einstein A coefficient** for the transition to distinguish it from the Einstein B coefficients for stimulated processes that will be discussed in Chapter 9.

Table 3.2 *Typical transition rates and radiative lifetimes for allowed and forbidden transitions at optical frequencies.*

Transition	Einstein A coefficient	Radiative lifetime
E1 allowed	$10^6 - 10^9$ s^{-1}	1 μs – 1 ns
E1 forbidden (M1 or E2)	$10^0 - 10^3$ s^{-1}	1 s – 1 ms

Eq. (3.24) has the following solution:

$$\begin{aligned} N_2(t) &= N_2(0)\exp(-At) \\ &= N_2(0)\exp(-t/\tau), \end{aligned} \tag{3.25}$$

where

$$\tau = \frac{1}{A}. \tag{3.26}$$

Equation (3.25) shows that the population of atoms in the upper level will decay by spontaneous emission with a time constant τ, which is called the **radiative lifetime** of the excited state.

The values of the Einstein A coefficient, and hence the radiative lifetime τ, vary considerably from transition to transition. Allowed E1 transitions have A coefficients in the range $10^6 - 10^9$ s^{-1} at optical frequencies, giving μs – ns radiative lifetimes. Forbidden transitions, on the other hand, are much slower because they are higher-order processes. The radiative lifetimes for M1 and E2 transitions are typically in the millisecond range or longer. This point is summarized in Table 3.2.

3.7 The Width and Shape of Spectral Lines

The radiation emitted in atomic transitions is not perfectly monochromatic. In this chapter we focus on the broadening mechanisms that affect the spectra of atomic gases. Section 11.1.2 in Chapter 11 considers how the principles developed here are adapted to the case where the atoms are embedded in solid-state environments.

The shape of the emission line is described by the **spectral line-shape function** $g(\nu)$. This is a function that peaks at the line center defined by

$$h\nu_0 = (E_2 - E_1), \tag{3.27}$$

and is normalized so that:

$$\int_0^\infty g(\nu)\,\mathrm{d}\nu = 1. \tag{3.28}$$

The most important parameter of the line-shape function is the **full width at half maximum** (FWHM) $\Delta\nu$, which quantifies the width of the spectral line. We shall see how the different types of line-broadening mechanisms give rise to two common line-shape functions, namely the **Lorentzian** and **Gaussian** functions.

In a gas of atoms, spectral lines are broadened by three main processes:

- natural broadening,
- collision broadening, and
- Doppler broadening.

We shall look at each of these processes separately below. A useful general division can be made at this stage by classifying the broadening as either **homogeneous** or **inhomogeneous**. Homogeneous processes affect all the individual atoms in the same way, with the natural and collision broadening mechanisms discussed in Section 3.8 and 3.9 being examples. All the atoms behave in the same way, and each atom produces the same emission spectrum. Inhomogeneous processes, by contrast, affect individual atoms in different ways. The Doppler broadening mechanism discussed in Section 3.10 is the standard example: the individual atoms are presumed to behave identically, but they are moving at different velocities, and one can associate different parts of the spectrum with the subset of atoms with the appropriate velocity. Inhomogeneous broadening is also found in solids, where different atoms may experience different local environments due to the inhomogeneity of the medium. (See Section 11.1.2.)

3.8 Natural Broadening

We have seen in Section 3.6 that the process of spontaneous emission causes the excited states of an atom to have a finite lifetime. Let us suppose that we somehow excite a number of atoms into level 2 at time $t = 0$. Equation (3.24) shows us that the rate of transitions is proportional to the instantaneous population of the upper level, and Eq. (3.25) shows that this population decays exponentially. Thus the rate of atomic transitions decays exponentially with time constant τ. For every transition from level 2 to level 1, a photon of angular frequency $\omega_0 = (E_2 - E_1)/\hbar$ is emitted. Therefore a burst of light with an exponentially decaying intensity will be emitted for $t > 0$:

$$I(t) = I(0)\exp(-t/\tau)\,. \tag{3.29}$$

This corresponds to a time-dependent electric field of the form:

$$\begin{aligned} t < 0: \quad & \mathcal{E}(t) = 0, \\ t \geq 0: \quad & \mathcal{E}(t) = \mathcal{E}_0 \, e^{-i\omega_0 t} \, e^{-t/2\tau}, \end{aligned} \tag{3.30}$$

where $\mathcal{E}_0$ is the amplitude at $t = 0$. The extra factor of 2 in the exponential in Eq. (3.30) compared to Eq. (3.29) arises because $I(t) \propto |\mathcal{E}(t)|^2$. The emission spectrum can be worked out by taking the Fourier transform of the electric field. (See Exercise 3.6.) The final result for the spectral line-shape function is:

$$g(\nu) = \frac{\Delta\nu}{2\pi} \frac{1}{(\nu - \nu_0)^2 + (\Delta\nu/2)^2}, \tag{3.31}$$

where the full width at half maximum is given by

$$\Delta\nu = \frac{1}{2\pi\tau}. \tag{3.32}$$

The spectrum described by Eq. (3.31) is called a **Lorentzian** line shape. This function is plotted in Figure 3.4.

It is interesting to rewrite Eq. (3.32) in the following form:

$$\Delta\nu \cdot \tau = \frac{1}{2\pi}. \tag{3.33}$$

By multiplying both sides by h, we can recast this as:

$$\Delta E \cdot \tau = h/2\pi \equiv \hbar. \tag{3.34}$$

If we realize that τ represents the average time the atom stays in the excited state (i.e., the uncertainty in the time), we can interpret this as the

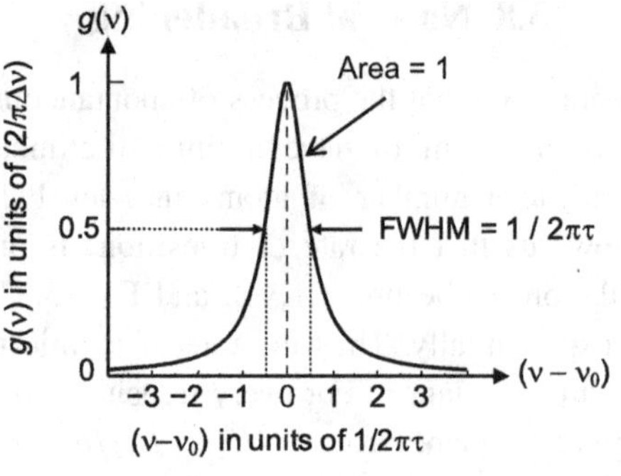

Figure 3.4 The Lorentzian line shape. The functional form is given in Eq. (3.31). The function peaks at the line center ν_0 and has an FWHM of $1/2\pi\tau$. The function is normalized so that the total area is unity.

energy–time uncertainty principle. Note that this has *nothing* to do with quantum mechanics: the derivation in Exercise 3.6 is completely classical.

3.9 Collision (Pressure) Broadening

The atoms in a gas jostle around randomly and frequently collide with each other and with the walls of the containing vessel. This interrupts the process of light emission and effectively shortens the lifetime of the excited state, resulting in additional broadening through Eq. (3.32), with τ replaced by τ_c, where τ_c is the mean time between collisions. It can be shown from the kinetic theory of gases that τ_c in an ideal gas is given by:

$$\tau_c \sim \frac{1}{\sigma_c P}\left(\frac{\pi m k_B T}{8}\right)^{1/2}, \tag{3.35}$$

where σ_c is the collision cross-section, m is the mass of the atom, T is the temperature, and P is the pressure. This shows that the collision rate τ_c^{-1} is proportional to P, which explains why collision broadening is sometimes called **pressure broadening**.

The collision cross section is an effective area that determines whether two atoms will collide or not. It will be approximately equal to the size of the atom. For example, for sodium atoms we have:

$$\sigma_c \sim \pi r_{\text{atom}}^2 \sim \pi \times (0.2\,\text{nm})^2 = 1.2 \times 10^{-19}\,\text{m}^2 .$$

At S.T.P., this gives $\tau_c \sim 6 \times 10^{-10}\,$s, which implies from Eq. (3.32) that $\Delta\nu \sim 0.3\,$GHz. Note that τ_c is much shorter than typical radiative lifetimes. For example, the strong yellow D-lines in sodium have a radiative lifetime of 16ns, which is nearly two orders of magnitude larger.

In conventional atomic discharge tubes, we reduce the effects of pressure broadening by working at low pressures. We see from Eq. (3.35) that this increases τ_c, and hence reduces the linewidth. This is why we tend to use low-pressure discharge lamps for spectroscopy.

3.10 Doppler Broadening

The spectrum emitted by a typical gas of atoms in a low-pressure discharge lamp is usually found to be much broader than the radiative lifetime would suggest, even when everything is done to avoid collisions. The reason for this

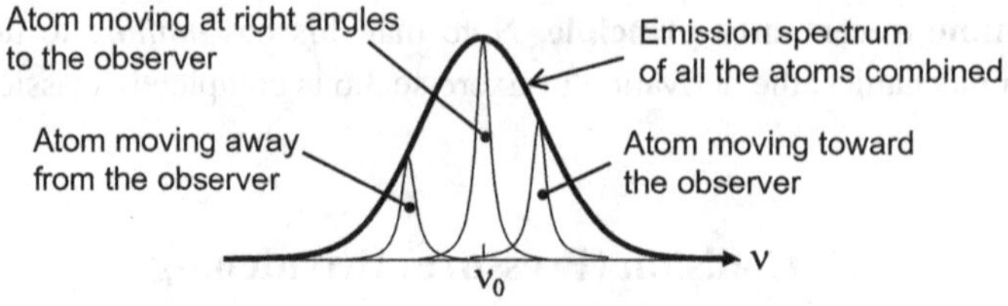

Figure 3.5 The Doppler broadening mechanism. Each individual atom is assumed to have a naturally broadened line, but the random thermal motion of the atoms causes their frequencies to be shifted by the Doppler effect.

discrepancy is the thermal motion of the atoms. The atoms in a gas move about randomly with a root-mean-square thermal velocity given by:

$$\frac{1}{2}mv_x^2 = \frac{1}{2}k_B T\,, \tag{3.36}$$

where k_B is Boltzmann's constant. At room temperature, the thermal velocities are quite large. For example, for sodium with an atomic weight of 23.0, we find $v_x \sim 330\,\text{ms}^{-1}$ at 300K. This random thermal motion gives rise to Doppler shifts in the observed frequencies, which then cause inhomogeneous line broadening, as illustrated in Figure 3.5.

Let us suppose that the atom is emitting light from a transition with center frequency ν_0. An atom moving with velocity v_x toward the observer will have its observed frequency shifted by the Doppler effect, according to:

$$\nu = \nu_0 \left(1 + \frac{v_x}{c}\right)\,. \tag{3.37}$$

The probability that an atom has velocity v_x is governed by the Boltzmann formula:

$$p(E) \propto \mathrm{e}^{-E/k_B T}\,. \tag{3.38}$$

On setting E equal to the kinetic energy, we find that the number of atoms with velocity v_x is given by the Maxwell–Boltzmann distribution:

$$N(v_x) \propto \exp\left(-\frac{mv_x^2}{2k_B T}\right)\,. \tag{3.39}$$

We can combine Eqs. (3.37) and (3.39) to find the number of atoms emitting at frequency ν:

$$N(\nu) \propto \exp\left(-\frac{mc^2(\nu-\nu_0)^2}{2k_B T \nu_0^2}\right)\,. \tag{3.40}$$

The frequency dependence of the light emitted is therefore given by:

$$I(\nu) \propto \exp\left(-\frac{mc^2(\nu-\nu_0)^2}{2k_B T \nu_0^2}\right). \tag{3.41}$$

This gives rise to a **Gaussian line shape** with $g(\nu)$ given by (see Exercise 3.7):

$$g(\nu) = \left(\frac{4\ln 2}{\pi}\right)^{1/2} \frac{1}{\Delta\nu_D} \exp\left(-\frac{mc^2(\nu-\nu_0)^2}{2k_B T \nu_0^2}\right), \tag{3.42}$$

where $\Delta\nu_D$ is the **Doppler linewidth** (i.e., the full width at half maximum):

$$\Delta\nu_D = 2\nu_0\left(\frac{(2\ln 2)k_B T}{mc^2}\right)^{1/2} = \frac{2}{\lambda}\left(\frac{(2\ln 2)k_B T}{m}\right)^{1/2}. \tag{3.43}$$

The value of $\Delta\nu_D$ at 300K is typically several orders of magnitude larger than the natural linewidth. (See, for example, Example 3.2.) The dominant broadening mechanism of low-pressure atomic gases at room temperature is therefore usually Doppler broadening, and the line shape is Gaussian rather than Lorentzian.

The advent of laser-cooling techniques have made it possible to reach extremely low temperatures in the micro-Kelvin range. (See Chapter 10.) Since $\Delta\nu_D$ is proportional to $\sqrt{T}$, this enables the natural Lorentzian shape of emission lines to be observed. (See Exercise 3.9.) Furthermore, cooling also reduces the collision broadening because $P \propto T$, and therefore $\tau_c \propto T^{-1/2}$. (See Eq. (3.35).)

Example 3.2 The radiative lifetime for the 632.8 nm line in neon is 2.95×10^{-7} s. What are the natural and Doppler linewidths at 300 K? What type of line shape would be expected at low pressure?

Solution: The natural linewidth is found from Eq. (3.32):

$$\Delta\nu = \frac{1}{2\pi \times 2.95 \times 10^{-7}} = 5.40 \times 10^5\,\text{Hz} \equiv 0.54\,\text{MHz}.$$

The Doppler linewidth can be worked out from Eq. (3.43) using the appropriate atomic weight of neon, namely 20.2:

$$\Delta\nu_D = \frac{2}{632.8 \times 10^{-9}}\left(\frac{(2\ln 2)\,k_B \times 300}{20.2 \times 1.67 \times 10^{-27}}\right)^{1/2} = 1.3\times10^9\,\text{Hz} \equiv 1.3\,\text{GHz}.$$

The Doppler linewidth is thus about three orders of magnitude larger than the natural linewidth. At low pressure, collisional broadening can be neglected. In these circumstances, the line shape would be Gaussian, which is appropriate for Doppler broadening, as opposed to Lorentzian.

3.11 Voigt Line Shapes

In the example just considered, the Doppler width was much larger than the natural linewidth, and so it was clear that the line shape would be Gaussian rather than Lorentzian. This will not always be the case, and there will some situations where the Lorentzian and Gaussian broadening mechanisms are of similar magnitude. One example of where this might occur is in a high-pressure lamp, where the collisional linewidth might be comparable to the Doppler linewidth. In these cases, the line shape is a convolution of the Gaussian and Lorentzian functions, and is called a **Voigt profile**.

3.12 Converting between Line Widths in Frequency and Wavelength Units

Spectral lines can be plotted against frequency, photon energy, wave number, or wavelength. Converting between linewidths for the first three of these examples presents no difficulty, since it just involves a linear scaling. (See Section 1.5.) However, converting to wavelengths is more complicated, because of the inverse relationship between wavelength and frequency.

Let us suppose that we have an atomic transition of center frequency ν_0 and FWHM $\Delta\nu$, where $\Delta\nu \ll \nu_0$. We convert to wavelengths through $\nu = c/\lambda$. This implies that:

$$\frac{\mathrm{d}\nu}{\mathrm{d}\lambda} = -\frac{c}{\lambda^2}, \tag{3.44}$$

and hence that the FWHM in wavelength units is given by:

$$\Delta\lambda = \left| -\frac{\lambda_0^2}{c}\Delta\nu \right| = \frac{\lambda_0^2}{c}\Delta\nu, \tag{3.45}$$

where $\lambda_0 = c/\nu_0$. A simple way of remembering this follows directly from Eq. (3.45), namely:

$$\frac{\Delta\lambda}{\lambda} = \frac{\Delta\nu}{\nu}, \tag{3.46}$$

where we have dropped the subscripts on the center frequency and wavelength.

Equations (3.45) and (3.46) work in the limit where $\Delta\nu \ll \nu_0$, or equivalently, $\Delta\lambda \ll \lambda_0$. In some cases (e.g., in molecular physics or solid-state physics), we might be considering a broad emission band rather than a narrow spectral line. In this situation, we have to go back to first principles to convert between frequency and wavelength units. Suppose that the emission band runs from frequency ν_1 to ν_2. The spectral width in wavelength units is then worked out from:

$$\Delta\lambda = |\lambda_2 - \lambda_1| = \left| \frac{c}{\nu_2} - \frac{c}{\nu_1} \right| . \tag{3.47}$$

Here, as in Eq. (3.45), the modulus is needed because an increase in frequency causes a decrease in wavelength, and vice versa. Note that Eq. (3.47) always works, and can be applied to the case of narrow spectral lines by putting $\nu_1 = \nu_0 - \Delta\nu/2$ and $\nu_2 = \nu_0 + \Delta\nu/2$, or, more easily, $\nu_1 = \nu_0$ and $\nu_2 = \nu_0 + \Delta\nu$. (See Exercise 3.12.)

Example 3.3 What is the spectral width in nm units of the neon 632.8 nm considered in Example 3.2?

Solution: The linewidth in frequency units was worked out in Example 3.2 to be 1.3×10^9 Hz. The linewidth in wavelength units is found from Eq. (3.45):

$$\Delta\lambda = \frac{(632.8 \times 10^{-9})^2}{c} \times 1.3 \times 10^9 = 1.7 \times 10^{-12}\,\text{m} = 0.0017\,\text{nm}.$$

This is much smaller than λ_0, and justifies the use of Eq. (3.45).

Exercises

3.1 In spherical polar coordinates, the parity operation $\boldsymbol{r} \to -\boldsymbol{r}$ corresponds $\theta \to (\pi - \theta)$ and $\phi \to (\pi + \phi)$; r is unchanged. Verify that the spherical harmonic functions listed in Table 2.2 have parity $(-1)^l$.

3.2 Explain why circularly polarized light can be written $\boldsymbol{\sigma}^{\pm} = \boldsymbol{x} \pm \mathrm{i}\boldsymbol{y}$. Show that the selection rules on m for σ^+ and σ^- light are $\Delta m = +1$ and $\Delta m = -1$, respectively. By writing linearly polarized light in terms of opposite circular polarizations, explain the selection rules for x or y linearly polarized light.

3.3 List the E1 transitions that can occur in emission from the 5d state of hydrogen. What are the wavelengths of the transitions?

3.4 What E1 absorption transitions can occur for a hydrogen atom in its ground state? What spectral series would be observed?

3.5 The Einstein A coefficient of the 589.6 nm transition in sodium is $6.14 \times 10^7\,\text{s}^{-1}$. A gas of sodium atoms is excited to the upper level of this transition at time $t = 0$. What fraction of the atoms are still in the upper level after 20 ns?

3.6 The spectrum of a time-varying source can be calculated by taking the Fourier transform of the electric field $\mathcal{E}(t)$ according to:

$$\mathcal{E}(\omega) = \frac{1}{\sqrt{2\pi}} \int_{-\infty}^{+\infty} \mathcal{E}(t)\, e^{i\omega t}\, dt\,.$$

Consider an optical source emitting a burst of radiation of angular frequency ω_0 that satisfies Eq. (3.30). The emission intensity $I(\omega)$ is proportional to $\mathcal{E}(\omega)^* \mathcal{E}(\omega)$.

(a) Show that $I(\omega)$ is given by:

$$I(\omega) = \frac{C}{(\omega - \omega_0)^2 + (1/2\tau)^2}\,,$$

where C is a constant.

(b) Rewrite $I(\omega)$ in terms of ν, and work out the frequencies at which $I(\nu)$ drops to half its maximum value. Derive Eq. (3.32).

(c) Apply the normalization condition given in Eq. (3.28) to derive Eq. (3.31).

3.7 Consider an atom emitting a Doppler-broadened line with $I(\nu)$ given by Eq. (3.41). This implies that the line-shape function $g(\nu)$ must be of the form:

$$g(\nu) = C \exp\left(-\frac{mc^2(\nu - \nu_0)^2}{2k_B T \nu_0^2}\right).$$

(a) Work out the frequencies at which the intensity drops to half its maximum value, and then derive Eq. (3.43).

(b) Find the value of the normalization constant C, to confirm that $g(\nu)$ in Eq. (3.42) satisfies Eq. (3.28).

3.8 Mercury (atomic weight 200.6) has a strong green line at 546.1 nm with an Einstein A coefficient of $4.87 \times 10^7\ s^{-1}$. What are the natural and Doppler linewidths at room temperature? Explain why low-pressure mercury lamps are popular for applications requiring narrow linewidths, as opposed to, say, neon or sodium lamps.

3.9 At what temperature would the Doppler linewidth of the neon 632.8 nm transition considered in Example 3.2 be equal to its natural linewidth?

3.10 Uranium (atomic weight 238.03) has a strong emission line at 424.4 nm with an Einstein A coefficient of $2.4 \times 10^7\ s^{-1}$. What is the width of this spectral line in wavelength units (nm):

(a) at very low temperatures, and

(b) at room temperature?

3.11 The hydrogen 2p $\rightarrow$ 1s transition at 121.57 nm consists of a fine-structure doublet separated by $0.366\,\mathrm{cm}^{-1}$. Up to what temperature would this doublet be resolvable ?

3.12 Put $\nu_2 = \nu_0 + \Delta\nu$ into Eq. (3.47) to derive Eq. (3.45) in the limit where $\nu_0 \gg \Delta\nu$.

3.13 A laser dye has a broad emission band from 550 to 650 nm. What is the spectral width of the band in Hz?

4
The Shell Model and Alkali Spectra

The solution of the Schrödinger equation in Chapter 2 allowed us to find the wave functions for hydrogenic atoms and to understand the meaning of the quantum numbers n, l, m_l, and m_s. Exact solutions were possible because we were dealing with a *two-body* system that consists of the nucleus and electron. This approach is not possible for the other elements, which are *many-body* systems that consist of one nucleus and many electrons. It is well known in classical physics that many-body problems have to be solved by approximation methods. For example, the Earth's orbit can be calculated exactly if we only consider the gravitational pull of the sun; but as soon as we throw in the other planets, then we have to use approximation techniques. The same applies to atoms. In this chapter, we consider the shell model, which is the first step to understanding the behavior of many-electron atoms.

4.1 The Central-Field Approximation

The Hamiltonian for an N-electron atom with nuclear charge $+Ze$ can be written in the form:

$$\hat{H} = \sum_{i=1}^{N} \left(-\frac{\hbar^2}{2m}\nabla_i^2 - \frac{Ze^2}{4\pi\epsilon_0 r_i} \right) + \sum_{i>j}^{N} \frac{e^2}{4\pi\epsilon_0 r_{ij}}, \tag{4.1}$$

where $N = Z$ for a neutral atom. The subscripts i and j refer to individual electrons and $r_{ij} = |\boldsymbol{r}_i - \boldsymbol{r}_j|$. The first summation accounts for the kinetic energy of the electrons and their Coulomb interaction with the nucleus, while the second accounts for the electron-electron repulsion.

It is not possible to find an exact solution to the Schrödinger equation with a Hamiltonian of the form given by Eq. (4.1) because the electron-electron repulsion term depends on the coordinates of two of the electrons, and so we cannot separate the wave function into a product of single-particle states. Furthermore, the electron-electron repulsion term is comparable in magnitude to the first summation, making it also impossible to use perturbation theory. The description of multi-electron atoms, therefore, usually starts with the **central-field approximation**.

A field is described as *central* if the potential energy has spherical symmetry about the origin, so that $V(\boldsymbol{r})$ only depends on r, and not on θ or ϕ. This means that the force is parallel to $\boldsymbol{r}$, i.e., it points centrally toward or away from the origin. In the case of a many-electron atom, we can get a good approximation to the net force that a specific electron experiences by treating all the other electrons as a charge cloud centered at the nucleus. In electromagnetism, a uniformly charged sphere can be considered as a point charge at the origin when outside the sphere. Similarly, the dominant repulsive force of the other electrons in the atom behaves like a point charge at the nucleus, leading to a radial force and a central field. In the central-field approximation, the off-central forces are assumed to be much smaller than the central ones, and so we determine the main structure of the atom by first solving a Schrödinger equation that only considers the central forces. A key concept that enters into the solution is **screening**, where we consider the effect of the other electrons on an individual electron as a negative charge cloud that screens the positive field produced by the nucleus.

In the central-field approximation, we rewrite the Hamiltonian of Eq. (4.1) in the form:

$$\hat{H} = \sum_{i=1}^{N} \left(-\frac{\hbar^2}{2m} \nabla_i^2 + V_{\text{central}}(r_i) \right) + V_{\text{residual}} \,, \tag{4.2}$$

where V_{central} is the central field, and V_{residual} is the **residual electrostatic interaction**. The residual electrostatic term accounts for the off-central forces, and the central-field approximation works in the limit where:

$$\left| \sum_{i=1}^{N} V_{\text{central}}(r_i) \right| \gg |V_{\text{residual}}| \,. \tag{4.3}$$

In this case, we can treat V_{residual} as a perturbation, and neglect it at this stage. We are then left with a Schrödinger equation in the form:

$$\hat{H}_{\text{central}} \Psi(\boldsymbol{r}_1, \boldsymbol{r}_2, \cdots \boldsymbol{r}_N) = E\, \Psi(\boldsymbol{r}_1, \boldsymbol{r}_2, \cdots \boldsymbol{r}_N) \,, \tag{4.4}$$

where

$$\hat{H}_{\text{central}} = \sum_{i=1}^{N} \hat{H}_i \,,$$

$$\hat{H}_i = -\frac{\hbar^2}{2m} \nabla_i^2 + V_{\text{central}}(r_i) \,. \tag{4.5}$$

Since the individual terms in Eq. (4.5) act on only one of the coordinates, they lead to N separate single-particle Schrödinger equations of the form:

$$\left(-\frac{\hbar^2}{2m}\nabla_i^2 + V_{\text{central}}(r_i)\right)\psi_i(\boldsymbol{r}_i) = E_i\,\psi_i(\boldsymbol{r}_i)\,. \qquad (4.6)$$

The solution of these single-particle Schrödinger equations is nontrivial. At this stage, however, all we need to know is that such solutions must exist. We can then form a many-particle wave function based on the single-particle wave functions by writing:

$$\Psi(\boldsymbol{r}_1, \boldsymbol{r}_2, \cdots \boldsymbol{r}_N) = \psi_1(\boldsymbol{r}_1)\,\psi_2(\boldsymbol{r}_2)\cdots\psi_N(\boldsymbol{r}_N)\,. \qquad (4.7)$$

Let us consider the effect of the first term in the Hamiltonian from Eq. (4.5) on this wave function:

$$\begin{aligned}\hat{H}_1\Psi(\boldsymbol{r}_1, \boldsymbol{r}_2, \cdots \boldsymbol{r}_N) &= \hat{H}_1\psi_1(\boldsymbol{r}_1)\,\psi_2(\boldsymbol{r}_2)\cdots\psi_N(\boldsymbol{r}_N)\,,\\ &= E_1\,\psi_1(\boldsymbol{r}_1)\,\psi_2(\boldsymbol{r}_2)\cdots\psi_N(\boldsymbol{r}_N),\\ &= E_1\,\Psi(\boldsymbol{r}_1, \boldsymbol{r}_2, \cdots \boldsymbol{r}_N)\,.\end{aligned}$$

The key step in the middle works because $\hat{H}_1$ only acts on the coordinates of electron 1. Similarly:

$$\begin{aligned}\hat{H}_2\Psi(\boldsymbol{r}_1, \boldsymbol{r}_2, \cdots \boldsymbol{r}_N) &= \psi_1(\boldsymbol{r}_1)\,\hat{H}_2\,\psi_2(\boldsymbol{r}_2)\cdots\psi_N(\boldsymbol{r}_N)\,,\\ &= \psi_1(\boldsymbol{r}_1)\,E_2\,\psi_2(\boldsymbol{r}_2)\cdots\psi_N(\boldsymbol{r}_N)\,,\\ &= E_2\,\Psi(\boldsymbol{r}_1, \boldsymbol{r}_2, \cdots \boldsymbol{r}_N)\,,\end{aligned}$$

and so on. We thus see that:

$$\begin{aligned}\hat{H}_{\text{central}}\Psi(\boldsymbol{r}_1, \boldsymbol{r}_2, \cdots \boldsymbol{r}_N) &= \hat{H}_1\Psi + \hat{H}_2\Psi + \cdots + \hat{H}_N\Psi\,,\\ &= E_1\Psi + E_2\Psi + \cdots + E_N\Psi\,,\\ &= (E_1 + E_2 + \cdots + E_N)\,\Psi\,,\\ &\equiv E\,\Psi(\boldsymbol{r}_1, \boldsymbol{r}_2, \cdots \boldsymbol{r}_N)\,,\end{aligned}$$

where

$$E = E_1 + E_2 \cdots E_N\,. \qquad (4.8)$$

This shows that the many-particle wave function in Eq. (4.7) is a solution of the many-particle, central-field Hamiltonian with total energy E equal to the sum of the single-particle energies. We might need a computer to solve any one of the single-particle Schrödinger equations of the type given in Eq. (4.6), but at least it is possible in principle, which leads to a tractable method to work out the energy states of the atom.

Before proceeding further, it is necessary to flag an issue that will become important later on in the book. The fact that electrons are indistinguishable particles means that we cannot distinguish physically between the case with electron 1 in state 1 and electron 2 in state 2, and vice versa. This means that the wave function in Eq. (4.7) is unphysical, and we should really write down a linear combination of all possibilities with the electron coordinates exchanged. We shall reconsider this point when discussing the helium atom in Chapter 6. At this stage, we just note the general point and move on.

The fact that the central-field potentials that appear in Eq. (4.6) only depend on the radial coordinate r_i (i.e., no dependence on the angles θ_i and ϕ_i) allows us to separate the wave function of each individual electron into a radial and angular part, just as we did for hydrogen. In analogy with Eq. (2.29), we then write:

$$\psi_i(\boldsymbol{r}_i) \equiv \psi_i(r_i, \theta_i, \phi_i) = R_i(r_i)\, Y_i(\theta_i, \phi_i)\,, \tag{4.9}$$

and proceed exactly as we did in Section 2.2. This leads to angular and radial equations for each electron, namely:

$$\hat{\boldsymbol{L}}_i^2 Y_{l_i m_i}(\theta_i, \phi_i) = \hbar^2 l_i(l_i+1) Y_{l_i m_i}(\theta_i, \phi_i)\,, \tag{4.10}$$

where $\hat{\boldsymbol{L}}_i$ is the angular momentum operator for the ith electron, and:

$$\left(-\frac{\hbar^2}{2m}\frac{1}{r_i^2}\frac{\mathrm{d}}{\mathrm{d}r_i}\left(r_i^2\frac{\mathrm{d}}{\mathrm{d}r_i}\right) + \frac{\hbar^2 l_i(l_i+1)}{2mr_i^2} + V_{\mathrm{central}}(r_i)\right) R_i(r_i) = E_i\, R_i(r_i)\,. \tag{4.11}$$

Equation (4.10) is exactly the same as Eq. (2.36), and the solution will therefore be a spherical harmonic function, $Y_{l_i m_i}(\theta_i, \phi_i)$, with orbital and magnetic quantum numbers l_i and m_i. It is worth noting that the central-field approximation naturally leads to states with well-defined orbital angular momentum. As noted in Section 2.2.3, the torque on the electron is zero if the force points centrally toward the nucleus, and this means that the orbital angular momentum must be constant.

Having established that the angular wave functions are spherical harmonics, and that l_i is an integer ≥ 0, we can then work out the quantized energies and radial wave functions by solving Eq. (4.11) for specific values of l_i. In doing so, we will inevitably have to introduce another quantum number n. The end result is that each electron in the atom has four quantum numbers:

- l and m_l: These drop out of the angular equation for each electron, namely Eq. (4.10).
- n: This arises from solving Eq. (4.11) for a given value of l with the appropriate form of $V_{\mathrm{central}}(r)$. Together, n and l determine the radial wave

function $R_{nl}(r)$ and the energy of the electron. The wave functions and energies are not expected to be the same as the hydrogenic ones given in Table 2.3 and Eq. (2.53) due to the different form of the central potential compared to the Coulombic $1/r$ dependence for hydrogen.

- m_s: Spin has not entered the argument. Each electron can therefore either have spin up ($m_s = +1/2$) or down ($m_s = -1/2$), as usual. We do not need to specify the spin quantum number s because it is always equal to 1/2.

The state of the many-electron atom is finally found by working out the wave functions of the individual electrons and finding the total energy according to Eq. (4.8), subject to the constraints imposed by the Pauli exclusion principle. This naturally leads to the shell model of the atom, which will be described in the following sections.

The details of how the central potential is worked out, and hence how the single-particle radial equation in Eq. (4.11) is solved, are beyond the scope of this book. In order to answer these questions, we need to know the electron probability density in order to work out the screening effect of the electrons on the nuclear field, and this requires prior knowledge of the wave functions. We thus have to find a self-consistent solution, which proceeds as follows:

(i) Make an initial guess of the wave functions of all the electrons in the atom.
(ii) Calculate the electron probability density from the wave functions, and use it to calculate their screening effect on the nuclear field.
(iii) Use this screened nuclear field as a first approximation for $V_{\text{central}}(r_i)$, and calculate $\psi_i(r_i)$ by solving Eq. (4.6) for each electron.
(iv) Use the revised wave functions to recalculate the screened nuclear field, and repeat the process.
(v) Keep iterating until the revised wave functions are the same as the original ones. At this point, a self-consistent solution has been obtained.

This is obviously a complicated process, and requires detailed numerical calculations – nowadays performed by computer. However, the key point is that the solutions do exist, and this process gives a methodology for understanding the atom. The shell model of the atom greatly simplifies the problem, as we shall now see.

4.2 The Shell Model and the Periodic Table

The shell model follows naturally from the central-field approximation, and its success in explaining the properties of atoms demonstrates that the approximation is a good one. The reason it works is based on the nature of the shells. An individual electron experiences the electrostatic potential due to the Coulomb repulsion from all the other electrons in the atom. Nearly all of the electrons in a many-electron atom are in filled core shells, which have spherically symmetric charge clouds. (The fact that filled shells are spherically symmetric is called **Unsöld's theorem**; see Exercise 4.1.) The off-radial forces from electrons in these closed shells cancel because of the spherical symmetry. Furthermore, the off-radial forces from electrons in unfilled shells are usually relatively small compared to the radial ones. We therefore expect the approximation given in Eq. (4.3) to be valid for most atoms. We can then proceed to find a self-consistent set of wave functions for the atom, and this will lead to a set of quantized energy states that have the following properties:

(i) The electronic states are specified by four quantum numbers: n, l, m_l, and m_s. The allowed values of these quantum numbers are given in Table 4.1.
(ii) The gross energy of the electron is determined by n and l – except in hydrogenic atoms, where E depends only on n. The energy levels increase with both n and l, with large jumps on moving to the next value of n for a given value of l, and smaller ones on increasing l for a given value of n.
(iii) In the absence of external magnetic fields, all the $\{m_l, m_s\}$ states with the same values of n and l are degenerate. Each (n, l) term, therefore, contains $2(2l + 1)$ degenerate levels. (See Table 2.4.) The degenerate states with the same values of n and l form a **shell**. The shells are labeled by the values of n and l, using spectroscopic notation for l; i.e., $l = 0$, 1, 2, 3, ... states are called $s, p, d, f, \ldots$ states. For example:

- the 1s shell has $n = 1$ and $l = 0$;
- the 2p shell has $n = 2$ and $l = 1$;
- the 4d shell has $n = 4$ and $l = 2$, etc.

The shell model of the atom is completed with the knowledge that electrons are indistinguishable, spin-1/2 fermionic particles. This means that they must obey the **Pauli exclusion principle**, which states that only one electron can

Table 4.1 *Quantum numbers for electrons in atoms.*

Quantum number	Symbol	Allowed values
Principal	n	Any integer > 0
Orbital	l	Integer from 0 to $(n - 1)$
Magnetic	m_l	Integer from $-l$ to $+l$
Spin	m_s	$\pm 1/2$

occupy a particular quantum state. If electrons were not fermions, they would all tend to go into the lowest energy shell to minimize the energy. However, the Pauli exclusion principle prevents this from happening. The 1s shell has the lowest energy, but only has two degenerate levels, and it can therefore only hold a maximum of two electrons. If the atom has more than two electrons, they have to go into higher energy shells. We then build up multi-electron atoms by adding electrons one by one, putting each electron into the lowest unfilled shell. Once the shell is full, the next electron has to go into the next unfilled shell at higher energy. The filling up of the shells in order of increasing energy in multi-electron atoms is sometimes called the **Aufbau principle**, from the German word *Aufbau*, meaning "building up."

The atomic shells are listed in order of increasing energy in Table 4.2. The ordering of the first few shells is straight forward, but it gets more complicated as n increases, as a shell with a large l value may have a higher energy than another one with a larger value of n but smaller value of l. For example, the 4s shell lies below the 3d shell, even though it has a larger value of n. The standard ordering can be remembered by following the scheme shown in Figure 4.1: the nl sub-shells are filled *diagonally* when laid out in rows determined by the principal quantum number n.

Table 4.2 *Atomic shells, listed in order of increasing energy.* N_{shell} *is equal to* $2(2l+1)$ *and is the number of electrons that can fit into the shell due to the degeneracy of the* m_l *and* m_s *levels. The last column gives the cumulative count of the number of electrons that can be held by the atom once the particular shell and all the lower ones have been filled.*

Shell	n	l	m_l	m_s	N_{shell}	N_{cum}
1s	1	0	0	$\pm1/2$	2	2
2s	2	0	0	$\pm1/2$	2	4
2p	2	1	$-1,0,+1$	$\pm1/2$	6	10
3s	3	0	0	$\pm1/2$	2	12
3p	3	1	$-1,0,+1$	$\pm1/2$	6	18
4s	4	0	0	$\pm1/2$	2	20
3d	3	2	$-2,-1,0,+1,+2$	$\pm1/2$	10	30
4p	4	1	$-1,0,+1$	$\pm1/2$	6	36
5s	5	0	0	$\pm1/2$	2	38
4d	4	2	$-2,-1,0,+1,+2$	$\pm1/2$	10	48
5p	5	1	$-1,0,+1$	$\pm1/2$	6	54
6s	6	0	0	$\pm1/2$	2	56
4f	4	3	$-3,-2,-1,0,+1,+2,+3$	$\pm1/2$	14	70
5d	5	2	$-2,-1,0,+1,+2$	$\pm1/2$	10	80
6p	6	1	$-1,0,+1$	$\pm1/2$	6	86
7s	7	0	0	$\pm1/2$	2	88

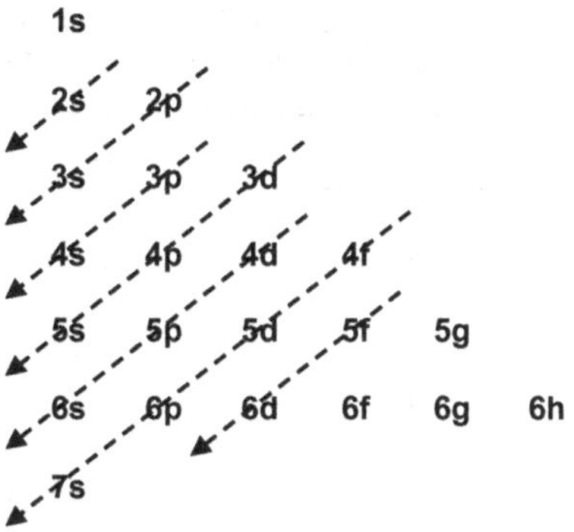

Figure 4.1 Atomic shells are filled in *diagonal* order when listed in rows, ordered according to the principal quantum number n.

Table 4.2 also shows the degeneracy ($\mathsf{N}_{\text{shell}}$) of the shell, together with the possible values of $\{m_l, m_s\}$. The final column shows the cumulative count N_{cum} of the total number of states with energy up to that particular shell. For example, N_{cum} is equal to 18 for the 3p shell, as the shell itself has 6 states and there are 12 more at lower energy in the 1s, 2s, 2p, and 3s shells, which have 2, 2, 6, and 2 states respectively. This value of N_{cum} determines the **electronic configuration** of the ground states of the elements, i.e., the quantum numbers of the electrons in the atom. This is done by filling up the shells according to the Aufbau principle until we run out of electrons.

The configurations of the first 11 elements are given in Table 4.3. The superscript attached to the shell tells us how many electrons are in a particular shell. Thus for sodium (Na) we have a configuration of $1s^2\ 2s^2\ 2p^6\ 3s^1$, i.e., two electrons in the 1s and 2s shells, six in the 2p shell, and one in the 3s shell. This gives a total of 11 electrons, as expected for a neutral atom with $Z = 11$.

The **periodic table** of the elements follows directly from their electronic configurations. A conventional periodic table showing the chemical symbols of the elements and their atomic numbers is given at the front of the book. Figure 4.2 shows a variation of the periodic table that focuses on the electronic configurations. The elements are arranged into 18 columns that identify specific groups. Each group is identified with a number from 1–18, as shown at the top. The line underneath shows the older convention with eight principal groups labeled 1A–8B, often written with Roman numerals: IA–VIIIB. In this older convention, the ten transition-metal groups in the middle are labelled IIIA–IIB. The **lanthanide** and **actinide** elements are shown at the bottom and fit in to the gaps marked with asterisks. Some groups of elements have special names. For example, the group 1 elements (with the exception

Table 4.3 *The electronic configuration of the first 11 elements of the periodic table. Z is the atomic number of the element.*

Element	*Z*	Electronic configuration
H	1	$1s^1$
He	2	$1s^2$
Li	3	$1s^2\ 2s^1$
Be	4	$1s^2\ 2s^2$
B	5	$1s^2\ 2s^2\ 2p^1$
C	6	$1s^2\ 2s^2\ 2p^2$
N	7	$1s^2\ 2s^2\ 2p^3$
O	8	$1s^2\ 2s^2\ 2p^4$
F	9	$1s^2\ 2s^2\ 2p^5$
Ne	10	$1s^2\ 2s^2\ 2p^6$
Na	11	$1s^2\ 2s^2\ 2p^6\ 3s^1$

Transition metals (groups 3–12)

1	2	3		4	5	6	7	8	9	10	11	12	13	14	15	16	17	18
1A	2A												3B	4B	5B	6B	7B	8B
$1s^1$																		$1s^2$
$2s^1$	$2s^2$												$2p^1$	$2p^2$	$2p^3$	$2p^4$	$2p^5$	$2p^6$
$3s^1$	$3s^2$												$3p^1$	$3p^2$	$3p^3$	$3p^4$	$3p^5$	$3p^6$
$4s^1$	$4s^2$	$3d^1$		$3d^2$	$3d^3$	**$4s^1$**	$3d^5$	$3d^6$	$3d^7$	$3d^8$	**$4s^1$**	$3d^{10}$	$4p^1$	$4p^2$	$4p^3$	$4p^4$	$4p^5$	$4p^6$
$5s^1$	$5s^2$	$4d^1$		$4d^2$	†	**$5s^1$**	$4d^5$	†	†	†	**$5s^1$**	$4d^{10}$	$5p^1$	$5p^2$	$5p^3$	$5p^4$	$5p^5$	$5p^6$
$6s^1$	$6s^2$	**$5d^1$**	*	$5d^2$	$5d^3$	$5d^4$	$5d^5$	$5d^6$	$5d^7$	†	**$6s^1$**	$5d^{10}$	$6p^1$	$6p^2$	$6p^3$	$6p^4$	$6p^5$	$6p^6$
$7s^1$	$7s^2$	**$6d^1$**	**															

*	†	$4f^3$	$4f^4$	$4f^5$	$4f^6$	$4f^7$	**$5d^1$**	$4f^9$	$4f^{10}$	$4f^{11}$	$4f^{12}$	$4f^{13}$	$4f^{14}$	$5d^1$	Lanthanides
**	$6d^2$	†	†	†	$5f^6$	$5f^7$	**$6d^1$**	$5f^9$	$5f^{10}$	$5f^{11}$					Actinides

Figure 4.2 Periodic table showing the electronic configurations of the valence electrons. The shells are filled according to the pattern given in Figure 4.1, apart from the cases shown in bold or marked by a † symbol. The modern chemical numbering of the groups is indicated at the top, together with the older convention underneath.

of element 1, namely hydrogen) are called **alkali metals**, while those in group 18 are called **noble gases**.

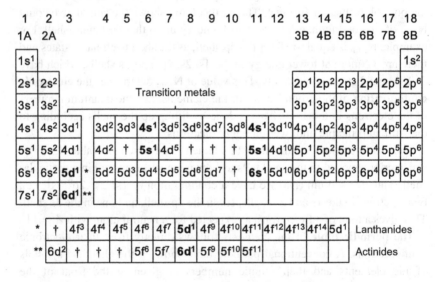

The electronic configurations given in Figure 4.2 correspond to the ground states of the elements. In each case, only the occupancy of the outermost shell is shown. All the other shells are occupied and have been filled according to the pattern shown in Figure 4.1. Thus, for example, the configuration of the first transition metal (i.e., scandium, $Z = 21$) is written $3d^1$. This assumes that the 1s, 2s, 2p, 3s, 3p, and 4s shells have been filled, so that the full configuration is $1s^2 2s^2 2p^6 3s^2 3p^6 4s^2 3d^1$. When such a configuration is written, the filled shells are often designated by the corresponding noble gas element. Thus, scandium could be written [Ar] $4s^2 3d^1$, where [Ar] indicates the configuration of argon ($Z = 18$) – namely, $1s^2 2s^2 2p^6 3s^2 3p^6$. Note that the number of electrons in the outermost shell determines the chemical valency of the element.

Close inspection of Figure 4.2 reveals that there are a few exceptions to the general shell-filling rule shown in Figure 4.1. The main exceptions are shown in bold typeface and occur when the filling order of the last two shells has been switched. An obvious example is group 11, which contains copper (Cu), silver (Ag), and gold (Au). Consider the case of copper, which has 29 electrons in a configuration of [Ar] $4s^1 3d^{10}$ instead of [Ar] $4s^2 3d^9$. The filled $3d^{10}$ shell is very stable, and so the [Ar] $4s^1 3d^{10}$ configuration actually has a lower energy than [Ar] $4s^2 3d^9$. The energy difference is not particularly large, which explains why copper sometimes behaves as though it is monovalent, and other times divalent. Silver and gold follow a similar pattern, with configurations of [Kr] $5s^1 4d^{10}$ and [Xe] $6s^1 5d^{10}$, respectively. Elements 24 and 42 in group 6 opt for a half-filled d-shell instead of a filled s-shell, giving them configurations of [Ar] $4s^1 3d^5$ and [Kr] $5s^1 4d^5$, respectively. The half-filled configuration has all the electron spins aligned, which can be energetically favorable. (See the discussion of Hund's rules in Section 5.9.) Gadolinium (Gd, $Z = 64$) and curium (Cm, $Z = 96$) opt similarly for half-filled f-shells, giving them configurations of [Xe] $6s^2 5d^1 4f^7$ and [Rn] $7s^2 6d^1 5f^7$, respectively.

The exceptions marked with the † symbol follow more complicated patterns. For example, niobium ($Z = 41$) has a configuration of [Kr] $5s^1 4d^4$, instead of [Kr] $5s^2 4d^3$. There is no simple reason that can be given to explain why this happens, other than to point out that the levels are sometimes close in energy, and so it is not surprising that there are occasional departures from the empirical rule shown in Figure 4.1.

Example 4.1 Write down the electronic configuration of yttrium, which has $Z = 39$.

Solution: Yttrium has 39 electrons, arranged in shells according to the sequence shown in Table 4.2. The value of N_{cum} is 38 for the 5s shell, and so this will be the last-filled shell, with the 39th electron going into the 4d shell. The configuration is thus $1s^2 2s^2 2p^6 3s^2 3p^6 4s^2 3d^{10} 4p^6 5s^2 4d^1$, or [Kr] $5s^2 4d^1$ for short.

4.3 Justification of the Shell Model

The theoretical justification for the shell model relies on the concept of **screening**. The idea is that the electrons in the inner shells screen the outer electrons from the potential of the nucleus. To see how this works, we take sodium as an example.

Sodium has an atomic number of 11, and therefore has a nuclear charge of $+11e$. The neutral atom has 11 electrons, and these are arranged in shells around the nucleus, as shown schematically in Figure 4.3. Rough values for the radii and energies of the electrons in their shells can be obtained by using the Bohr formula:

$$r_n = \frac{n^2}{Z} a_0 , \tag{4.12}$$

$$E_n = -\left(\frac{Z}{n}\right)^2 R_{\mathrm{H}} , \tag{4.13}$$

where $a_0 = 5.29 \times 10^{-11}$ m is the Bohr radius of hydrogen, $R_{\mathrm{H}} = 13.6$ eV is the Rydberg energy, and Z is the atomic number.

The first two electrons go into the $n = 1$ shell. These electrons see the full nuclear charge of $+11e$. With $n = 1$ and $Z = 11$, we find $r_1 \sim 1^2/11 \times a_0 = 0.05$ Å and $E_1 \sim -11^2 R_{\mathrm{H}} = -1650$ eV. The next eight electrons go into the $n = 2$ shell. There are actually two sub-shells, namely 2s and 2p, but we ignore this level of detail at this stage. The $n = 2$ electrons are presumed to orbit outside the $n = 1$ shell, and the two inner electrons partly screen the nuclear charge. The $n = 2$ electrons therefore see an effective nuclear charge $Z_{\mathrm{eff}} \sim (+11 - 2)\, e = +9e$. The radius is therefore $r_2 \sim (2^2/9) \times a_0 = 0.24$ Å and the energy is $E_2 \sim -(9/2)^2 R_{\mathrm{H}} = -275$ eV. Finally, the outermost electron in the $n = 3$ shell orbits outside the filled $n = 1$ and $n = 2$ shells, and it therefore sees $Z_{\mathrm{eff}} \sim (+11 - 2 - 8)\, e = +1e$. With $Z_{\mathrm{eff}} \sim 1$ and $n = 3$ we

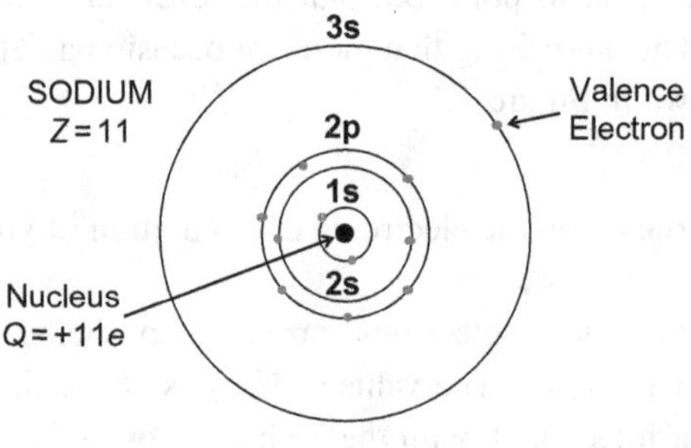

Figure 4.3 Schematic representation of the electronic configuration of the sodium atom according to the shell model. The orbital radii are not to scale, and should not be taken literally.

Table 4.4 *Radii and energies of the principal atomic shells of sodium according to the Bohr model. The unit of 1 Ångstrom* (Å) = 10^{-10} *m. The final column gives the energies deduced from experimental data.*

Shell	n	Z_{eff}	radius (Å)	Energy (eV)	Experimental energy (eV)
1s	1	11	0.05	−1650	−1072
2s, 2p	2	9	0.24	−275	−64, −31
3s	3	1	4.8	−1.5	−5.1

find $r_3 \sim 4.8$ Å and $E_3 \sim -1.5$ eV. These values are summarized in Table 4.4. Note the large jump in energy and radius in moving from one shell to the next.

The treatment of the screening based on Bohr-type orbits is clearly oversimplified, as it does not allow for electron probability distributions, and therfore does not treat the electron-electron repulsion properly. This is evident from the final column in Table 4.4, which shows the actual values of the energies deduced from X-ray and optical data. The experimental values differ significantly from the Bohr-orbit ones – which is not surprising, given the simplicity of the Bohr model. Nevertheless, the basic point stands. The inner shells screen the outer ones, and this leads to big jumps in energy on moving from one shell to the next. The model is therefore reasonably self-consistent, with electrons layered in shells of increasing radius and decreasing binding energy around the nucleus.

4.4 Experimental Evidence for the Shell Model

There is a wealth of experimental evidence to confirm that the shell model is a good one. The main points are discussed briefly here.

4.4.1 The Periodic Table of Elements

The periodic table of elements follows from the electronic configurations of the elements, which is derived from the shell structure of atoms. The periodic table underpins the chemical activity of the elements. It can thus be argued that the whole subject of chemistry can be regarded as experimental proof of the shell structure of atoms.

4.4.2 Ionization Potentials and Atomic Radii

The ionization potentials of the noble gas elements are the highest within a particular period of the atomic table, while those of the alkali metals are the

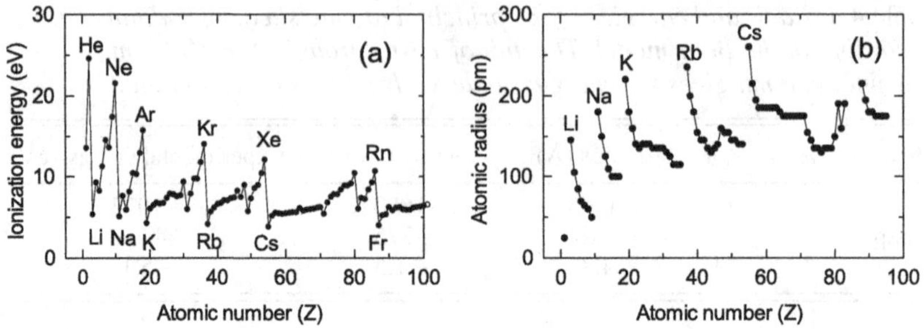

Figure 4.4 (a) First ionization potentials of the elements up to $Z = 100$. The noble gas elements (He, Ne, Ar, Kr, Xe, Rn) have highly stable, fully filled shells with large ionization potentials. The alkali metals (Li, Na, K, Rb, Cs, Fr) have one weakly bound valence electron outside fully filled shells. (b) Atomic radius versus Z. Data from Kramida et al. (2016) and Slater (1964).

lowest. This can be seen by looking at the data in Figure 4.4(a). The ionization potential gradually increases as a shell is being filled across a row of the periodic table due to the increase in Z, and hence the nuclear charge. Once the shell is filled, and the outermost electron shifts to the next shell, the ionization energy drops abruptly. This shows that the filled shells are very stable, and that the valence electrons go in larger, less tightly bound orbits. The results correlate with the chemical activity of the elements. The noble gases require large amounts of energy to liberate their outermost electrons, and they are therefore chemically inert. The alkali metals, on the other hand, need much less energy, and are therefore highly reactive.

An opposite trend is observed in the atomic radius, as shown in Figure 4.4(b). These data show that the radius is largest for the alkali metals, and then drops as a shell is being filled up. The decrease in radius across a shell is caused by the increase in Z, with the larger nuclear charge pulling the electrons more closely to the nucleus. Once a shell is full, the electron has to go to the next shell, causing a jump in its radius due to the larger value of n. The jump in radius when the outermost electron is pushed into a higher shell shows the screening effect of the inner shell electrons, and indicates that we have weakly bound valence electrons outside strongly bound, small-radius inner shells.

4.4.3 X-Ray Spectra

Measurements of X-ray spectra allow the energies of the inner shells to be determined directly. The energy of an electron in an inner shell with principal quantum number n is given by:

$$E_n = -\frac{Z_n^{\text{eff}2}}{n^2} R_{\text{H}}, \tag{4.14}$$

where Z_n^{eff} is the effective nuclear charge and $R_{\text{H}} = 13.6\,\text{eV}$. The difference between Z and Z_n^{eff} is caused by the screening effect of the other electrons. This is conveniently expressed by writing:

$$Z_n^{\text{eff}} = Z - \sigma_n, \tag{4.15}$$

where σ_n is a phenomenological **screening parameter** that accounts for the screening of the nucleus by the other electrons. In X-ray notation, it is customary to indicate the value of n by a letter (K, L, M, N, ...) corresponding to $n = 1, 2, 3, 4, \ldots$, respectively. This old spectroscopic notation dates back to the early work on X-ray spectra.

As an example of how this works, consider the $n = 1$ shell of sodium (i.e., the K shell). If we take Z_n^{eff} equal to Z (i.e., 11), we find $E_1 = -1650\,\text{eV}$, as discussed in Section 4.3. However, the actual value of E_1 is $-1072\,\text{eV}$ (see Table 4.4), which implies $Z_1^{\text{eff}} = 8.88$, and hence $\sigma_1 = 2.12$. This value of σ_1 includes the screening effect of the other electron in the K-shell, plus the remaining nine electrons in higher shells. The calculation of the screening parameter σ requires knowledge of the wave functions. The majority of the probability density of the L- and M-shell electrons will reside outside the K-shell, but there will be *some* probability that they actually lie closer to the nucleus. Hence the outer-shell electrons can have a small screening effect on an inner shell, and if there lots of them, this can amount to a significant effect. The value of $\sigma_1 = 2.12$ for sodium includes a contribution of $\sim$1 from the other K-shell electron, and $\sim$1 for the eight electrons in the L-shell. (The screening effect of the single electron in the M-shell would be very small.)

The values of σ_n for the various elements can be deduced from analysis of X-ray spectra. Here, we consider how this is done first for emission spectra, and then for absorption spectra. The values that are obtained can then be compared to detailed self-consistent theoretical calculations of the wave functions. In this way, a detailed picture of the inner-shell wave functions of a multi-electron atom can be developed.

The experimental arrangement for observing an X-ray emission spectrum is shown in Figure 4.5(a). Electrons are accelerated across a potential drop of several kV and then impact on a target. This ejects core electrons from the inner shells of the target, as shown in Figure 4.5(b). X-ray photons are emitted as the higher energy electrons drop down to fill the empty level (or **hole**) in the lower shell. Each target emits a series of characteristic lines, with each series designated by the lower shell. Thus, for example, the X-rays emitted when an

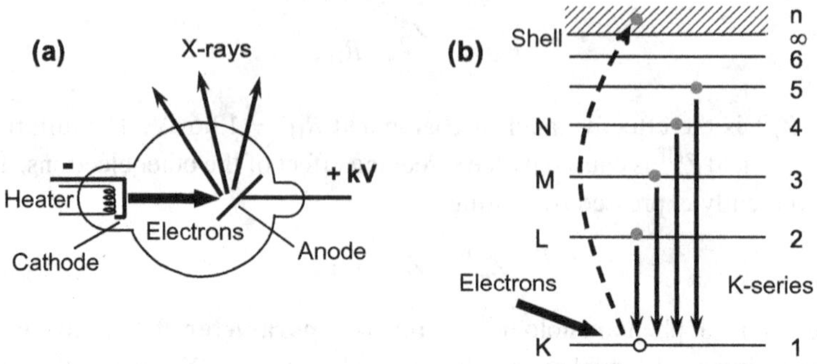

Figure 4.5 (a) A typical X-ray tube. Electrons are accelerated with a voltage of several kV and impact on a target, causing it to emit X-rays. (b) Transitions occurring in the K-series emission lines. An electron from the discharge tube ejects one of the K-shell electrons of the target, leaving an empty level in the K-shell. X-ray photons are emitted as electrons from the higher shells drop down to fill the hole in the K-shell. Note that the vertical energy scale is *not* linear.

electron jumps from a higher shell to fill a hole in the K-shell ($n = 1$) would be called the K-series, as illustrated in Figure 4.5(b). Similarly, the L- and M-series correspond to transitions ending at the L-shell ($n = 2$) or M-shell ($n = 3$), respectively. It is important to realize that the vertical energy scale in Figure 4.5(b) is only schematic. The energy gap from K to L is *much* larger than the gap from L to M.

Figure 4.6(a) shows a typical X-ray emission spectrum. The spectrum consists of a series of sharp lines on top of a continuous spectrum. The groups of sharp lines are generated by radiative transitions following the ejection of an inner-shell electron. A particular set of lines is only observed if the tube voltage is high enough to eject the relevant electron. Figure 4.6(a) shows the spectrum of gold at 20 kV and 30 kV. Neither of these voltages is sufficient to eject a K-shell electron, and so the lines correspond to the L-series, as shown in the inset. (We shall see when discussing Figure 4.6(b) that we would need more than 81 kV to eject a K-shell electron.) Once an L-series transition has occurred, the hole in the L-shell moves to the upper level, e.g., to the M-shell after an M $\rightarrow$ L transition. We should then see M-series X-ray transitions, which would occur at lower energies (~2 keV). These M-series lines are not observed in Figure 4.6(a), as the air absorbs very strongly at lower energies due to X-ray absorption by the oxygen and nitrogen atoms.

The energy of an X-ray transition from $n \rightarrow n'$ is given by:

$$h\nu = |E_{n'} - E_n| = \left| -\frac{Z_{n'}^{\mathrm{eff}\,2}}{n'^2} + \frac{Z_n^{\mathrm{eff}\,2}}{n^2} \right| R_{\mathrm{H}} . \quad (4.16)$$

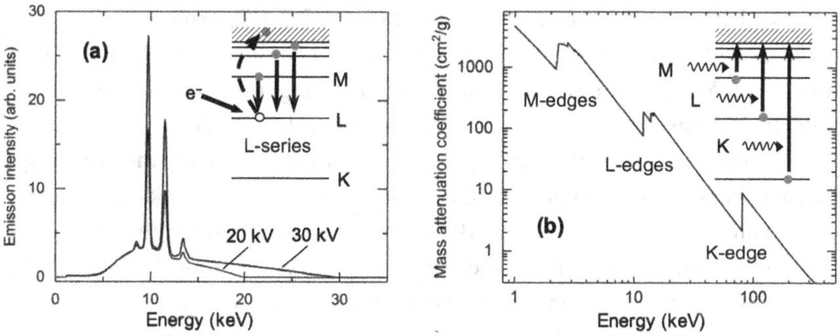

Figure 4.6 X-ray spectra of gold ($Z = 79$). (a) Emission spectra at two different electron voltages. The sharp lines are caused by L-series transitions after an L-shell electron has been ejected, as shown in the inset. The continuum is caused by bremsstrahlung. (b) Mass attenuation coefficient, which is proportional to the absorption coefficient. The inset indicates the transitions that occur at the edges. In both insets, the vertical energy scale is not linear. Data in (a) courtesy of Amptek, Inc. Data in (b) from Hubbell and Seltzer (2004).

In practice, the wavelengths of the various series of emission lines are found to obey **Moseley's law**, where we make the approximation $Z_n^{\text{eff}} = Z_{n'}^{\text{eff}}$ and write both as $(Z - \sigma_n)$. For example, the K-shell lines are given by:

$$\frac{hc}{\lambda} \approx (Z - \sigma_K)^2 R_H \left(\frac{1}{1^2} - \frac{1}{n^2} \right), \tag{4.17}$$

where $n > 1$ and $\sigma_K \sim 3$ for heavy atoms. Similarly, the L-shell spectra obey:

$$\frac{hc}{\lambda} \approx (Z - \sigma_L)^2 R_H \left(\frac{1}{2^2} - \frac{1}{n^2} \right), \tag{4.18}$$

where $n > 2$, and $\sigma_L \sim 10$. We can see that these are just the expected wavelengths predicted by the Bohr model, except that we have an effective charge of $(Z - \sigma_n)$ instead of Z. There is no real scientific justification for the approximation $Z_n^{\text{eff}} = Z_{n'}^{\text{eff}}$ in Moseley's law. The law is an empirical one and reflects the fact that the transition wavelength is mainly dominated by the energy of the lower shell.

In applying Moseley's law to the L-series of gold ($Z = 79$) with $\sigma_L = 10$, we find $h\nu = 9.0\,\text{keV}$, $12\,\text{keV}$, and $14\,\text{keV}$, for the $n = 3 \to 2$, $4 \to 2$, and $5 \to 2$ transitions respectively. These energies are in the right spectral range as the lines observed in Figure 4.6(a), but careful analysis requires a more detailed approach. This is because all shells above K are split into sub-shells: the L-shell has two (2s and 2p), the M-shell has three (3s, 3p, and 3d), etc. The spectra are further complicated by spin-orbit coupling, which splits sub-shells with $l \geq 1$ into doublets. (See Chapter 7, Section 7.7.) The spin-orbit splitting increases with Z, and so can be quite large for heavy atoms. (See Section 7.3.3.) In practice, Moseley's law is normally sufficient to get the rough wavelengths of the transitions correct – in this case in the range 1.0–1.3 Å.

The continuous spectrum in Figure 4.6(a) is caused by **bremsstrahlung**. This name comes from combining the German words *brems* (braking, or deceleration) and *strahlung* (radiation). Bremsstrahlung occurs when the electron is scattered by the atoms without ejecting a core electron from the target. The acceleration of the electron associated with its change of direction causes it to radiate. Conservation of energy demands that the photon energy must be less than eV, where V is the voltage across the tube, which corresponds to a minimum wavelength of hc/eV. The increase of the maximum bremsstrahlung energy with voltage is clearly shown in Figure 4.6(a). The cutoff at lower energies is caused by atmospheric absorption.

Figure 4.6(b) shows the X-ray absorption spectrum of gold up to 300 keV. The data actually plots the mass attenuation coefficient, which is equal to α/ρ, where α is the absorption coefficient and ρ is the mass density. The spectrum consists of a series of **absorption edges** followed by decreasing absorption. The edges occur whenever the incoming photon has enough energy to promote an electron from an inner shell to empty states above the highest occupied shell, as sketched in the inset. The final state for the electron after the absorption transition could either be one of the excited states of the valence electrons or a continuum state above the ionization limit. The absorption probability decreases above the edge due to the reduced overlap between the initial localized electron wave function and the delocalized states in the continuum.

The energy of the absorption edge can be worked out from Eq. (4.16). Since the binding energy of the valence electrons is negligible on the scale of X-ray energies, we can effectively put $E_{n'} = 0$ in Eq. (4.16). The edge therefore occurs at:

$$h\nu^{\text{edge}} = \frac{Z_n^{\text{eff}\,2}}{n^2} R_{\text{H}} \equiv \frac{(Z - \sigma_n)^2}{n^2} R_{\text{H}} \,. \tag{4.19}$$

The absorption edge thus gives a direct measurement to the inner-shell energy. Figure 4.6(b) shows that the K-shell energy of gold is −80.7 keV. This implies that a voltage ≥ 80.7 kV must be applied to eject a K-shell electron, which explains why no K-shell emission lines were observed in Figure 4.6(a).

It is apparent in Figure 4.6(b) that there is substructure in the L-edge, but not in the K-edge. As discussed earlier, the L-shell is split into the 2s and 2p states, corresponding to $l = 0$ and 1, respectively. These have slightly different screening parameters, and hence slightly different energies, on account of the different shape of their radial wave functions. We therefore get separate L-shell edges for these two states. The K-shell, by contrast, can only have $l = 0$, and thus consists of a unique state, namely the 1s level. Close inspection of the data reveals that the 2p sub-shell at higher energy is split into a doublet by the spin-orbit effect, as discussed earlier in this section and in Section 7.7. Hence the L-edge has three sub-edges at 11.9, 13.7, and 14.4 keV. The M-shell also has multiple sub-edges, due to the 3s, 3p, and 3d states, the latter two being further split into doublets by spin-orbit coupling.

Detailed lists of X-ray transition energies may be found online on the NIST X-Ray Transition Energies Database. (See Deslattes et al., 2005.) Other useful X-ray information can be found in the Lawrence Berkeley National Laboratory's X-ray data book. (See Thompson, 2009.)

Example 4.2 An X-ray tube has a molybdenum target ($Z = 42$). (a) Estimate the wavelength of the longest wavelength K-shell line. (b) Estimate the tube voltage that would have to be applied to observe the line.

Solution: (a) The longest wavelength K-shell transition corresponds to $n = 2 \rightarrow 1$. On substituting into Moseley's law (Eq. [4.17]) with $\sigma_K = 3$, we find $h\nu = 16\,\text{keV}$. This corresponds to a wavelength of 0.80 Å.
(b) The K-shell lines will only be observed if the tube voltage is sufficient to eject a K-shell electron. This requires that $eV \geq |E_K|$. On substituting into Eq. (4.14) with $Z_K^{\text{eff}} = 39$, as appropriate for $\sigma_k = 3$, we find $V \geq 21\,\text{kV}$.

The experimental values for (a) and (b) are 0.71 Å and 20.0 kV, respectively. The Mo K_α ($n = 2 \rightarrow 1$) line is widely used in medical X-ray imaging. The absorption edge implies that the actual value of σ_K is 3.65.

Example 4.3 The L_1, L_2, and L_3 absorption edges of gold occur at 14.4, 13.7, and 11.9 keV, respectively. (See Figure 4.6(b).) What are the screening parameters of these sub-shells?

Solution: The screening parameters can be worked out using eqn 4.19. For an L-shell we have $n = 2$, and for gold we have $Z = 79$. Hence we obtain screening parameters of 13.9, 15.5, and 19.8 for the L_1, L_2 and L_3 sub-shells, respectively. Note that these are larger than the value of $\sigma_L \sim 10$ quoted in the discussion of Moseley's law, highlighting its empirical nature.

4.5 Alkali Metals

After considering the inner shells of a multi-electron atom in the previous section, we now consider the outermost shells. The electrons in these outermost electrons are called the **valence electrons** of the atom. They are responsible for

Table 4.5 *Alkali metals and their electronic configurations.*

Element	Z	Electronic configuration
Lithium	3	[He] $2s^1$
Sodium	11	[Ne] $3s^1$
Potassium	19	[Ar] $4s^1$
Rubidium	37	[Kr] $5s^1$
Cesium	55	[Xe] $6s^1$
Francium	87	[Rn] $7s^1$

the chemical activity of the elements, and also for their optical spectra. Hence, they are the main focus of this book. As a specific example, we consider the **alkali metals** such as lithium, sodium, and potassium, which come from group 1 of the periodic table, and are extremely important in atomic physics.

Alkalis have one single valence electron outside filled inner shells, as indicated in Table 4.5. They are therefore approximately one-electron systems. The energy levels of the valence electron can be worked out by solving the N-electron Schrödinger equation given in Eq. (4.1). Within the central-field approximation, the valence electron satisfies a Schrödinger equation of the type given in eqn 4.6, which can be written in the form:

$$\left(-\frac{\hbar^2}{2m}\nabla^2 + V_{\text{eff}}^l(r)\right)\psi = E\,\psi\,. \tag{4.20}$$

The Coulomb repulsion from the core electrons is lumped into the effective potential $V_{\text{eff}}^l(r)$. This is only an approximation to the real behavior, but it can be reasonably good, depending on how well we work out $V_{\text{eff}}^l(r)$. Note that the effective potential depends on l. This arises from the term in l that appears in Eq. (4.11), and has important consequences, as we shall see below.

The overall dependence of $V_{\text{eff}}(r)$ on r must look something like Figure 4.7(a). At very large values of r, the valence electron will be well outside any filled shells, and will thus only see an attractive potential equivalent to a charge of $+e$. On the other hand, if r is very small, the electron will see the full nuclear charge of $+Ze$. The potential at intermediate values of r must lie somewhere between these two limits – hence the generic form of $V_{\text{eff}}(r)$ shown in Figure 4.7(a). The task of calculating $V_{\text{eff}}^l(r)$ from first principles keeps theoretical atomic physicists busy. Here we adopt a simpler, phenomenological approach to describe the energies. To see how this works, we consider the sodium atom.

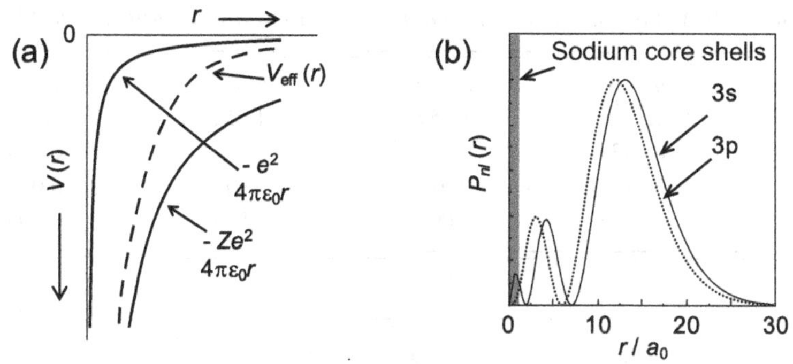

Figure 4.7 (a) Typical effective potential $V_{\text{eff}}(r)$ for the valence electrons of an atom with atomic number Z. (b) Radial probability densities for hydrogenic 3s and 3p wave functions. a_0 is the Bohr radius (0.529 Å). The shaded region near $r = 0$ represents the inner core shells for the case of sodium with $Z = 11$.

The shell model picture of sodium is shown in Figure 4.3. As discussed in Section 4.3, the simplistic Bohr picture predicts an energy of -1.51 eV for the valence electron in the $n = 3$ shell. However, the actual energy of the 3s ground state is -5.14 eV. The Bohr picture thus gives the right order or magnitude, but is clearly inaccurate in the detail. We can get a better model by writing the energy of each (n, l) term as:

$$E_{nl} = -\frac{R_H}{[n - \delta(l)]^2}, \tag{4.21}$$

where $\delta(l)$ is the phenomenological **quantum defect**. The quantum defect allows for the penetration of the inner shells by the valence electron. Since this formula only applies to the valence electron, the value of n cannot correspond to a filled shell. Hence for sodium we must have $n \geq 3$.

The dependence of the quantum defect on l can be understood with reference to Figure 4.7(b). This shows the radial probability densities $P_{nl}(r) = r^2|R(r)|^2$ for the 3s and 3p orbitals of a hydrogenic atom with $Z = 1$, which might be expected to be a reasonable approximation for the single valence electron of sodium. The shaded region near $r = 0$ represents the inner $n = 1$ and $n = 2$ shells with radii of $\sim 0.09a_0$ and $\sim 0.44a_0$, respectively. (See Section 4.3.) We see that both the 3s and 3p orbitals penetrate the inner shells, and that this penetration is much greater for the 3s electron. The electron will therefore see a larger effective nuclear charge for part of its orbit, and this will have the effect of reducing the energies. The energy reduction is largest for the 3s electron due to its larger core penetration.

Table 4.6 *Values of the quantum defect $\delta(l)$ for sodium against n and l.*

l	$n = 3$	$n = 4$	$n = 5$	$n = 6$
0	1.373	1.357	1.352	1.349
1	0.883	0.867	0.862	0.859
2	0.010	0.011	0.013	0.011
3	–	0.000	–0.001	–0.008

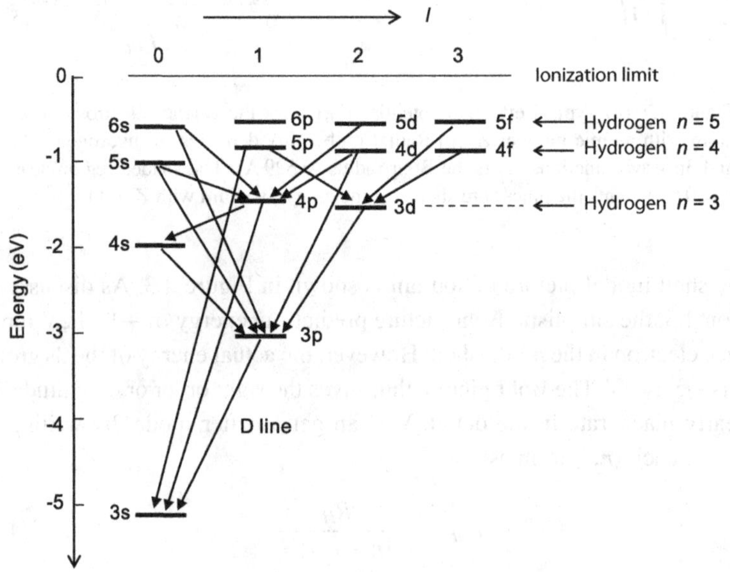

Figure 4.8 Approximate energy-level diagram for sodium, showing the ordering of the levels and the main E1-allowed optical transitions. The hydrogen energy levels are shown for comparison at the right. The first transition from the ground state, namely the D-line 3s ↔ 3p transition, is labeled.

The quantum defect $\delta(l)$ was introduced empirically to account for the optical spectra of alkalis. In principle it should depend on both n and l, but it was found experimentally to depend mainly on l. This can be seen from the values of the quantum defect for sodium tabulated in Table 4.6. The corresponding energy spectrum is shown schematically in Figure 4.8. As noted previously in Example 3.1, the representation of the energy states of an atom with energy on the y-axis and the angular momentum states on the x-axis is called a **Grotrian diagram**. This representation was introduced by Walter Grotrian (1890–1954) in his 1928 book, and is named after him.

Two aspects of the energy spectrum in Figure 4.8 are particularly noteworthy:

- The energy depends on both n and l. This contrasts with hydrogen, where the energy depends only on n. (See Eq. [2.53].) The degeneracy of the l states in hydrogen is called an "accidental" degeneracy, and follows from the exact Coulombic form of the potential in one-electron atoms. It is called accidental because l appears in the Schrödinger equation, but not in the energy. In all other atoms, the energy depends on both n and l.
- The energy levels converge to the hydrogenic limit for the d- and f-states on account of their small quantum defects. For large enough n, the quantum defect also becomes negligible for the p- and s-states, and these also converge to the hydrogenic limit. This is a consequence of the fact that the penetration of the inner shells is negligible for shells with large n, and the Rydberg atom picture of Figure 2.6 applies. In fact, this argument applies to the highly excited states of *any* atom, as these involve promotion of just one of the valence electrons to high lying shells, as discussed in Section 1.3.

The optical spectra for sodium and other alkalis can be worked out by applying the selection rules discussed in Section 3.4 to transitions between the energy levels of the valence electron. The key selection rule is $\Delta l = \pm 1$. Note that there is no selection rule on n, so transitions like 3p $\rightarrow$ 3s are perfectly acceptable. (We did not consider these in Example 3.1 because they have no energy in hydrogen!) The transition energy for the $nl \rightarrow n'l'$ transition is worked out from Eq. (4.21) as:

$$h\nu \equiv \frac{hc}{\lambda} = R_H \left| \frac{1}{[n-\delta(l)]^2} - \frac{1}{[n'-\delta(l')]^2} \right| . \qquad (4.22)$$

A particularly important transition for an alkali atom is the **D-line**. The D-line nomenclature dates back to Fraunhofer's catalogue of spectral lines. (See Section 12.2.) It originally applied just to sodium, but is now extended to all alkali atoms and refers to the transition between the first excited state and the ground state, i.e., the np $\leftrightarrow$ ns transition, where ns is the ground state. In the case of sodium, the D-line is thus the 3p $\rightarrow$ 3s transition. Its wavelength is considered in Example 4.4 below. This line is in fact split into a doublet by spin-orbit coupling, as will be discussed in Section 7.5.

The arguments developed here about alkalis apply equally well to other atoms with just one valence electron outside filled shells. In particular, the singly charged ions of the divalent alkaline-earth metals in group 2 of the periodic table (i.e., Be^+, Mg^+, Ca^+, etc.) are iso-electronic to alkalis. These ions are extensively studied in atomic physics.

Example 4.4 Estimate the wavelength of the 3p → 3s transition in sodium.

Solution: The wavelength is given by $h\nu = hc/\lambda = E_{3p} - E_{3s}$. This can be worked out from Eq. (4.22) using the values of δ given in Table 4.6. This gives:

$$\frac{1}{\lambda} = \frac{R_H}{hc}\left(\frac{1}{[3-\delta(3\mathrm{s})]^2} - \frac{1}{[3-\delta(3\mathrm{p})]^2}\right),$$
$$= (1.10 \times 10^5\,\mathrm{cm}^{-1}) \times \left(\frac{1}{1.627^2} - \frac{1}{2.117^2}\right).$$

The wave number $\overline{\nu} \equiv 1/\lambda$ of the transition is thus $1.70 \times 10^4\,\mathrm{cm}^{-1}$, and so λ is equal to 590 nm. This corresponds to the yellow-orange part of the spectrum.

Exercises

4.1 Unsöld's theorem states that a filled or half-filled sub-shell of atomic orbitals is spherically symmetrical and thus contributes an orbital angular momentum of zero. Verify that this is true for s-, p-, and d-shells by working out $\sum_{m=-l}^{m=+l} |Y_{l,m}(\theta,\phi)|^2$ for the appropriate spherical harmonic functions given in Table 2.2 .

4.2 Write down the electronic configurations of (a) sulfur ($Z = 16$), (b) nickel ($Z = 28$), (c) zirconium ($Z = 40$), (d) praseodymium ($Z = 59$).

4.3 Discuss qualitiatively how a graph of the *second* ionization potential of the elements against Z would compare to Figure 4.4(a).

4.4 The three ionization potentials of lithium ($Z = 3$) are: Li I, 5.392 eV; Li II, 75.64 eV; Li III, 122.5 eV. Explain the trends in this data.

4.5 (a) Estimate the energy of the L-shell absorption edge in lead ($Z = 82$).
(b) Estimate the wavelength of the M → L emission line.

4.6 (a) The K-shell absorption edge of copper ($Z = 29$) occurs at 8.99 keV. What is the effective charge experienced by the K-shell electrons?
(b) Use Moseley's law and your answer to part (a) to estimate the energy of the L → K transition.
(c) The actual energy of the KL_3 line is 8.05 keV. Account for the discrepancy with your answer to part (b).
(d) Deduce the screening parameter of the L_3 shell.

4.7 The K and L absorption edges of scandium ($Z = 21$) occur at 4.49 and 0.498 keV, respectively. Deduce values for the screening parameters of the K- and L-shell electrons. Explain why the L-edge has sub-structure, but the K-edge does not.

4.8 The K absorption edge of molybdenum ($Z = 42$) occurs at 20.00 keV, while the L-shell has three edges labeled L_1, L_2, and L_3 at 2.88, 2.63, and 2.52 keV, respectively.

(a) Deduce the wavelengths of the X-ray lines that would be observed in the emission spectrum for the L $\rightarrow$ K transitions.
(b) Explain why K $\rightarrow$ L transitions are not observed in the absorption spectrum.

4.9 The ground state electronic configuration of rubidium is [Kr]5s.

(a) The first ionization energy of rubidium is 4.177eV. What is the quantum defect of the 5s electron?
(b) The quantum defect of the 6p electron is 2.68. What is the wavelength of the 6p $\rightarrow$ 5s transition?

4.10 Francium is an alkali metal with a ground-state configuration of [Rn] 7s. Its first ionization potential is equal to 4.07 eV, and the wavelength of the 8p $\rightarrow$ 7s transition is 428 nm.

(a) Find the values of $\delta(0)$ and $\delta(1)$ for francium.
(b) Estimate the wavelength of the 7p $\rightarrow$ 7s transition.
(c) Find the maximum value of $\delta(2)$ that would allow the 8p $\rightarrow$ 6d transition to occur in emission.

4.11 The ground-state electronic configuration of the alkali metal cesium is [Xe] 6s. The approximate values of the quantum defects are as follows: $\delta(0) \approx 4.11$, $\delta(1) \approx 3.61$, $\delta(2) \approx 2.45$, $\delta(3) \approx 0.02$. An electron is promoted to the 7p shell. What wavelengths could be observed in emission?

5
Angular Momentum

The treatment of angular momentum is very important to understand the properties of atoms. It is now time to explore these effects in detail, and to see how this leads to the classification of the quantized states of atoms by their angular momentum.

5.1 Conservation of Angular Momentum

In the sections that follow, we will consider several different types of angular momentum and the ways in which they are coupled together. Before going into the details, it is useful to stress one very important point related to conservation of angular momentum. In an isolated atom, there are many forces (and hence torques) acting inside the atom. These *internal* forces cannot change the *total* angular momentum of the atom, since conservation of angular momentum demands that the angular momentum of the atom as a whole must be constant in the absence of any *external* torques. The total angular momentum of the atom is normally determined by its electrons. The total electronic angular momentum is written $\boldsymbol{J}$ and is specified by the quantum number J. The principle of conservation of angular momentum therefore requires that isolated atoms always have well-defined J-states. It is this J-value that determines, for example, the magnetic dipole moment of the atom.

It should be noted in passing that the nucleus can possess angular momentum and that electrons can exchange angular momentum with the nucleus through hyperfine interactions. (See Section 7.8.2.) These interactions are very weak, and can usually be neglected except when explicitly considering nuclear effects. With this caveat, we can safely regard the total electronic angular momentum $\boldsymbol{J}$ as a conserved quantity.

The principle of conservation of angular momentum does not apply, of course, when external perturbations are applied. The most obvious example is the perturbation caused by the emission or absorption of a photon. In this case the angular momentum of the atom must change because the photon itself carries angular momentum, and the angular momentum of the whole system (atom + photon) has to be conserved. The change in J is then governed by selection rules, as discussed, for example, in Section 5.8. Another obvious example is the effect of a strong external DC magnetic field. In this case, it is possible for the magnetic field to produce states where the component of angular momentum along the direction of the field is well-defined, but not the total angular momentum. (See the discussion of the Paschen-Back effect in Section 8.1.3.)

5.2 Types of Angular Momentum

The electrons in atoms possess two different types of angular momentum, namely orbital and spin. These are discussed separately below.

5.2.1 Orbital Angular Momentum

The electrons in atoms orbit around the nucleus, and therefore possess orbital angular momentum. In classical mechanics, we define the orbital angular momentum of a particle by:

$$\boldsymbol{L} = \boldsymbol{r} \times \boldsymbol{p}\,, \tag{5.1}$$

where $\boldsymbol{r}$ is the radial position, and $\boldsymbol{p}$ is the linear momentum. The components of $\boldsymbol{L}$ are given by:

$$\begin{pmatrix} L_x \\ L_y \\ L_z \end{pmatrix} = \begin{pmatrix} x \\ y \\ z \end{pmatrix} \times \begin{pmatrix} p_x \\ p_y \\ p_z \end{pmatrix} = \begin{pmatrix} yp_z - zp_y \\ zp_x - xp_z \\ xp_y - yp_x \end{pmatrix} . \tag{5.2}$$

In quantum mechanics, we represent the linear momentum by differential operators of the type:

$$\hat{p}_x = -i\hbar \frac{\partial}{\partial x} \,. \tag{5.3}$$

Therefore, the quantum mechanical operators for the Cartesian components of the orbital angular momentum are given by:

$$\hat{L}_x = -i\hbar \left(y\frac{\partial}{\partial z} - z\frac{\partial}{\partial y} \right) , \tag{5.4}$$

$$\hat{L}_y = -i\hbar\left(z\frac{\partial}{\partial x} - x\frac{\partial}{\partial z}\right), \tag{5.5}$$

$$\hat{L}_z = -i\hbar\left(x\frac{\partial}{\partial y} - y\frac{\partial}{\partial x}\right). \tag{5.6}$$

Note that the "hat" symbol indicates that we are representing an operator and not just a number.

In classical mechanics, the magnitude of the angular momentum is given by:

$$L^2 = L_x^2 + L_y^2 + L_z^2 .$$

We therefore define the quantum-mechanical operator for the magnitude of the angular momentum by:

$$\hat{\boldsymbol{L}}^2 = \hat{L}_x^2 + \hat{L}_y^2 + \hat{L}_z^2 . \tag{5.7}$$

The operators like $\hat{L}_x^2$ that appear here should be understood in terms of repeated operations:

$$\begin{aligned}\hat{L}_x^2\psi &= -\hbar^2\left(y\frac{\partial}{\partial z} - z\frac{\partial}{\partial y}\right)\left(y\frac{\partial\psi}{\partial z} - z\frac{\partial\psi}{\partial y}\right)\\ &= -\hbar^2\left(y^2\frac{\partial^2\psi}{\partial z^2} - y\frac{\partial\psi}{\partial y} - z\frac{\partial\psi}{\partial z} - 2yz\frac{\partial^2\psi}{\partial y\partial z} + z^2\frac{\partial^2\psi}{\partial y^2}\right).\end{aligned}$$

Note that we have already met the $\hat{\boldsymbol{L}}^2$ and $\hat{L}_z$ operators when we solved the Schrödinger equation for hydrogen in Section 2.2. (See Eqs. [2.31] and [2.40].)

When considering hydrogen, the spherical symmetry of the atom made it convenient to work in spherical polar rather than Cartesian coordinates. The two approaches are, of course, completely equivalent, and the operators are physically identical whether expressed in their spherical polar or Cartesian forms. For example, we can show that the two forms of the $\hat{L}_z$ operator given in Eqs. (2.40) and (5.6) are equivalent as follows: We have two sets of coordinates (x, y, z) and (r, θ, ϕ), with $x = r\sin\theta\cos\phi$, $y = r\sin\theta\sin\phi$, and $z = r\cos\theta$. We can then write:

$$\begin{aligned}\frac{\partial}{\partial\phi} &= \frac{\partial x}{\partial\phi}\frac{\partial}{\partial x} + \frac{\partial y}{\partial\phi}\frac{\partial}{\partial y} + \frac{\partial z}{\partial\phi}\frac{\partial}{\partial z},\\ &= -r\sin\theta\sin\phi\frac{\partial}{\partial x} + r\sin\theta\cos\phi\frac{\partial}{\partial y} + 0,\\ &= -y\frac{\partial}{\partial x} + x\frac{\partial}{\partial y}.\end{aligned}$$

Hence:

$$\hat{L}_z = -i\hbar\frac{\partial}{\partial\phi} = -i\hbar\left(x\frac{\partial}{\partial y} - y\frac{\partial}{\partial x}\right), \tag{5.8}$$

which is the same as Eq. (5.6). The proof that Eq. (5.7) is equivalent to Eq. (2.31) is considered in Exercise 5.1.

The **commutator** of two quantum-mechanical operators $\hat{A}$ and $\hat{B}$ is defined by:

$$[\hat{A}, \hat{B}] = \hat{A}\hat{B} - \hat{B}\hat{A}\,. \tag{5.9}$$

In simple terms, the commutator tells us whether the order in which the operators are applied matters or not. If $[\hat{A}, \hat{B}] = 0$, the operators are said to commute, and the order does not matter. A key property of the orbital angular momentum operator is that its components do not commute with each other, but they do commute with $\hat{\boldsymbol{L}}^2$. We can summarize this by writing:

$$\begin{aligned} [\hat{L}_x, \hat{L}_y] &= \hat{L}_x\hat{L}_y - \hat{L}_y\hat{L}_x \neq 0\,, \\ [\hat{\boldsymbol{L}}^2, L_z] &= \hat{\boldsymbol{L}}^2\hat{L}_z - \hat{L}_z\hat{\boldsymbol{L}}^2 = 0\,. \end{aligned} \tag{5.10}$$

The non-commutation of the components can be proved as follows:

$$\begin{aligned} \hat{L}_x\hat{L}_y\psi &= (-i\hbar)^2 \left(y\frac{\partial}{\partial z} - z\frac{\partial}{\partial y}\right)\left(z\frac{\partial\psi}{\partial x} - x\frac{\partial\psi}{\partial z}\right), \\ &= -\hbar^2 \left(yz\frac{\partial^2\psi}{\partial z\partial x} + y\frac{\partial\psi}{\partial x} - yx\frac{\partial^2\psi}{\partial z^2} - z^2\frac{\partial^2\psi}{\partial y\partial x} + zx\frac{\partial^2\psi}{\partial y\partial z}\right). \end{aligned}$$

On the other hand, we have:

$$\begin{aligned} \hat{L}_y\hat{L}_x\psi &= (-i\hbar)^2 \left(z\frac{\partial}{\partial x} - x\frac{\partial}{\partial z}\right)\left(y\frac{\partial\psi}{\partial z} - z\frac{\partial\psi}{\partial y}\right), \\ &= -\hbar^2 \left(zy\frac{\partial^2\psi}{\partial x\partial z} - z^2\frac{\partial^2\psi}{\partial x\partial y} - xy\frac{\partial^2\psi}{\partial z^2} + xz\frac{\partial^2\psi}{\partial z\partial y} + x\frac{\partial\psi}{\partial y}\right). \end{aligned}$$

On recalling that $\partial^2\psi/\partial x\partial y = \partial^2\psi/\partial y\partial x$, we find:

$$\begin{aligned} \hat{L}_x\hat{L}_y\psi - \hat{L}_y\hat{L}_x\psi \equiv [\hat{L}_x, \hat{L}_y]\psi &= -\hbar^2 \left(y\frac{\partial\psi}{\partial x} - x\frac{\partial\psi}{\partial y}\right), \\ &= i\hbar \times -i\hbar \left(x\frac{\partial\psi}{\partial y} - y\frac{\partial\psi}{\partial x}\right), \\ &= i\hbar\hat{L}_z\psi\,. \end{aligned}$$

We therefore conclude that:

$$[\hat{L}_x, \hat{L}_y] = i\hbar\hat{L}_z\,. \tag{5.11}$$

The other commutators of the angular momentum operators, namely $[\hat{L}_y, \hat{L}_z]$ and $[\hat{L}_z, \hat{L}_x]$, are obtained by cyclic permutation of the indices in Eq. (5.11): $x \to y$, $y \to z$, $z \to x$.

The commutation of $\hat{\boldsymbol{L}}^2$ with $\hat{L}_z$ (i.e., $[\hat{\boldsymbol{L}}^2, \hat{L}_z] = 0$) can be proven in a number of ways. Here is one: It can be shown (see Exercise 5.2) that the commutators of two arbitrary operators satisfy:

$$[\hat{A}^2, \hat{B}] = \hat{A}[\hat{A}, \hat{B}] + [\hat{A}, \hat{B}]\hat{A}\,. \tag{5.12}$$

We combine this result with Eq. (5.7) and the cyclic permutations of Eq. (5.11) to write:

$$\begin{aligned}
[\hat{\boldsymbol{L}}^2, \hat{L}_z] &= [\hat{L}_x^2, \hat{L}_z] + [\hat{L}_y^2, \hat{L}_z] + [\hat{L}_z^2, \hat{L}_z]\,,\\
&= [\hat{L}_x^2, \hat{L}_z] + [\hat{L}_y^2, \hat{L}_z] + 0\,,\\
&= \hat{L}_x[\hat{L}_x, \hat{L}_z] + [\hat{L}_x, \hat{L}_z]\hat{L}_x + \hat{L}_y[\hat{L}_y, \hat{L}_z] + [\hat{L}_y, \hat{L}_z]\hat{L}_y\,,\\
&= -i\hbar\hat{L}_x\hat{L}_y - i\hbar\hat{L}_y\hat{L}_x + i\hbar\hat{L}_y\hat{L}_x + i\hbar\hat{L}_x\hat{L}_y\,,\\
&= 0\,.
\end{aligned}$$

It can be shown that the measurable quantities corresponding to two quantum-mechanical operators that do not commute must obey an uncertainty principle. The general result for operators $\hat{A}$ and $\hat{B}$ is:

$$\Delta A^2 \Delta B^2 \geq \frac{1}{4}\left|\langle[\hat{A}, \hat{B}]\rangle\right|^2\,. \tag{5.13}$$

As a specific example, consider the case of the position and momentum operators $\hat{x}$ and $\hat{p}$:

$$[\hat{x}, \hat{p}]\psi = (\hat{x}\hat{p} - \hat{p}\hat{x})\psi = -i\hbar x\left(\frac{\mathrm{d}\psi}{\mathrm{d}x}\right) + i\hbar\frac{\mathrm{d}(x\psi)}{\mathrm{d}x} = i\hbar\psi\,.$$

Hence $[\hat{x}, \hat{p}] = i\hbar$. It then follows from Eq. (5.13) that $\Delta x \Delta p \geq \hbar/2$, which is the Heisenberg uncertainty principle.

The non-commutation of the components of $\boldsymbol{L}$ thus implies that it is not possible to know the values of L_x, L_y, and L_z simultaneously: we can only know *one* of them (usually L_z) at any time. Once L_z is known, we cannot know L_x and L_y as well. On the other hand, the fact that $\hat{L}_z$ commutes with $\hat{\boldsymbol{L}}^2$ (cf, Eq. [5.10]) means that we *can* know the modulus squared of the angular momentum vector and its z-component simultaneously. In summary:

- We can know $|L^2|$ and the magnitude of one of the components of the angular momentum.
- For mathematical convenience, we usually take the component we know to be L_z.
- We cannot know the values of all three components of the angular momentum simultaneously.

The eigenvalues of $\hat{\boldsymbol{L}}^2$ and $\hat{L}_z$ were discussed in Section 2.2.3. The orbital angular momentum is specified by two quantum numbers: l and m. The latter is sometimes given an extra subscript (i.e., m_l) to distinguish it from the spin quantum number m_s considered below. The magnitude of $\boldsymbol{l}$ is given by:

$$|\boldsymbol{l}|^2 = l(l+1)\hbar^2, \tag{5.14}$$

and the component along the z-axis by

$$l_z = m\hbar \,. \tag{5.15}$$

Note that we have switched to a lowercase notation here because we are referring to a *single* electron. (See Section 5.7.) The quantum number l can take positive integer values (including 0), and m can take values in integer steps from $-l$ to $+l$. The number of m states that each l-state is therefore equal to $(2l+1)$. These m-states are degenerate in isolated atoms, but can be split by external perturbations (e.g., magnetic or electric fields.)

The quantization of the angular momentum can be represented pictorially in the **vector model**, as shown previously in Figure 2.3. In this model, the angular momentum is represented as a vector of length $\sqrt{l(l+1)}\hbar$ angled so that its component along the z-axis is equal to $m\hbar$. The x- and y-components of the angular momentum are not known.

In classical mechanics, the orbital angular momentum is conserved when the force $\boldsymbol{F}$ is radial: i.e., $\boldsymbol{F} \equiv F\hat{\boldsymbol{r}}$, where $\hat{\boldsymbol{r}}$ is a unit vector parallel to $\boldsymbol{r}$. This follows from the equation of motion:

$$\frac{\mathrm{d}\boldsymbol{l}}{\mathrm{d}t} = \boldsymbol{\Gamma} = \boldsymbol{r} \times \boldsymbol{F} = \boldsymbol{r} \times F\hat{\boldsymbol{r}} = 0\,, \tag{5.16}$$

where $\boldsymbol{\Gamma}$ is the torque. In the hydrogen atom, the Coulomb force on the electron acts toward the nucleus, and hence $\boldsymbol{l}$ is conserved, which leads to well-defined quantized values. It is also the case that the individual electrons of many-electron atoms have well-defined l states. This follows from the central-field approximation (see Section 4.1), where the dominant resultant force on the electron is radial (i.e., central). Note, however, that the inclusion of noncentral forces via the residual electrostatic interaction leads to some mixing of the orbital angular momentum states. This can explain why transitions that are apparently forbidden by selection rules can sometimes be observed, albeit with low transition probabilities.

5.2.2 Spin Angular Momentum

A wealth of data derived from the optical, magnetic, and chemical properties of atoms points to the fact that electrons possess an additional type of angular

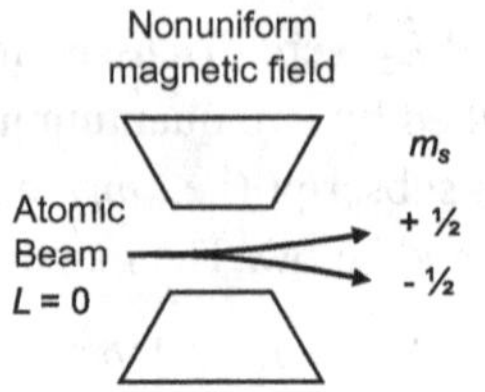

Figure 5.1 The Stern–Gerlach experiment. A beam of monovalent atoms with $L = 0$ (i.e., zero orbital angular momentum and hence zero orbital magnetic dipole moment) is deflected in two discrete ways by a nonuniform magnetic field. The force on the atoms arises from the interaction between the field and the magnetic moment due to the electron spin.

momentum called **spin**. The electron behaves as if it spins around its own internal axis, but this analogy should not be taken literally; the electron is, as far as we know, a point particle, and so cannot be spinning in any classical way. In fact, spin is a purely quantum effect with no classical explanation. Paul Dirac at Cambridge successfully accounted for electron spin when he produced the relativistic wave equation that bears his name in 1928.

The discovery of spin goes back to the Stern–Gerlach experiment, in which a beam of atoms is deflected by a nonuniform magnetic field. (See Figure 5.1). The force on a magnetic dipole in a nonuniform magnetic field is given by:

$$F_z = \mu_z \frac{\mathrm{d}B}{\mathrm{d}z}, \tag{5.17}$$

where $\mathrm{d}B/\mathrm{d}z$ is the field gradient, which is assumed to point along the z-direction, and μ_z is the z-component of the magnetic dipole of the atom. In Chapter 7 we shall explore the origin of magnetic dipoles in detail. At this stage, all we need to know is that the magnetic dipole is directly proportional to the angular momentum of the atom. (See Section 7.1.)

In the original Stern–Gerlach experiment, silver atoms were used. These have a ground-state electronic configuration of [Kr] $4\mathrm{d}^{10}\,5\mathrm{s}^1$. Filled shells have no net orbital angular momentum because there are as many positive m_l states occupied as negative ones. The 5s electron has $l = 0$, and therefore the total orbital angular momentum of the atom is zero. This implies that the orbital magnetic dipole of the atom is also zero, and hence we expect no deflection. However, the experiment showed that the atoms were deflected either up or down, as indicated in Figure 5.1.

In order to explain the up/down deflection of atoms with no orbital angular momentum, we have to assume that each electron possesses an additional type of magnetic dipole moment. This magnetic dipole is attributed to the spin angular momentum. In analogy with orbital angular momentum, spin angular momentum is described by two quantum numbers s and m_s, where m_s can take the $(2s+1)$ values in integer steps from $-s$ to $+s$. The magnitude of the spin angular momentum is given by:

$$|\boldsymbol{s}|^2 = s(s+1)\hbar^2 \tag{5.18}$$

and the component along the z-axis is given by:

$$s_z = m_s\hbar\,. \tag{5.19}$$

The fact that atoms with a single s-shell valence electron (e.g., silver) are only deflected in two directions (i.e., up or down) implies that $(2s+1) = 2$, and hence that $s = 1/2$. The spin quantum numbers of the electron therefore have the following values:

$$s = 1/2\,,$$
$$m_s = \pm 1/2\,.$$

The Stern–Gerlach experiment is just one of many pieces of evidence that support the hypothesis of electron spin. Here is an incomplete list of other evidence based on atomic physics:

- The periodic table of elements cannot be explained unless we assume that the electrons possess spin.
- High-resolution spectroscopy of atomic spectral lines shows that they frequently consist of closely spaced multiplets. This fine structure is caused by spin–orbit coupling, which can only be explained by postulating that electrons possess spin. See Chapter 7.
- If we ignore spin, we expect to observe the *normal* Zeeman effect in an external magnetic field. However, most atoms display the *anomalous* Zeeman effect, which is a consequence of spin. See Chapter 8.
- The ratio of the magnetic dipole moment to the angular momentum is called the gyromagnetic ratio. (See Section 7.1.) The gyromagnetic ratio can be measured directly by a number of methods. In 1915, Einstein and de Haas measured the gyromagnetic ratio of iron and came up with a value twice as large as expected. They rejected this result, attributing it to experimental errors. However, we now know that the magnetism in iron is caused by the spin rather than the orbital angular momentum, so the experimental value was correct. (The electron spin g-factor is 2; see Section 7.2.) This is a salutary lesson from history that even great physicists like Einstein and de Haas can get their error analysis wrong!

5.3 Addition of Angular Momentum

Having discovered that electrons have different types of angular momentum, now we need to know how add them together. Let us suppose that $\boldsymbol{C}$ is the resultant of two angular momentum vectors $\boldsymbol{A}$ and $\boldsymbol{B}$, as shown in Figure 5.2(a), so that:

$$\boldsymbol{C} = \boldsymbol{A} + \boldsymbol{B}. \tag{5.20}$$

We assume for the sake of simplicity that $|\boldsymbol{A}| > |\boldsymbol{B}|$. (The argument is unaffected if $|\boldsymbol{A}| < |\boldsymbol{B}|$.) We define θ as the angle between the two vectors, as shown in Figure 5.2(a).

In *classical* mechanics, the angle θ can take any value from 0° to 180°. Therefore, $|\boldsymbol{C}|$ can take any value from $(|\boldsymbol{A}| + |\boldsymbol{B}|)$ to $(|\boldsymbol{A}| - |\boldsymbol{B}|)$. This is *not* the case in *quantum* mechanics because the lengths of the angular momentum vectors must be quantized according to:

$$\begin{aligned} |\boldsymbol{A}| &= \sqrt{A(A+1)}\hbar \\ |\boldsymbol{B}| &= \sqrt{B(B+1)}\hbar \\ |\boldsymbol{C}| &= \sqrt{C(C+1)}\hbar, \end{aligned} \tag{5.21}$$

where A, B, and C are quantum numbers, which must be either integers or half-integers. This makes it apparent that θ can only take specific values. The rule for working out the allowed values of C from the known values of A and B is that C takes values in integer steps from $(A + B)$ down to $|A - B|$, for example:

$$C = A \oplus B = (A+B),\ (A+B-1),\ \cdots,\ |A-B|. \tag{5.22}$$

The $\oplus$ symbol here indicates that we are adding together angular momentum quantum numbers.

Example 5.1 What are the possible values of the quantum numbers of the resultant for the following cases:

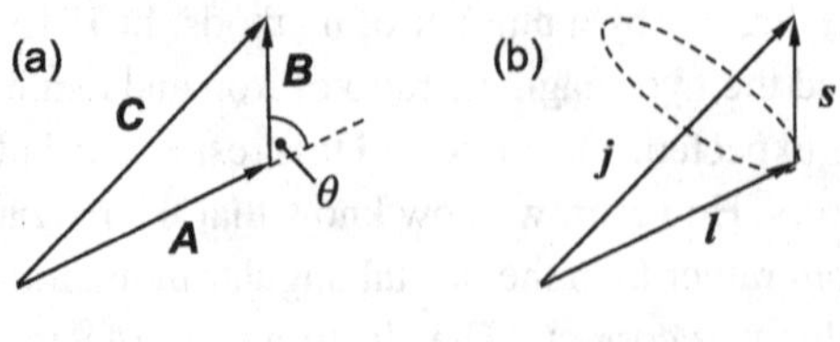

Figure 5.2 (a) Vector addition of two angular momentum vectors $\boldsymbol{A}$ and $\boldsymbol{B}$ to form the resultant $\boldsymbol{C}$. (b) Vector model of the atom. The spin-orbit interaction couples $\boldsymbol{l}$ and $\boldsymbol{s}$ together to form the resultant $\boldsymbol{j}$.

(a) $\boldsymbol{J} = \boldsymbol{L} + \boldsymbol{S}$, and $L = 3$, $S = 1$.
(b) $\boldsymbol{L} = \boldsymbol{l}_1 + \boldsymbol{l}_2$, and $l_1 = 2$, $l_2 = 0$.
(c) $\boldsymbol{S} = \boldsymbol{s}_1 + \boldsymbol{s}_2$, and $s_1 = 1/2$, $s_2 = 1/2$.
(d) $\boldsymbol{J} = \boldsymbol{j}_1 + \boldsymbol{j}_2$, and $j_1 = 5/2$, $j_2 = 3/2$.

Solution:

(a) $J = 3 \oplus 1$, $3 + 1 = 4$, $|3 - 1| = 2$, therefore $J = 4, 3, 2$.
(b) $L = 2 \oplus 0$, $2 + 0 = 2$, $|2 - 0| = 2$, therefore $L = 2$.
(c) $S = 1/2 \oplus 1/2$, $1/2 + 1/2 = 1$, $|1/2 - 1/2| = 0$, therefore $S = 1, 0$.
(d) $J = 5/2 \oplus 3/2$, $5/2 + 3/2 = 4$, $|5/2 - 3/2| = 1$, therefore $J = 4, 3, 2, 1$.

5.4 Spin-Orbit Coupling

The orbital and spin angular momenta of electrons in atoms are not totally independent of each other, but interact through the **spin-orbit interaction**. Spin-orbit coupling and its effects are considered in detail in Chapter 7, and at this stage we just need to know two basic things:

(i) Spin-orbit coupling derives from the interaction between the magnetic dipole due to spin and the magnetic field that the electron experiences due to its orbital motion. We can thus write the spin-orbit interaction in the form (see Eq. [7.31]):

$$\hat{H} = -\boldsymbol{\mu}_{\text{spin}} \cdot \boldsymbol{B}_{\text{orbital}} \propto \boldsymbol{l} \cdot \boldsymbol{s}\,, \tag{5.23}$$

since $\boldsymbol{\mu}_{\text{spin}} \propto \boldsymbol{s}$ and $\boldsymbol{B}_{\text{orbital}} \propto \boldsymbol{l}$.

(ii) The spin-orbit interaction increases with Z. (See Section 7.3.3.) It is therefore weak in light atoms, and stronger in heavy atoms.

We introduce the spin-orbit interaction here because it is one of the mechanisms that is important in determining the angular momentum coupling schemes that apply to different atoms.

5.5 Angular Momentum Coupling in Single-Electron Atoms

If an atom has just a single electron, the addition of the orbital and spin angular momenta is relatively straightforward. The physical mechanism that couples the orbital and spin angular momenta together is the spin-orbit interaction, and the resultant total angular momentum vector $\boldsymbol{j}$ is defined by:

$$\boldsymbol{j} = \boldsymbol{l} + \boldsymbol{s}\,. \tag{5.24}$$

Vector $\boldsymbol{j}$ is described by the quantum numbers j and m_j, which denote its magnitude and z-component according to the usual rules for quantum mechanical angular momenta, namely:

$$|\boldsymbol{j}|^2 = j(j+1)\hbar^2, \tag{5.25}$$

and

$$j_z = m_j\hbar, \tag{5.26}$$

where m_j takes values of $j, (j-1), \cdots, -j$. The spin-orbit coupling of $\boldsymbol{l}$ and $\boldsymbol{s}$ to form the resultant $\boldsymbol{j}$ is illustrated by Figure 5.2(b). The magnitudes of the vectors shown in Figure 5.2(b) are, respectively: $|\boldsymbol{j}| = \sqrt{j(j+1)}\hbar$, $|\boldsymbol{l}| = \sqrt{l(l+1)}\hbar$, and $|\boldsymbol{s}| = \sqrt{s(s+1)}\hbar$.

The allowed values of j are worked out by applying Eq. (5.22), with the knowledge that the spin quantum number s is always equal to 1/2. If the electron is in a state with orbital quantum number l, we then find $j = l \oplus s = (l \pm 1/2)$, except when $l = 0$, in which case we just have $j = 1/2$. In the second case, the angular momentum of the atom arises purely from the electron spin.

5.6 Angular Momentum Coupling in Multi-Electron Atoms

The Hamiltonian for an N-electron atom can be written in the form:

$$\hat{H} = \hat{H}_0 + \hat{H}_1 + \hat{H}_2, \tag{5.27}$$

where:

$$\hat{H}_0 = \sum_{i=1}^{N}\left(-\frac{\hbar^2}{2m}\nabla_i^2 + V_{\text{central}}(r_i)\right), \tag{5.28}$$

$$\hat{H}_1 = -\sum_{i=1}^{N}\frac{Ze^2}{4\pi\epsilon_0 r_i} + \sum_{i>j}^{N}\frac{e^2}{4\pi\epsilon_0|\boldsymbol{r}_i - \boldsymbol{r}_j|} - \sum_{i=1}^{N} V_{\text{central}}(r_i), \tag{5.29}$$

$$\hat{H}_2 = \sum_{i=1}^{N}\xi(r_i)\boldsymbol{l}_i\cdot\boldsymbol{s}_i. \tag{5.30}$$

As discussed in Section 4.1, $\hat{H}_0$ is the central-field Hamiltonian and $\hat{H}_1$ is the residual electrostatic potential. $\hat{H}_2$ is the spin-orbit interaction summed over the electrons of the atom.

In Chapter 4 we neglected both $\hat{H}_1$ and $\hat{H}_2$, and just concentrated on $\hat{H}_0$. This led to the conclusion that each electron occupies a state in a shell defined by the quantum numbers n and l. The reason why we neglected $\hat{H}_1$ is that the off-radial forces due to the electron-electron repulsion are smaller than the radial ones, while $\hat{H}_2$ was neglected because the spin-orbit effects are much smaller than the main terms in the Hamiltonian. It is now time to study what happens when these two terms are included. In doing so, there are two obvious limits to consider:

- **LS coupling**: $\hat{H}_1 \gg \hat{H}_2$.
- **jj coupling**: $\hat{H}_2 \gg \hat{H}_1$.

Since the spin-orbit interaction increases with Z, LS coupling mainly occurs in atoms with small to medium Z, while jj coupling occurs in some atoms with large Z. In the sections that follow, we will focus on the LS coupling limit. The less common case of jj coupling is considered briefly in Section 5.10. There are, inevitably, a small number of atoms with medium-large Z (e.g., germanium $Z = 32$) in which neither limit applies. We then have **intermediate coupling**, which is not considered further in this book.

5.7 LS Coupling

In the **LS coupling** limit (alternatively called **Russell–Saunders coupling**), the residual electrostatic interaction is much stronger than the spin-orbit interaction. We therefore deal with the effect of the residual electrostatic interaction on the states defined by the central field Hamiltonian first. This produces coupled states in which the resultant L and S values are known. We then apply the spin-orbit interaction as a second, smaller perturbation on these LS states. The LS coupling regime applies to most atoms of small and medium atomic number.

Let us first discuss some issues of notation. We shall need to distinguish between the quantum numbers that refer to the individual electrons within an atom and the state of the atom as a whole. The convention is:

- Lowercase quantum numbers (j, l, and s) refer to *individual electrons*.
- Uppercase quantum numbers (J, L, and S) refer to the resultant angular momentum states of the *whole atom*.

For single-electron atoms like hydrogen, there is no difference. The same applies to alkalis, which are quasi one-electron atoms. For all other atoms with two or more valence electrons, we have to distinguish carefully.

We can use this notation to determine the angular momentum states that the LS coupling scheme produces. The residual electrostatic interaction has the effect of coupling the orbital and spin angular momenta of the individual electrons together, so that we find their resultants according to:

$$\boldsymbol{L} = \sum_i \boldsymbol{l}_i\,, \tag{5.31}$$

$$\boldsymbol{S} = \sum_i \boldsymbol{s}_i\,. \tag{5.32}$$

Filled shells of electrons have no net angular momentum, and so the summation only needs to be carried out over the valence electrons. In a many-electron atom, the rule given in Eq. (5.22) usually allows several possible values of the quantum numbers L and S for a particular electronic configuration. Their energies will differ due to the residual electrostatic interaction. The atomic states defined by the values of L and S are called **terms**.

For each atomic term, we can find the total angular momentum of the whole atom from:

$$\boldsymbol{J} = \boldsymbol{L} + \boldsymbol{S}\,. \tag{5.33}$$

The values of J, the quantum number corresponding to $\boldsymbol{J}$, are found from L and S, according to Eq. (5.22). The states of different J for each LS-term have different energies due to the spin-orbit interaction. In analogy with Eq. (5.23), the spin-orbit interaction of the whole atom is written:

$$\Delta E_{\text{so}} \propto -\boldsymbol{\mu}_{\text{spin}}^{\text{atom}} \cdot \boldsymbol{B}_{\text{orbital}}^{\text{atom}} \propto \boldsymbol{L} \cdot \boldsymbol{S}\,, \tag{5.34}$$

where the superscript atom indicates that we take the resultant values for the whole atom. The details of the spin-orbit interaction in the LS-coupling limit are considered in Section 7.6. At this stage, all we need to know is that the spin-orbit interaction splits the LS terms into **levels** labeled by J, and that the splitting is much smaller than the energy difference between different LS terms.

It is convenient to introduce a shorthand notation to label the energy levels that occur in the LS coupling regime. Each level is labeled by the quantum numbers J, L, and S, and is represented in the form:

$$^{(2S+1)}\mathrm{L}_J\,.$$

The superscript $(2S + 1)$ and subscript J appear as numbers, whereas L is a capital letter that follows the usual rule: S, P, D, F correspond, respectively, to $L = 0, 1, 2, 3$. Thus, for example, a $^2\mathrm{P}_{1/2}$ level has quantum numbers $S = 1/2$, $L = 1$, and $J = 1/2$, while a $^3\mathrm{D}_3$ level has $S = 1$, $L = 2$, and $J = 3$. For $L > 3$, the letters increment alphabetically, with the exception that the letter J is omitted in order to avoid confusion with the angular momentum quantum number J. Hence $L = 6$ is designated by I, but $L = 7$ is designated by K.

The factor of $(2S+1)$ in the top left is called the **multiplicity**. It indicates the degeneracy of the level due to the spin – i.e., the number of M_S states available.

If $S = 0$, the multiplicity is 1, and the terms are called **singlets**. If $S = 1/2$, the multiplicity is 2, and we have **doublet** terms. If $S = 1$, we have **triplet** terms, and so on.

Example 5.2 Magnesium ($Z = 12$) is a divalent metal with ground-state electronic configuration of [Ne] $3s^2$.

(a) What is the angular momentum level of the ground state?
(b) What are the allowed angular momentum levels of the (3s,3p) excited state?

Solution:

(a) The $3s^2$ ground state is a filled shell and has no net angular momentum. Both L and S are zero, and hence $J = 0$. The ground state is thus a ${}^1S_0$ level.
(b) For the (3s,3p) excited state, we have one valence electrons in an s-shell with $l = 0$ and the other in a p-shell with $l = 1$. We first work out the possible values of L and S from Eqs. (5.31) and (5.32) using Eq. (5.22):

- $L = l_1 \oplus l_2 = 0 \oplus 1 = 1$.
- $S = s_1 \oplus s_2 = 1/2 \oplus 1/2 = 1$ or 0.

We thus have two terms: a ${}^3P$ triplet and a ${}^1P$ singlet. The allowed J-levels for each term are then worked out from Eq. (5.33):

- For the ${}^3P$ triplet, we have $J = L \oplus S = 1 \oplus 1 = 2$, 1, or 0. We thus have three levels: ${}^3P_2$, ${}^3P_1$, and ${}^3P_0$.
- For the ${}^1P$ singlet, we have $J = L \oplus S = 1 \oplus 0 = 1$. We thus have a single ${}^1P_1$ level.

These levels are illustrated in Figure 5.3. The ordering of the energy states should not concern us at this stage. The main point to realize is the general way the states split as the new interactions are turned on, and the terminology used to designate the states.

5.8 Electric-Dipole Selection Rules in the LS Coupling Limit

In an electric-dipole transition within an atom with LS coupling, a single electron makes a jump from one atomic shell to a new one. The rules that apply to this electron are the same as the ones discussed in Section 3.4. However, we also have to think about the angular momentum state of the whole atom

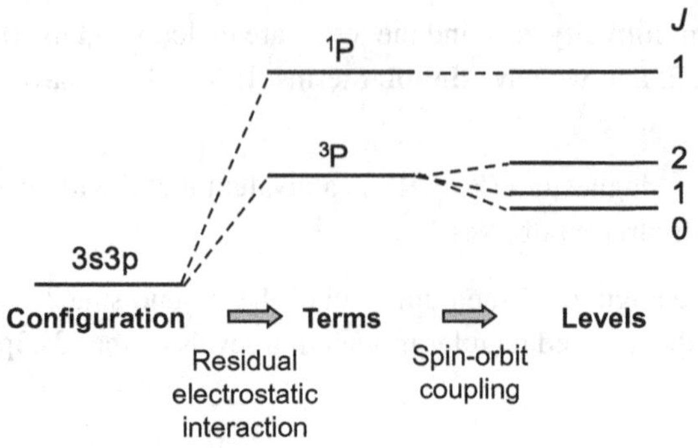

Figure 5.3 Splitting of the energy levels for the (3s,3p) configuration of magnesium in the LS-coupling regime.

as specified by the quantum numbers (L, S, J). The rules that emerge are as follows:

(1) The parity of the wave function must change.
(2) $\Delta l = \pm 1$ for the electron that jumps between shells.
(3) $\Delta L = 0, \pm 1$, but $L = 0 \to 0$ is forbidden.
(4) $\Delta J = 0, \pm 1$, but $J = 0 \to 0$ is forbidden.
(5) $\Delta S = 0$.

Rule 1 follows from the odd parity of the dipole operator. Rule 2 applies the $\Delta l = \pm 1$ single-electron rule to the individual electron that makes the jump in the transition, while rule 3 applies rule 2 to the resultant orbital angular momentum of the whole atom according to the rules for addition of angular momenta. $\Delta L = 0$ transitions are obviously forbidden in one-electron atoms, because $L = l$ and l must change. However, in atoms with more than one valence electron, it is possible to get transitions between different configurations that satisfy rule 2, but have the same value of L. An example is the allowed $3p3p\,^3P_1 \to 3p4s\,^3P_2$ transition in silicon at 250.6 nm. Rule 4 follows from the fact that the total angular momentum must be conserved in the transition, allowing us to write:

$$\boldsymbol{J}^{\text{initial}} = \boldsymbol{J}^{\text{final}} + \boldsymbol{J}^{\text{photon}} \,. \tag{5.35}$$

The photon carries one unit of angular momentum, and so we conclude from Eq. (5.22) that $\Delta J = -1,\ 0$, or $+1$. However, the $\Delta J = 0$ rule cannot be applied to $J = 0 \to 0$ transitions because it is not possible to satisfy Eq. (5.35) in these circumstances. Finally, rule 5 is a consequence of the fact that the photon does not interact with the spin.

5.9 Hund's Rules

We have seen that there are many levels in the energy spectrum of a multi-electron atom. Of these, one will have the lowest energy, and will form the **ground state**. All the others are excited states. Each atom has a *unique* ground state, which is determined by minimizing the energy of its valence electrons with the residual electrostatic and spin-orbit interactions included. In principle, this is a very complicated calculation. Fortunately, however, **Hund's rules** allow us to determine which level is the ground state for atoms that have LS coupling without lengthy calculation. The rules are:

(i) The term with the largest multiplicity (i.e., largest S) has the lowest energy.
(ii) For a given multiplicity, the term with the largest L has the lowest energy.
(iii) The level with $J = |L - S|$ or $J = L + S$ has the lowest energy depending, respectively, on whether the shell is less than, or more than, half-full.

The first of these rules basically tells us that the electrons try to align themselves with their spins parallel in order to minimize the exchange interaction. (See Chapter 6.) The other two rules follow from minimizing the spin-orbit interaction.

Let us have a look at carbon as an example. Carbon has an atomic number $Z = 6$ with two valence electrons in the outermost 2p shell. Each valence electron therefore has $l = 1$ and $s = 1/2$. Consider first the (2p,np) excited state configuration with one electron in the 2p shell and the other in the np shell, where $n \geq 3$. We have from Eq. (5.22) that $L = 1 \oplus 1 = 0$, 1 or 2, and $S = 1/2 \oplus 1/2 = 0$ or 1. We thus have three singlet terms (^{1}S, ^{1}P, ^{1}D), and three triplet terms (^{3}S, ^{3}P, ^{3}D). This gives rise to three singlet levels:

$$^1\mathrm{S}_0, \quad ^1\mathrm{P}_1, \quad ^1\mathrm{D}_2\,,$$

and seven triplet levels:

$$^3\mathrm{S}_1, \quad ^3\mathrm{P}_0, \quad ^3\mathrm{P}_1, \quad ^3\mathrm{P}_2, \quad ^3\mathrm{D}_1, \quad ^3\mathrm{D}_2, \quad ^3\mathrm{D}_3\,.$$

We thus have a confusing array of *ten* levels in the energy spectrum for the (2p,np) configuration.

The situation in the ground-state configuration (2p,2p) is simplified by the fact that the electrons are **equivalent**, i.e., in the *same* shell. The Pauli exclusion principle forbids the possibility that two or more electrons should have the same set of quantum numbers, and in the case of an atom with two valence electrons, it can be shown that this implies that $L + S$ must be equal to an even number. There is no easy explanation for this rule, but the simplest example of its application, namely to two electrons in the same s-shell, is considered in Section 6.3. For these two s-electrons, we have $L = 0 \oplus 0 = 0$ and $S = 1/2 \oplus 1/2 = 0$ or 1, giving rise to two terms: ^{1}S and ^{3}S. Both terms are allowed when the electrons are in different s-shells, but the ($L+S =$ even) rule tells us that only the singlet ^{1}S term is allowed if the electrons are in the same s-shell. The proof that the triplet term does not exist for the (1s,1s) ground-state configuration of helium is given in Section 6.3.

On applying the rule that $L+S$ must be even to the equivalent 2p electrons in the carbon ground state, we find that only the ^{1}S, ^{1}D, and ^{3}P terms are allowed, which means that only five of the ten levels listed above are possible:[1]

$$^1\mathrm{S}_0,\ \ ^1\mathrm{D}_2,\ \ ^3\mathrm{P}_0,\ \ ^3\mathrm{P}_1,\ \ ^3\mathrm{P}_2\,.$$

We can now apply Hund's rules to find out which of these is the ground state. The first rule states that the triplet levels have the lower energy. Since these all have $L = 1$, we do not need to consider the second rule. The shell is less than half full, and so we have $J = |L - S| = 0$. The ground state is thus the $^3\mathrm{P}_0$ level. All the other levels are excited states. Note that Hund's rules *cannot* be used to find the energy ordering of excited states with reliability. For example, consider the ten possible levels of the (2p,3p) excited-state configuration of carbon, listed previously. Hund's rules predict that the $^3\mathrm{D}_1$ level has the lowest energy, but the lowest state is actually the $^1\mathrm{P}_1$ level.

It is important to notice that if we had forgotten the rule that $L + S$ must be even, we would have incorrectly concluded that the ground state of carbon is a $^3\mathrm{D}_1$ term, which does not exist for the $2\mathrm{p}^2$ configuration. The situation can get even more complicated if we have more than two equivalent valence electrons, as, for example, in nitrogen or many transition metals. It is therefore safer to use a different version of Hund's rules, based on the allowed combinations of $\{m_s, m_l\}$ sublevels:

(i) Maximize the spin, and set $S = \sum m_s$.
(ii) Maximize the orbital angular momentum, subject to rule 1, and set $L = \sum m_l$.
(iii) $J = |L - S|$ if the shell is less than half-full; otherwise $J = |L + S|$.

These rules should work in all cases, since they incorporate the Pauli exclusion principle properly.

The ground-state levels for the first 11 elements, as worked out from Hund's rules, are listed in Table 5.1. Experimental results confirm these predictions.

[1] The full derivation of the allowed states for the $n\mathrm{p}^2$ configuration of a group IV (14) atom is considered, for example, in Woodgate (1980), §7.2.

Table 5.1 *Electronic configurations and ground-state terms of the first 11 elements in the periodic table.*

Z	Element	Configuration	Ground state
1	H	$1s^1$	$^2S_{1/2}$
2	He	$1s^2$	$^1S_0$
3	Li	$1s^2\ 2s^1$	$^2S_{1/2}$
4	Be	$1s^2\ 2s^2$	$^1S_0$
5	B	$1s^2\ 2s^2\ 2p^1$	$^2P_{1/2}$
6	C	$1s^2\ 2s^2\ 2p^2$	$^3P_0$
7	N	$1s^2\ 2s^2\ 2p^3$	$^4S_{3/2}$
8	O	$1s^2\ 2s^2\ 2p^4$	$^3P_2$
9	F	$1s^2\ 2s^2\ 2p^5$	$^2P_{3/2}$
10	Ne	$1s^2\ 2s^2\ 2p^6$	$^1S_0$
11	Na	$1s^2\ 2s^2\ 2p^6\ 3s^1$	$^2S_{1/2}$

	m_l		
m_s	-1	0	+1
+1/2		↑	↑
-1/2			

Figure 5.4 Distribution of the two valence electrons of the carbon ground state within the m_s and m_l states of the 2p shell.

Note that full shells always give $^1S_0$ levels with no net angular momentum: $S = L = J = 0$.

Example 5.3 Apply the second version of Hund's rules to deduce the ground-state level of carbon.

Solution In the ground state of carbon, we have two electrons in p-shells. The two electrons can go into the six possible $\{m_s, m_l\}$ sub-levels of the 2p shell, as shown in Figure 5.4.

(i) To get the largest value of the spin, we must have both electron spins aligned with $m_s = +1/2$. This gives $S = 1/2 + 1/2 = 1$.
(ii) Having put both electrons into spin-up states, we cannot now put both electrons into $m_l = +1$ states because of Pauli's exclusion principle. The best we can do is to put one into an $m_l = 1$ state and the other into an $m_l = 0$ state, as illustrated in Figure 5.4. This gives $L = 1 + 0 = 1$.
(iii) The shell is less than half-full, and so we have $J = |L - S| = 0$.

	m_l				
m_s	-2	-1	0	+1	+2
+1/2	↑	↑	↑	↑	↑
-1/2				↓	↓

Figure 5.5 Distribution of the seven d electrons in the ground state of the Co^{2+} ion among the available $\{m_s, m_l\}$ states.

We thus deduce that the ground state is the $^3P_0$ level, as before.

Example 5.4 Find the ground-state angular momentum level of the Co^{2+} ion.

Solution The electronic configuration of the Co^{2+} ion is [Ar] $3d^7$. We thus have seven valence electrons in a d-shell. The available states are shown in Figure 5.5. Hund's first rule says that we maximize the spin. This is achieved by putting the maximum number of electrons, five, in the top row, and the remaining two in the bottom row. This implies that:

$$S = \sum m_s = 5 \times (+1/2) + 2 \times (-1/2) = 3/2\,.$$

Hund's second rule says that we now maximize $\sum m_l$, subject to rule 1. This is done by putting the two electrons in the bottom row as far to the right as possible, in the $m_l = +2$ and $m_l = +1$ states, giving:

$$L = \sum m_l = 2 + 1 + 0 = 3\,.$$

The 0 here represents $\sum m_l$ for the top row: $\sum m_l = -2 + -1 + 0 + 1 + 2 = 0$. The shell is more than half full, so $J = L + S = 9/2$. The ground state is thus the $^4F_{9/2}$ level.

5.10 jj Coupling

The spin-orbit interaction gets larger as Z increases. (See, e.g., Eq. [7.43].) This means that in some atoms with large Z (e.g., tin with $Z = 50$), we can have a situation in which the spin-orbit interaction is stronger than the residual electrostatic interaction. In this regime, **jj coupling** occurs. The spin-orbit interaction couples the orbital and spin angular momenta of the individual electrons together first, and we then find the resultant $\boldsymbol{J}$ for the whole atom by adding together the individual $\boldsymbol{j}$s:

$$j_i = l_i + s_i$$
$$J = \sum_{i=1}^{N} j_i. \tag{5.36}$$

These J-states are then split by the weaker residual electrostatic potential, which acts as a perturbation.

Note that the values of L and S are *not* known in the jj coupling scheme, just as the values of j for the individual electrons are not known when LS coupling occurs. This is an example of the concept of 'good quantum numbers', and is discussed further in Section 8.2. The LS and jj coupling regimes are, ultimately, both simplifying approximations that derive from the central-field approximation, and their validity depends on whether the residual electrostatic or the spin-orbit interaction is the dominant perturbation.

Exercises

5.1 In this exercise, we show that the forms of the $\hat{\boldsymbol{L}}^2$ operator in Cartesian and spherical polar coordinates (i.e., Eqs. [5.7] and [2.31]) are completely equivalent to each other.

(a) Verify that the forms of $\hat{L}_x$ and $\hat{L}_y$ in spherical polar coordinates are:

$$\hat{L}_x = -i\hbar\left(-\sin\phi\frac{\partial}{\partial\theta} - \cot\theta\cos\phi\frac{\partial}{\partial\phi}\right),$$
$$\hat{L}_y = -i\hbar\left(\cos\phi\frac{\partial}{\partial\theta} - \cot\theta\sin\phi\frac{\partial}{\partial\phi}\right).$$

(b) Substitute these results, together with $\hat{L}_z$ from Eq. (2.40) into Eq. (5.7) to show that:

$$\hat{\boldsymbol{L}}^2 = -\hbar^2\left[\frac{1}{\sin\theta}\frac{\partial}{\partial\theta}\left(\sin\theta\frac{\partial}{\partial\theta}\right) + \frac{1}{\sin^2\theta}\frac{\partial^2}{\partial\phi^2}\right].$$

5.2 Show that $[\hat{A}^2, \hat{B}] = \hat{A}[\hat{A}, \hat{B}] + [\hat{A}, \hat{B}]\hat{A}$. (Hint: work backwards.)

5.3 Explain why it is necessary to apply a *nonuniform* magnetic field to deflect the atom in the Stern–Gerlach experiment. Justify the form of Eq. (5.17), starting from the energy of a magnetic dipole in a magnetic field, namely $U = -\boldsymbol{\mu}\cdot\boldsymbol{B}$.

5.4 Consider an np n'p configuration of a divalent atom in either the LS- or jj coupling regimes.

(a) What is the total number of combinations of $\{m_s, m_l\}$ states for the two electrons before they are coupled?
(b) The possible levels that can occur in the LS coupling regime were considered in Section 5.9. Verify that the total number of M_J states is the same as for part (a).
(c) Work out the possible J states that can occur in the jj coupling regime. Verify that the total number of M_J states is the same as for parts (a) and (b).

5.5 For each of the following transitions among LS-coupled levels, state whether it is allowed or forbidden for electric–dipole transitions. For forbidden transitions, state which selection rule(s) is/are violated:

$^3P_1 \rightarrow {}^1S_0$
$^2P_{3/2} \rightarrow {}^2S_{1/2}$
$^2D_{3/2} \rightarrow {}^2S_{1/2}$
$^4D_{3/2} \rightarrow {}^2P_{1/2}$
$^3F_4 \rightarrow {}^3G_4$
$^3G_4 \rightarrow {}^3H_5$
$^3H_5 \rightarrow {}^3G_3$
$^4I_{11/2} \rightarrow {}^4K_{11/2}$

5.6 The 3d and 4s shells of the Sc^+ ion (scandium, $Z = 21$) are close in energy, so that the following three configurations have to be considered in finding its lowest energy levels: $4s^2$, 3d4s, and $3d^2$.

(a) Deduce all the possible angular momentum levels that are possible for these three configurations.
(b) The ground state is actually one of the 3d4s levels. Explain why no electric–dipole transitions are possible from the ground state to any of the levels you have just listed.
(c) Use Hund's rules to deduce which of the angular momentum states of the 3d4s configuration is the ground state.
(d) The first strong transition from the ground state has an energy of 3.45 eV. Explain, with reasoning, whether you would expect the equivalent transition in neutral calcium ($Z = 20$) to have a larger or smaller energy.

5.7 Cadmium is a divalent metal with its two valence electrons in the 5s shell. The lowest energy transition in absorption occurs by promoting one of the 5s electrons to the 5p shell. Write down the angular momentum quantum numbers L, S, and J for the levels involved in the transition.

5.8 The Cr^{4+} ion has an electronic configuration of [Ar] $3d^2$. Write down the quantum numbers L, S, and J for the allowed angular momentum states of the 3d4p excited state configuration of the Cr^{4+} ion.

5.9 The ground state of silicon has an electronic configuration of [Ne] $3s^2 3p^2$.

(a) Use Hund's rules to determine the values of L, S, and J for the ground state of silicon.
(b) Write down the quantum numbers of L, S, and J for the 3p4s excited state configuration, and deduce which transitions are possible between the 3p4s excited state levels and the ground state.
(c) Explain why there are more L, S, J levels for the 3p4p configuration than in the $3p^2$ configuration.

5.10 Use Hund's rules to determine the ground state angular momentum quantum numbers L, S and J for the following atoms: (i) beryllium ($Z = 4$); (ii) phosphorus ($Z = 15$); (iii) manganese ($Z = 25$); (iv) iron ($Z = 26$); (v) neodymium ($Z = 60$).

6
Helium and Exchange Symmetry

The wave functions of a many-electron atom were first considered in Section 4.1. We noted there that a wave function of the type given in Eq. (4.7) is unphysical, as it does not take account of the fact that electrons are indistinguishable particles. In this chapter, we return to explore the consequences of the particle indistinguishability. We focus on two-electron systems, since this is the simplest case in which the issue occurs, using helium as the main example. We shall then briefly look at other atoms with two valence electrons, such as those in group 2 of the periodic table. The discussion will inevitably lead to the idea of exchange energy, and also to the Pauli exclusion principle.

6.1 Exchange Symmetry

Consider a many-electron atom with N electrons, as illustrated in Figure 6.1(a). The wave function of the atom will be a function of the coordinates of the individual electrons:

$$\Psi \equiv \Psi(\boldsymbol{r}_1, \boldsymbol{r}_2, \cdots, \boldsymbol{r}_K, \boldsymbol{r}_L, \cdots \boldsymbol{r}_N).$$

However, electrons are **indistinguishable particles**. It is not physically possible to stick labels on the individual electrons and then keep tabs on them as they move around their orbits. This means that the many-electron wave function must have **exchange symmetry**: i.e., no physically measurable property can be affected by an interchange of the particle labels. The most fundamental measurable property is probability density. On applying the principle of exchange symmetry to the probability density, we conclude that:

$$|\Psi(\boldsymbol{r}_1, \boldsymbol{r}_2, \cdots, \boldsymbol{r}_K, \boldsymbol{r}_L, \cdots \boldsymbol{r}_N)|^2 = |\Psi(\boldsymbol{r}_1, \boldsymbol{r}_2, \cdots, \boldsymbol{r}_L, \boldsymbol{r}_K, \cdots \boldsymbol{r}_N)|^2 . \tag{6.1}$$

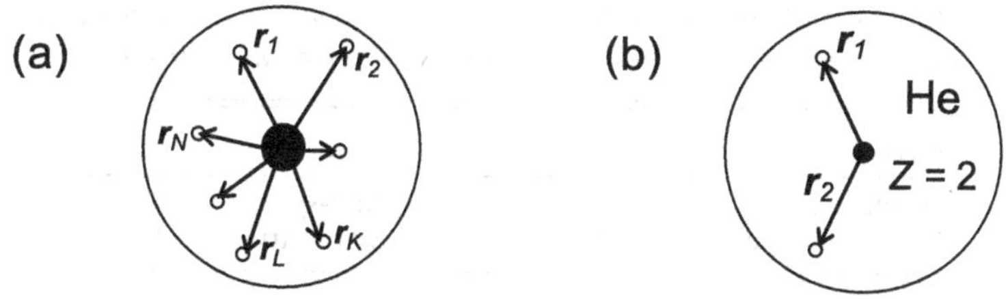

Figure 6.1 (a) A multi-electron atom with N electrons. (b) The helium atom.

Note that two of the labels have been switched on the right-hand side. Equation 6.1 will be satisfied if:

$$\Psi(\boldsymbol{r}_1,\boldsymbol{r}_2,\cdots,\boldsymbol{r}_K,\boldsymbol{r}_L,\cdots\boldsymbol{r}_N) = \pm\,\Psi(\boldsymbol{r}_1,\boldsymbol{r}_2,\cdots,\boldsymbol{r}_L,\boldsymbol{r}_K,\cdots\boldsymbol{r}_N)\,. \tag{6.2}$$

The + sign applies if the particles are **bosons**. These are said to be *symmetric* with respect to particle exchange. The − sign applies to **fermions**, which are *anti-symmetric* with respect to particle exchange.

Electrons are spin-1/2 particles, and are therefore fermions. Hence the wave function of a multi-electron atom must be *anti-symmetric* with respect to particle exchange. The discussion of exchange symmetry gets quite complicated when there are lots of electrons, and so we shall just concentrate here on the case of helium, which is the simplest example in which exchange effects become important.

6.2 Helium Wave Functions

Figure 6.1(b) shows a schematic diagram of a helium atom. It consists of one nucleus with $Z = 2$ and two electrons. The position coordinates of the electrons are written $\boldsymbol{r}_1$ and $\boldsymbol{r}_2$, respectively.

The Hamiltonian of helium is given in Eq. (6.8) and only acts on the spatial coordinates of the electrons. There is no interaction between the spatial coordinates and the spin, which means that, to first order, the overall wave function can be written as a product of a spatial wave function and a spin wave function:

$$\Psi = \psi_{\text{spatial}}(\boldsymbol{r}_1,\boldsymbol{r}_2)\,\psi_{\text{spin}}\,. \tag{6.3}$$

As we have seen above, the fact that electrons are indistinguishable fermions requires that Ψ must be anti-symmetric with respect to exchange of electrons 1

Table 6.1 *Allowed combinations of the exchange symmetries of the spatial and spin wave functions of fermionic particles.*

ψ_{spatial}	ψ_{spin}
Symmetric	Anti-symmetric ($S = 0$)
Anti-symmetric	Symmetric ($S = 1$)

Table 6.2 *Spin wave functions for a two-electron system. The arrows indicate whether the spin of the individual electrons is up or down (i.e., $m_s = +1/2$ or $-1/2$). The $+$ or $-$ sign in the symmetry column gives the symmetry with respect to particle exchange. M_S is obtained by adding the m_s values of the two electrons together.*

Spin wave function	symmetry	M_S
$\uparrow_1 \uparrow_2$	$+$	$+1$
$\frac{1}{\sqrt{2}}(\uparrow_1 \downarrow_2 + \downarrow_1 \uparrow_2)$	$+$	0
$\frac{1}{\sqrt{2}}(\uparrow_1 \downarrow_2 - \downarrow_1 \uparrow_2)$	$-$	0
$\downarrow_1 \downarrow_2$	$+$	-1

and 2. Table 6.1 lists the two possible combinations of wave function symmetries that can produce an antisymmetric total wave function.

Let us first consider the spin wave function. We have two spin 1/2 electrons, and so the total spin quantum number S is given by $S = 1/2 \oplus 1/2 = 1$ or 0. $S = 0$ states are called **singlets** because they only have one possible M_S value, namely 0, while $S = 1$ states are called **triplets** because they have three possible M_S values, namely $+1$, 0, and -1.

There are four possible ways of combining the spins of the two electrons so that the total wave function has exchange symmetry. These are listed in Table 6.2. The component of $\mathbf{S}$ along the z-axis is obtained by adding together the s_z values of the individual electrons. This gives the S_z value of the whole helium atom, and hence the spin quantum number M_S. For the states with $M_S = 0$, one electron must be spin-up, and the other down. In order for the wave function to have exchange symmetry, we must allow for both possibilities: electron 1 up and electron 2 down, and vice versa. The factor of $1/\sqrt{2}$ is required for normalization. (See Exercise 6.1).

The exchange symmetries of the four M_S states in Table 6.2 are found by swapping the labels on the electrons. It is immediately obvious that the $M_S = +1$ and $M_S = -1$ states are symmetric, since the wave functions are unchanged when swapping the labels. The symmetry of the $M_S = 0$ states is found as follows:

$$\Psi^+ = \frac{1}{\sqrt{2}}(\uparrow_1\downarrow_2 + \downarrow_1\uparrow_2) \overset{\text{swap labels}}{\longrightarrow} \frac{1}{\sqrt{2}}(\uparrow_2\downarrow_1 + \downarrow_2\uparrow_1) = +\Psi^+ .$$
$$\Psi^- = \frac{1}{\sqrt{2}}(\uparrow_1\downarrow_2 - \downarrow_1\uparrow_2) \overset{\text{swap labels}}{\longrightarrow} \frac{1}{\sqrt{2}}(\uparrow_2\downarrow_1 - \downarrow_2\uparrow_1) = -\Psi^- . \quad (6.4)$$

This shows that Ψ^+ is symmetric, and Ψ^- is anti-symmetric. We therefore have three symmetric M_S wave functions, namely $+1$, -1, and Ψ^+, and one anti-symmetric one, namely Ψ^-. Since the two S-states must have well-defined symmetries, and the symmetric $M_S = +1$ and $M_S = -1$ are derived unambiguously from the triplet state, we deduce that Ψ^+ must be the $M_S = 0$ state of the triplet. This then implies that Ψ^- is the singlet state. We thus conclude that triplet spin states have positive exchange symmetry, while singlets have negative symmetry, as noted in Table 6.1.

Now let us consider the spatial wave functions. The state of the atom will be specified by the configuration of the two electrons. In the ground state, both electrons are in the 1s shell, and so we have a configuration of $1s^2$. In the excited states, one of the electrons will be in a higher shell. The configuration is thus given by the n, l values of the two electrons, and we write the configuration as $(n_1 l_1,\ n_2 l_2)$, where $(n_1 l_1)$ is normally (1 0), i.e., 1s. This means that the spatial part of the helium wave function must contain terms of the type $u_A(\boldsymbol{r}_1)\, u_B(\boldsymbol{r}_2)$, where $u_{nl}(\boldsymbol{r})$ is the wave function for an electron with quantum numbers n and l, and the subscripts A and B stand for the quantum numbers n, l of the two electrons.

As with the spin wave functions, we must take account of the fact that the electrons are indistinguishable: we cannot distinguish between the state with electron 1 in state A and electron 2 in state B, and vice versa. $u_B(\boldsymbol{r}_1)\, u_A(\boldsymbol{r}_2)$ is therefore an equally valid wave function for the particular electronic configuration. The wave function for the configuration A, B must therefore take the form:

$$\psi_{AB}(\boldsymbol{r}_1,\boldsymbol{r}_2) = \frac{1}{\sqrt{2}}\Big(u_A(\boldsymbol{r}_1)\, u_B(\boldsymbol{r}_2) \pm u_B(\boldsymbol{r}_1)\, u_A(\boldsymbol{r}_2)\Big) . \quad (6.5)$$

The $1/\sqrt{2}$ factor ensures that $\psi_{AB}(\boldsymbol{r}_1,\boldsymbol{r}_2)$ is correctly normalized. (See Exercise 6.2.) Following the same sort of reasoning as in Eq. (6.4), it is easy to verify that the wave function with the plus sign is symmetric with respect to particle exchange, while the wave function with the minus sign is antisymmetric.

We have seen above that spin singlet and triplet states are, respectively, antisymmetric and symmetric under exchange symmetry. The fact that the overall symmetry must be negative then implies that spin singlets and triplets must be paired off with symmetric and antisymmetric spatial wave functions respectively. This leads to the pairing of spin and spatial wave functions shown

Table 6.3 *Spin and spatial wave functions for a two-electron atom with electronic configuration designated by the labels* A *and* B.

S	M_S	ψ_{spin}	ψ_{spatial}
0	0	$\frac{1}{\sqrt{2}}(\uparrow_1\downarrow_2 - \downarrow_1\uparrow_2)$	$\frac{1}{\sqrt{2}}\Big(u_A(\boldsymbol{r}_1)\,u_B(\boldsymbol{r}_2) + u_B(\boldsymbol{r}_1)\,u_A(\boldsymbol{r}_2)\Big)$
	+1	$\uparrow_1\uparrow_2$	
1	0	$\frac{1}{\sqrt{2}}(\uparrow_1\downarrow_2 + \downarrow_1\uparrow_2)$	$\frac{1}{\sqrt{2}}\Big(u_A(\boldsymbol{r}_1)\,u_B(\boldsymbol{r}_2) - u_B(\boldsymbol{r}_1)\,u_A(\boldsymbol{r}_2)\Big)$
	−1	$\downarrow_1\downarrow_2$	

in Table 6.3. In Section 6.4 we shall calculate the energies of the symmetric and anti-symmetric spatial wave functions and see that they are different. The opposite pairing of the spatial and spin symmetries then links spin singlets and triplets to different energy states. This is a surprising result when you consider that spin does not directly enter the Hamiltonian given in Eq. (6.8).

6.3 The Pauli Exclusion Principle

Let us consider what happens if we try to put the two electrons in the same atomic shell, as, for example, in the ground-state $1s^2$ configuration of helium. The spatial wave functions will be given by Eq. (6.5) with $A = B$. The antisymmetric combination with the minus sign in the middle is zero in this case. From Table 6.3 we see that this implies that there are no triplet $S = 1$ states if both electrons are in the same shell.

The absence of the triplet state for the $1s^2$ configuration is equivalent to the **Pauli exclusion principle**. We are trying to put two electrons in the same state as defined by the n, l, m_l quantum numbers. This is only possible if the two electrons have different m_s values. In other words, their spins must be aligned antiparallel. The $S = 1$ state contains terms with both spins pointing in the same direction, and is therefore not allowed. The analysis of the symmetry of the wave function discussed here shows us that the Pauli exclusion principle is a consequence of the fact that electrons are indistinguishable fermions.

The fact that the triplet state does not exist for the helium ground state is a demonstration of the rule that $L + S$ must be even for a two-electron atom with both electrons in the same shell. This rule was introduced without any justification in Section 5.9. In the case of the $1s^2$ configuration, we have $L = 0$, and therefore $S = 1$ is not allowed. The general justification of the rule is beyond the scope of this book, but the example of the helium ground state at least demonstrates that the rule is true for the simplest case.

6.3.1 Slater Determinants

The antisymmetric wave function given in Eq. (6.5) can be written as a determinant:

$$\psi_{\text{spatial}} = \frac{1}{\sqrt{2}} \begin{vmatrix} u_A(\boldsymbol{r}_1) & u_A(\boldsymbol{r}_2) \\ u_B(\boldsymbol{r}_1) & u_B(\boldsymbol{r}_2) \end{vmatrix} . \tag{6.6}$$

This can be generalized to give the correct anti-symmetric wave function when we have more than two electrons:

$$\Psi = \frac{1}{\sqrt{N!}} \begin{vmatrix} u_\alpha(1) & u_\alpha(2) & \cdots & u_\alpha(N) \\ u_\beta(1) & u_\beta(2) & \cdots & u_\beta(N) \\ \vdots & \vdots & \ddots & \vdots \\ u_\nu(1) & u_\nu(2) & \cdots & u_\nu(N) \end{vmatrix} , \tag{6.7}$$

where $\{\alpha, \beta, \cdots, \nu\}$ each represent a set of quantum numbers $\{n, l, m_l, m_s\}$ for the individual electrons, and $\{1, 2, \cdots, N\}$ are the electron labels. Determinants of this type are called **Slater determinants**. A determinant is zero if any two rows are equal, which tells us that each electron in the atom must have a unique set of quantum numbers, as required by the Pauli exclusion principle. We shall not make further use of Slater determinants in this book. They are mentioned here for completeness.

6.4 The Hamiltonian for Helium

The gross-structure Hamiltonian for the helium atom is given by:

$$\hat{H} = \left(-\frac{\hbar^2}{2m}\nabla_1^2 - \frac{2e^2}{4\pi\epsilon_0 r_1}\right) + \left(-\frac{\hbar^2}{2m}\nabla_2^2 - \frac{2e^2}{4\pi\epsilon_0 r_2}\right) + \frac{e^2}{4\pi\epsilon_0 r_{12}}, \tag{6.8}$$

where $r_{12} = |\boldsymbol{r}_1 - \boldsymbol{r}_2|$. The first two bracketed terms account for the kinetic energy of the two electrons and their attraction toward the nucleus, which has a charge of $+2e$. The final term is the electron-electron repulsion. It is this repulsion that makes the equations difficult to deal with.

The Hamiltonian in Eq. (6.8) only acts on the spatial coordinates, which explains why the wave function separates into a product of spatial and spin wave functions as in Eq. (6.3). The Hamiltonian is unaffected by exchange of the electron labels 1 and 2, which implies that the states must have clear symmetry under particle exchange, as discussed in Section 6.2. Spin does not enter the Hamiltonian, and so the energy will be determined only by the spatial part of the wave function. However, the pairing of spin singlets and triplets with different spatial wave functions implies that the spin states end up with different energies.

In Section 4.1 and following we described how to deal with a many-electron Hamiltonian by splitting it into a central field and a residual electrostatic interaction. In the case of helium, we just have one Coulomb repulsion term and it is easier to go back to first principles. We can then use the correctly symmeterized wave functions to calculate the energies for specific electronic configurations.

The energy of the electronic configuration $(n_1 l_1,\ n_2 l_2)$ is found by computing the expectation value of the Hamiltonian:

$$\langle E \rangle = \iint \psi^*_{\text{spatial}}\, \hat{H}\, \psi_{\text{spatial}}\, \mathrm{d}^3\boldsymbol{r}_1 \mathrm{d}^3\boldsymbol{r}_2\,. \tag{6.9}$$

The spin wave functions do not appear here because the Hamiltonian does not affect the spin directly, and so the spin wave functions just integrate out to unity.

We start by rewriting the Hamiltonian given in Eq. (6.8) in the following form:

$$\hat{H} = \hat{H}_1 + \hat{H}_2 + \hat{H}_{12}\,, \tag{6.10}$$

where

$$\hat{H}_i = -\frac{\hbar^2}{2m}\nabla_i^2 - \frac{2e^2}{4\pi\varepsilon_0 r_i}\,, \tag{6.11}$$

$$\hat{H}_{12} = \frac{e^2}{4\pi\varepsilon_0\, |\boldsymbol{r}_1 - \boldsymbol{r}_2|}\,. \tag{6.12}$$

The energy can be split into three parts:

$$E = E_1 + E_2 + E_{12}\,, \tag{6.13}$$

where:

$$E_i = \iint \psi^*_{\text{spatial}} \hat{H}_i \psi_{\text{spatial}} \mathrm{d}^3\boldsymbol{r}_1 \mathrm{d}^3\boldsymbol{r}_2\,, \tag{6.14}$$

and:

$$E_{12} = \iint \psi^*_{\text{spatial}} \hat{H}_{12} \psi_{\text{spatial}} \mathrm{d}^3\boldsymbol{r}_1 \mathrm{d}^3\boldsymbol{r}_2\,. \tag{6.15}$$

The first two terms in Eq. (6.13) represent the energies of the two electrons in the absence of the electron-electron repulsion. If we use hydrogenic wave functions as our starting point,[1] then the energies will just be given by:

$$E_1 + E_2 = -\frac{4R_H}{n_1^2} - \frac{4R_H}{n_2^2}, \tag{6.16}$$

where the factor of $4 \equiv Z^2$ accounts for the nuclear charge. (See Appendix C for the evaluation of the integrals.) The third term is the electron-electron Coulomb repulsion energy:

$$E_{12} = \iint \psi^*_{\text{spatial}} \frac{e^2}{4\pi\epsilon_0 r_{12}} \psi_{\text{spatial}} \, \mathrm{d}^3\boldsymbol{r}_1 \mathrm{d}^3\boldsymbol{r}_2 . \tag{6.17}$$

As shown in Appendix C, the end result for the correctly symmeterized wave functions given in Eq. (6.5) is:

$$E_{12} = D_{AB} \pm J_{AB}, \tag{6.18}$$

where the plus sign is for singlets and the minus sign is for triplets. D_{AB} is the **direct** Coulomb energy given by:

$$D_{AB} = \frac{e^2}{4\pi\epsilon_0} \iint u_A^*(\boldsymbol{r}_1)\, u_B^*(\boldsymbol{r}_2) \,\frac{1}{r_{12}}\, u_A(\boldsymbol{r}_1)\, u_B(\boldsymbol{r}_2)\, \mathrm{d}^3\boldsymbol{r}_1\, \mathrm{d}^3\boldsymbol{r}_2 , \tag{6.19}$$

and J_{AB} is the **exchange** Coulomb energy given by:

$$J_{AB} = \frac{e^2}{4\pi\epsilon_0} \iint u_A^*(\boldsymbol{r}_1)\, u_B^*(\boldsymbol{r}_2) \,\frac{1}{r_{12}}\, u_B(\boldsymbol{r}_1)\, u_A(\boldsymbol{r}_2)\, \mathrm{d}^3\boldsymbol{r}_1\, \mathrm{d}^3\boldsymbol{r}_2 . \tag{6.20}$$

Note that in the exchange integral, we are integrating the expectation value of $1/r_{12}$ with each electron in a different shell. This is why it is called the exchange energy. The total energy of the configuration $(n_1 l_1,\ n_2 l_2)$ is then given by:

$$E = E_1 + E_2 + D_{AB} \pm J_{AB}, \tag{6.21}$$

where the plus and minus signs apply to singlet ($S = 0$) and triplet ($S = 1$) states, respectively. In the hydrogenic wave function approximation, this becomes:

$$E(n_1 l_1,\ n_2 l_2) = -\frac{4R_H}{n_1^2} - \frac{4R_H}{n_2^2} + D_{AB} \pm J_{AB} . \tag{6.22}$$

[1] It is natural to use hydrogenic wave functions as our initial guess for $u_A(\boldsymbol{r}_i)$ and $u_B(\boldsymbol{r}_i)$. However, the electron-electron repulsion term is not a *small* perturbation, and the *actual* wave functions will be significantly different due to the departure of the effective potential (see Eq. [4.20]) from the strict $1/r$ dependence of hydrogen. This means that the discussion in the main text is only approximately true. Nevertheless, it does correctly identify the exchange terms. A more detailed discussion of the helium wave functions may be found, for example, in Woodgate (1980), Chapter 5.

The key point is that the energies of the singlet and triplet states differ by $2J_{AB}$.

Here are a few points of special note that emerge from this discussion of the exchange energy:

- The exchange splitting is *not* a small energy. It is part of the *gross structure* of the atom. This contrasts with other spin-dependent effects, such as the spin-orbit interaction (see Chapter 7), that only contribute to the *fine structure*. The value of $2J_{AB}$ for the first excited state of helium, namely the 1s2s configuration, is 0.80 eV.
- We can give a simple physical reason why the symmetry of the spatial wave function affects the energy so much. If we put $\boldsymbol{r}_1 = \boldsymbol{r}_2$ into Eq. (6.5), we see that we get $\psi_{\text{spatial}} = 0$ for the antisymmetric state. This means that there is zero probability that the two electrons can sit on top of each other at the same point in space in the triplet state, which reduces the Coulomb repulsion energy. On the other hand, $\psi_{\text{spatial}}(\boldsymbol{r}_1 = \boldsymbol{r}_2) \neq 0$ for singlet states with symmetric spatial wave functions. They therefore have a larger Coulomb repulsion energy.
- The exchange energy is sometimes written in the form:

$$\Delta E_{\text{exchange}} \propto -J\,\boldsymbol{s}_1 \cdot \boldsymbol{s}_2\,. \tag{6.23}$$

  This emphasizes the point that the change of energy correlates with the relative alignment of the electron spins. If both spins are aligned, as they are in the triplet states, the energy is lower. If the spins are antiparallel, the energy is higher.
- The notation given in Eq. (6.23) is extensively used when explaining the phenomenon of **ferromagnetism**. The interaction that induces the spins in iron to align parallel to each other is the spin-dependent change of the Coulomb repulsion energy caused by symmeterizing the wave function. In antiferromagntic materials, the exchange constant J is negative, and the spins align antiparallel to their nearest neighbors.

6.5 The Helium Term Diagram

The energy-level diagram for helium can be worked out if we can evaluate the direct and exchange Coulomb energies. The total energy for each configuration is given by Eq. (6.21).

Table 6.4 *Electron configurations for the ground state and excited states of helium.*

Ground state	1s 1s ($\equiv$ 1s^2)
First excited state	1s 2s
Second excited state	1s 2p
Third excited state	1s 3s
Fourth excited state	1s 3p
Ionization limit	1s ∞l

The Ground State

In the ground state, both electrons are in the 1s shell, and we have a configuration of 1s^2. We have seen above that we can only have $S = 0$ for this configuration. The energy is thus given by:

$$E(1\text{s}^2) = 2E_{1\text{s}} + \left(D_{1\text{s}^2} + J_{1\text{s}^2}\right), \tag{6.24}$$

where $E_{1\text{s}}$ is the energy of the 1s electrons with the electron-electron repulsion neglected. The computation of $E(1\text{s}^2)$ from first principles is nontrivial, and involves finding self-consistent wave functions in a screened nuclear field as in Eq. (4.11). If we assume hydrogenic wave functions, then $E_{1\text{s}} = -4R_{\text{H}}$, and $D+J \approx 34$ eV (See Woodgate (1980), §5.2), implying $E(1\text{s}^2) = 2\times(-54.4)+34 = -75$ eV. Analysis of the ionization potentials enables us to deduce that the actual value of $E(1\text{s}^2)$ is -79.0 eV. This shows that the hydrogenic wave function approximation is a reasonable starting point, but does not give the exact energy as it neglects the effect of the Coulomb repulsion on the wave functions.

Ionization Potentials

The excited states of helium are made by promoting one of the electrons to higher shells, according to the scheme shown in Table 6.4. We do not need to consider two-electron jump excited states, such as the 2s 2s configuration here, as the energy of such states is above the single-electron ionization limit. This can be seen from a simplistic Bohr model picture neglecting electron-electron repulsion, where an energy of $2\times 3/4 \times 4R_H = 6R_{\text{H}}$ is required to promote two electrons to the $n = 2$ shell, which is larger the energy to move one electron to infinity, namely $4R_{\text{H}}$.

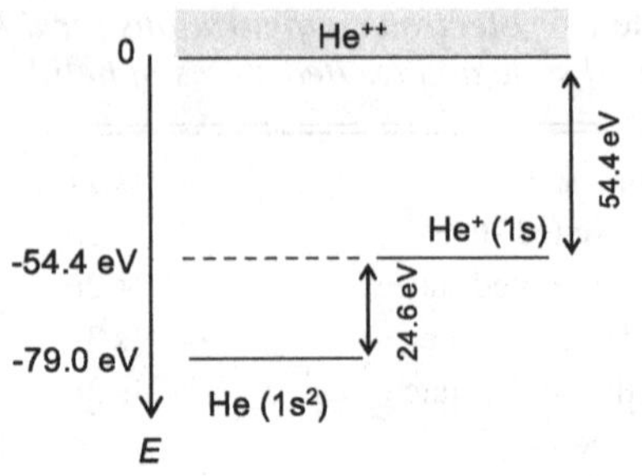

Figure 6.2 The ionization energies of helium.

When the second electron has been promoted into the energy continuum at $n_2 = \infty$, we are left with a singly ionized helium atom: He^+. This is now a hydrogenic system. We have one electron in the 1s shell orbiting around a nucleus with charge $+2e$, and the energy is just $-Z^2R_H = -54.4\,\text{eV}$. This means that the *second* ionization potential is equal to 54.4 eV.

The experimental value of the first ionization potential is 24.6 eV, which means that the $1s^2$ ground state lies 24.6 eV below the He^+ ground state, as illustrated in Figure 6.2. The absolute energy of the helium ground state is thus $-54.4 - 24.6 = -79.0\,\text{eV}$. Note that this is an example of the point made in the discussion of Figure 1.5 in Section 1.3, namely that the ionization limit of the neutral He atom corresponds to the ground state of the He^+ ion.

Optical Spectra

The first few excited-state configurations of helium are listed in Table 6.4. For each configuration, we have two spin states corresponding to $S = 0$ or 1. The triplet $S = 1$ terms are at lower energy than the singlets. (See Eq. [6.21].) The Grotrian diagram for helium is shown in Figure 6.3. Note that the singlet and triplet states are separated. This is because the $\Delta S = 0$ selection rule tells us that we cannot get optical transitions between singlets and triplets.

It is clear from Figure 6.3. that the energy of the (1s, nl) state approaches the hydrogenic energy $-R_H/n^2$ when n is large. This was also the case for alkali atoms, as mentioned in Section 4.5. The excited electron in a high n state is well outside the 1s shell, which just partly screens the nuclear potential. The outer electron then sees $Z_{\text{eff}} = 1$, and we have a purely hydrogenic potential.

Excited states such as the 1s 2s configuration are said to be **metastable**. They cannot relax easily to the ground state, and therefore have very long lifetimes. The relaxation would involve a 2s → 1s transition, which is forbidden by the

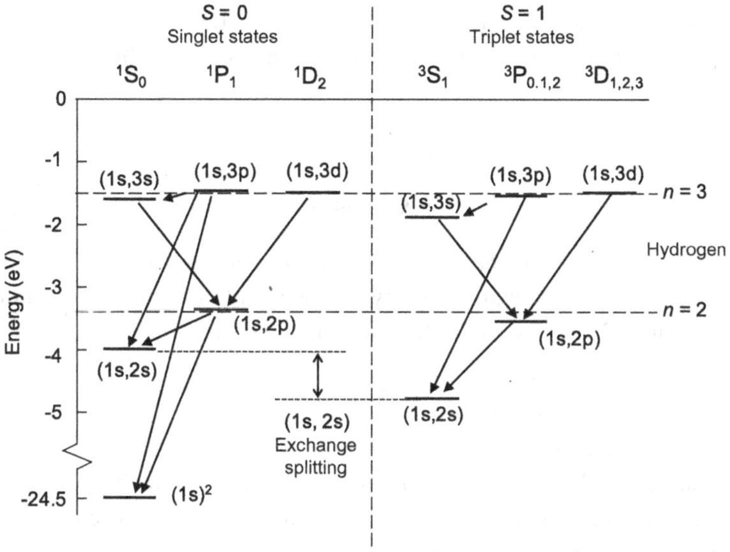

Figure 6.3 Approximate energy-level diagram for helium. The diagram is split into singlet and triplet states since only $\Delta S = 0$ transitions are allowed by the E1 selection rules. The energy difference between the singlet and triplet terms for the same configuration is caused by the exchange energy, as identified for the 1s 2s configuration. The dashed line shows the hydrogen energies for comparison. The arrows indicate the allowed E1 transitions.

$\Delta l = \pm 1$ selection rule. In the absence of collisions, the metastable states decay by forbidden or two-photon transitions, as discussed in Section 12.5.

The fact that helium has a series of transitions between states with $S = 0$ has important consequences for the Zeeman effect. This point will be developed in detail in Chapter 8.

6.6 Optical Spectra of Divalent Metals

The principles that we have been discussing here with respect to helium apply equally well to other two-electron atoms. In particular, they apply to the alkaline-earth metals in group 2 of the periodic table (i.e., Be, Mg, Ca, etc.), and also those in group 12 (i.e., Zn, Cd, Hg). Similar principles would also apply to divalent ions such as Al^{+}.

The key feature of all these elements is that they have two valence electrons in an s-shell outside a filled shell. The term diagram for group 2 elements would appear generically similar to Figure 6.3, and the optical spectra would follow similar rules, with singlet and triplet transitions split by the exchange energy.

Exercises

6.1 Show that $\Psi^{\pm} = C(\uparrow_1 \downarrow_2 \pm \downarrow_1 \uparrow_2)$ is normalized when $C = 1/\sqrt{2}$.

6.2 Use the orthonormality property of the eigenfunctions of the Hamiltonian to show that the wave function of Eq. (6.5) satisfies:

$$\iint \psi_{AB}(\boldsymbol{r}_1, \boldsymbol{r}_2)^* \psi_{AB}(\boldsymbol{r}_1, \boldsymbol{r}_2)\, \mathrm{d}^3\boldsymbol{r}_1\, \mathrm{d}^3\boldsymbol{r}_2 = 1\,.$$

6.3 The first ionization potential of helium is 24.6 eV. What is the effective charge seen by the 1s electron? Justify the answer you obtain.

6.4 The first ionization potential of beryllium ($Z = 4$) is 9.33 eV. Why is it smaller than that of helium?

6.5 The first ionization potential of magnesium ($Z = 12$) is 7.646 eV. Estimate the wavelength of the 3s8p $^1\mathrm{P}_1 \rightarrow$ 3s3s $^1\mathrm{S}_0$ transition.

6.6 The 1s4p $\rightarrow$ 1s2s transition in helium consists of two lines with wavelengths of 316 nm and 397 nm.

(a) Identify the angular momentum quantum numbers L and S of the levels involved in both transitions.

(b) Deduce the exchange splitting of the 1s 2s configuration. (You may assume that the exchange splitting of the 1s 4p level is negligible.)

(c) Calculate the wavelength of the 4p $\rightarrow$ 2s transition in the He^+ ion.

6.7 The 1s3d $\rightarrow$ 1s2p transitions of helium occur at 668 and 588 nm, while the 1s2p $\rightarrow$ 1s2s transitions occur at 2058 and 1083 nm. Use this information to deduce the exchange splittings of the 1s 2p and 1s 2s configurations, stating your answer in cm^{-1} units. (Assume that the exchange splitting of the 1s 3d configuration is negligible.)

6.8 Cadmium is a group 12 element with a ground-state electronic configuration of [Kr] $4\mathrm{d}^{10}5\mathrm{s}^2$. The 508.6 nm emission line corresponds to the 5s6s $^3\mathrm{S}_1 \rightarrow$ 5s5p $^3\mathrm{P}_2$ transition, and is one of a triplet of lines; the other two have wavelengths of 480.0 nm and 467.8 nm.

(a) Give two reasons why you would not expect to observe the 508.6 nm line in an absorption spectrum.

(b) Deduce the atomic levels involved in the other two transitions. (The magnitude of the splittings is considered in Exercise 7.9.)

7
Fine Structure and Nuclear Effects

The **gross structure** of atoms, which has been our focus up to this stage in the book, only includes the largest interaction terms in the Hamiltonian: the electron kinetic energy, the electron-nuclear attraction, and the electron-electron repulsion. It is now time to discuss the smaller interactions that arise from magnetic effects. In this chapter, we consider the effects associated with *internal* magnetic fields, which cause **fine structure** and **hyperfine structure** in atomic spectra. The discussion of the effects produced by *external* fields is given in the next chapter.

7.1 Orbital Magnetic Dipoles

The argument for the concept of spin in the context of the Stern–Gerlach experiment in Section 5.2.2 was based on the assertion that there is a connection between angular momentum and magnetic dipoles. This relationship is easy to understand for the orbital angular momentum. Let us first consider an electron in a circular Bohr orbit, as illustrated in Figure 7.1(a). The electron orbit is equivalent to a current loop, which generates a magnetic dipole of magnitude μ given by:

$$\mu = i \times \text{Area} = -(e/T) \times (\pi r^2)\,, \tag{7.1}$$

where T is the period of the orbit. Now $T = 2\pi r/v$, and so we obtain:

$$\mu = -\frac{ev}{2\pi r}\pi r^2 = -\frac{e}{2m_\mathrm{e}}m_\mathrm{e}vr = -\frac{e}{2m_\mathrm{e}}L, \tag{7.2}$$

where we have substituted L for the orbital angular momentum $m_\mathrm{e}vr$.

The generalization of Eq. (7.2) to noncircular orbits proceeds as follows: Consider an electron at position vector $\boldsymbol{r}$ in a noncircular orbit as shown in

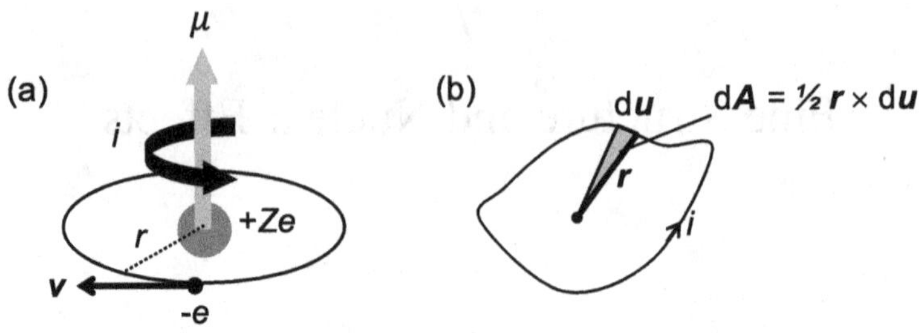

Figure 7.1 (a) The orbital motion of the electron around the nucleus in a circular Bohr orbit is equivalent to a current loop, which generates a magnetic dipole moment. (b) Magnetic dipole moment of an electron in a noncircular orbit.

Figure 7.1(b). The magnetic dipole moment is given by:

$$\boldsymbol{\mu} = \oint i\,\mathrm{d}\boldsymbol{A}\,, \tag{7.3}$$

where i is the current in the loop and $\mathrm{d}\boldsymbol{A}$ is the incremental area swept out by the electron as it performs its orbit. It is apparent from Figure 7.1(b) that $\mathrm{d}\boldsymbol{A}$ is related to the path element $\mathrm{d}\boldsymbol{u}$ by:

$$\mathrm{d}\boldsymbol{A} = \frac{1}{2}\boldsymbol{r} \times \mathrm{d}\boldsymbol{u}\,, \tag{7.4}$$

and so Eq. (7.3) becomes:

$$\boldsymbol{\mu} = \frac{1}{2}\oint i\boldsymbol{r} \times \mathrm{d}\boldsymbol{u}\,. \tag{7.5}$$

We can write the current as $i = \mathrm{d}q/\mathrm{d}t$, where q is the charge, which implies:

$$\begin{aligned}\boldsymbol{\mu} &= \frac{1}{2}\oint \frac{\mathrm{d}q}{\mathrm{d}t}\boldsymbol{r} \times \mathrm{d}\boldsymbol{u}\,,\\ &= \frac{1}{2}\oint \mathrm{d}q\,\boldsymbol{r} \times \frac{\mathrm{d}\boldsymbol{u}}{\mathrm{d}t}\,,\\ &= \frac{1}{2}\oint \mathrm{d}q\,\boldsymbol{r} \times \boldsymbol{v}\,,\\ &= \frac{1}{2m_\mathrm{e}}\oint \mathrm{d}q\,\boldsymbol{r} \times \boldsymbol{p}\,,\end{aligned} \tag{7.6}$$

where $\boldsymbol{v}$ is the velocity and $\boldsymbol{p}$ the momentum. The angular momentum is defined as usual by:

$$\boldsymbol{L} = \boldsymbol{r} \times \boldsymbol{p} \tag{7.7}$$

and so we finally obtain:

$$\boldsymbol{\mu} = \frac{1}{2m_{\mathrm{e}}} \oint \boldsymbol{L} \mathrm{d}q = \frac{1}{2m_{\mathrm{e}}} \boldsymbol{L} \oint \mathrm{d}q = \frac{1}{2m_{\mathrm{e}}} \boldsymbol{L}(-e) \equiv -\frac{e}{2m_{\mathrm{e}}} \boldsymbol{L}, \tag{7.8}$$

as in Eq. (7.2). Note that the result works because the angular momentum $\boldsymbol{L}$ is a constant of the motion in the central-field approximation (see Section 5.2.1), and so can be taken out of the integral.

Equation (7.8) shows us that the orbital angular momentum is directly proportional to the magnetic dipole moment. The proportionality constant $e/2m_{\mathrm{e}}$ is called the **gyromagnetic ratio**. We can recall that the orbital angular momentum in an atom is quantized with magnitude and z-component given by (see Eqs. [2.44] and [2.45]):

$$|\boldsymbol{L}|^2 = l(l+1)\hbar^2 \,, \tag{7.9}$$

$$L_z = m_l \hbar \,, \tag{7.10}$$

where l is an integer ≥ 0, and m_l runs in integer steps from $-l$ to $+l$. It is then apparent that the magnitude and z-component of an atomic magnetic dipole are given, respectively, by:

$$|\boldsymbol{\mu}| = \frac{e}{2m_{\mathrm{e}}} |\boldsymbol{L}| = \frac{e}{2m_{\mathrm{e}}} \hbar\sqrt{l(l+1)} \equiv \mu_{\mathrm{B}}\sqrt{l(l+1)} \,, \tag{7.11}$$

$$\mu_z = -\frac{e}{2m_{\mathrm{e}}} L_z = -\frac{e}{2m_{\mathrm{e}}} m_l \hbar \equiv -\mu_{\mathrm{B}}\, m_l \,, \tag{7.12}$$

where μ_{B} is the **Bohr magneton** defined by:

$$\mu_{\mathrm{B}} = \frac{e\hbar}{2m_{\mathrm{e}}} = 9.27 \times 10^{-24}\, \mathrm{JT}^{-1}. \tag{7.13}$$

This shows that the size of atomic dipoles is of order μ_{B}.

7.2 Spin Magnetism

We have seen in Section 5.2.2 that electrons also have spin angular momentum. The deflections measured in Stern–Gerlach experiments (see Figure 5.1) and other experimental measurements enable the magnitude of the magnetic moment due to spin to be determined. The component along the z-axis is found to obey:

$$\mu_z = -g_{\mathrm{s}} \mu_{\mathrm{B}} m_s \,, \tag{7.14}$$

where g_{s} is the **g factor** of the electron, and $m_s = \pm 1/2$ is the spin magnetic quantum number. This is identical in form to Eq. (7.12) apart from the factor of g_s. The experimental value of g_s was found to be close to 2. The Dirac equation predicts that g_s should be exactly 2, and more recent calculations based on quantum electrodynamics (QED) give a value of $2.0023192\cdots$, which agrees very accurately with the most precise experimental data.

It should be noted that other branches of physics sometimes use a different sign convention in which the electron spin g factor is negative. The negative charge of the electron is factored into the g factor, which is defined by:

$$\boldsymbol{\mu}_{\text{spin}} = g_e \frac{\mu_B}{\hbar} \boldsymbol{s}, \tag{7.15}$$

where $\boldsymbol{s}$ is the spin angular momentum, and $\mu_B/\hbar = e/2m_e$ is the magnitude of the electron gyromagnetic ratio. This implies:

$$\mu_z = g_e \frac{\mu_B}{\hbar} s_z = g_e \mu_B m_s . \tag{7.16}$$

On comparing to Eq. (7.14), it is apparent that g_s and g_e are related to each other through:

$$g_s = |g_e| = -g_e . \tag{7.17}$$

The convention in which the sign of the g factor relates to the charge of the particle is frequently used in tables of fundamental constants. However, in atomic physics we are almost always dealing with electrons, and so it is more convenient to use the the positive value g_s rather than the negative one g_e.

7.3 Spin-Orbit Coupling

The fact that electrons in atoms have both orbital and spin angular momentum leads to a new interaction term in the Hamiltonian called **spin-orbit coupling**. Sophisticated theories (e.g., those based on the Dirac equation) indicate that this is actually a relativistic effect. At the level of this book, however, it is more intuitive to consider spin-orbit coupling as the interaction between the magnetic field due to the orbital motion of the electron and its magnetic moment due to spin.

7.3.1 Spin-Orbit Coupling in the Bohr Model

We can derive a simple estimate of the magnitude of spin-orbit coupling by considering a single electron in a Bohr-like circular orbit around the nucleus as shown in Figure 7.2. Just as the sun appears to orbit Earth when observed from Earth, the nucleus moves in a circular orbit of radius r_n around the electron

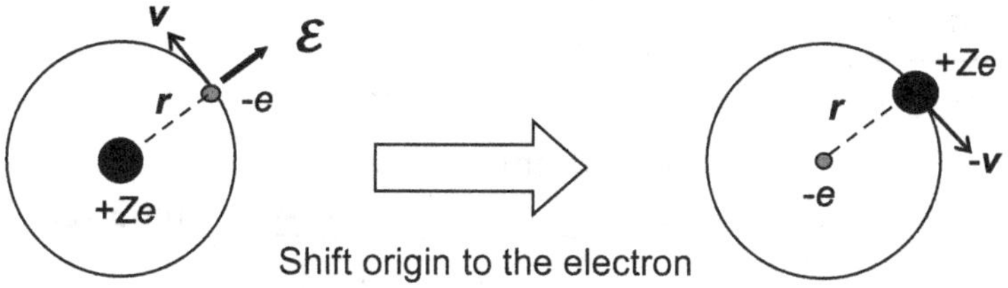

Figure 7.2 An electron moving with velocity $\boldsymbol{v}$ through the electric field $\boldsymbol{\mathcal{E}}$ of the nucleus experiences a magnetic field equal to $(\boldsymbol{\mathcal{E}} \times \boldsymbol{v})/c^2$. The magnetic field can be understood by shifting the origin to the electron and calculating the magnetic field due to the orbital motion of the nucleus around the electron. The velocity of the nucleus in this frame is equal to $-\boldsymbol{v}$.

in the rest frame of the electron. The orbit of the nucleus is equivalent to a current loop, which produces a magnetic field at the origin. Now the magnetic field produced by a circular loop of radius r carrying a current i is given by:

$$B_z = \frac{\mu_0 i}{2r}, \tag{7.18}$$

where z is taken to be the direction perpendicular to the loop. As in Section 7.1, the current is given by the charge Ze divided by the orbital period $T = 2\pi r/v$. On substituting for v and r using Eqs. (2.16) and (2.17) from the Bohr model, we find:

$$B_z = \frac{\mu_0 Zev_n}{4\pi r_n^2} = \left(\frac{Z^4}{n^5}\right)\frac{\mu_0 \alpha ce}{4\pi a_0^2}, \tag{7.19}$$

where $\alpha = e^2/2\epsilon_0 hc \approx 1/137$ is the **fine-structure constant** defined in Eq. (2.19). For hydrogen with $Z = n = 1$, this gives $B_z \approx 12$ Tesla, which is a large field.

The electron at the origin experiences this orbital field, and we thus have a magnetic interaction of the form:

$$\Delta E_{\text{so}} = -\boldsymbol{\mu}_{\text{spin}} \cdot \boldsymbol{B}_{\text{orbital}}, \tag{7.20}$$

which, from Eq. (7.14), becomes:

$$\Delta E_{\text{so}} = g_{\text{s}}\mu_{\text{B}} m_s B_z = \pm \mu_{\text{B}} B_z, \tag{7.21}$$

where we have used $g_{\text{s}} = 2$ and $m_s = \pm 1/2$ in the last equality. By substituting from Eq. (7.19) and making use of Eq. (7.13), we find:

$$|\Delta E_{\text{so}}| = \left(\frac{Z^4}{n^5}\right)\frac{\mu_0 \alpha ce^2 \hbar}{8\pi m_{\text{e}} a_0^2} \equiv \alpha^2 \frac{Z^2}{n^3}|E_n|, \tag{7.22}$$

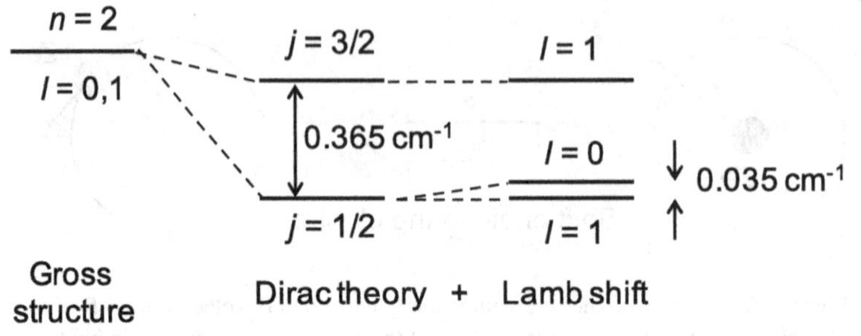

Figure 7.3 Fine structure in the $n = 2$ level of hydrogen.

where E_n is the quantized energy given by Eq. (2.8). For the $n = 2$ orbit of hydrogen, this gives:

$$|\Delta E_{\mathrm{so}}| = \alpha^2 R_H / 2^5 = 13.6\,\mathrm{eV}/(32 \times 137^2) = 0.02\,\mathrm{meV} \equiv 0.2\,\mathrm{cm}^{-1},$$

which is the right of order of magnitude as the actual spin-orbit splitting. (See Figure 7.3.) The key point is that the spin-orbit interaction is about 10^5 times smaller than the gross-structure energy in hydrogen. The equivalent Bohr-model value for the $n = 1$ orbit is 0.7 meV ($6\,\mathrm{cm}^{-1}$), but this is not very meaningful as the $n = 1$ level only has $l = 0$, and so the spin-orbit interaction is, in fact, zero – as we shall see.

A connection with relativistic theories can be made by noting that Eq. (7.22) can be rewritten using Eq. (2.16) as:

$$|\Delta E_{\mathrm{so}}| = \left(\frac{v_n}{c}\right)^2 \frac{|E_n|}{n}. \tag{7.23}$$

This shows that the spin-orbit interaction energy depends on v^2/c^2, just as we would expect for a relativistic correction to the Bohr model. This is hardly surprising, given that Dirac tells us that we should really think of spin-orbit coupling as a relativistic effect.

7.3.2 Spin-Orbit Coupling Beyond the Bohr Model

The semiclassical Bohr-model picture is sufficient to introduce the concept of spin-orbit coupling, but a more detailed analysis is required to get good agreement with experimental data. It is a general result of relativistic electrodynamics that a charged particle moving through a static electric field $\mathcal{E}$ with velocity $\boldsymbol{v}$ sees a magnetic field in its own rest frame given by:

$$\boldsymbol{B} = \frac{1}{c^2}\,\boldsymbol{\mathcal{E}} \times \boldsymbol{v}\,. \tag{7.24}$$

The demonstration of the validity of this formula for the special case of circular orbits and a Coulomb field is given in Exercise 7.1. The general proof for noncircular orbits and non-Coulombic fields may be found, for example, in Jackson (1998).[1]

The spin-orbit coupling is found from the interaction of the magnetic field due to the orbital motion and magnetic dipole due to the spin:

$$\Delta E_{\text{so}} = -\boldsymbol{\mu}_{\text{spin}}\cdot\boldsymbol{B}_{\text{orbital}}\,, \tag{7.25}$$

where $\boldsymbol{\mu}_{\text{spin}}$ is given by (see Eq. [7.15] with $-g_{\text{s}} = g_{\text{e}}$):

$$\boldsymbol{\mu}_{\text{spin}} = -g_{\text{s}}\frac{e}{2m_{\text{e}}}\boldsymbol{s} = -g_{\text{s}}\frac{\mu_{\text{B}}}{\hbar}\boldsymbol{s}\,. \tag{7.26}$$

On substituting Eqs. (7.24) and (7.26) into Eq. (7.25), we obtain:

$$\Delta E_{\text{so}} = \frac{g_{\text{s}}\mu_{\text{B}}}{\hbar c^2}\boldsymbol{s}\cdot(\boldsymbol{\mathcal{E}} \times \boldsymbol{v})\,. \tag{7.27}$$

As explained in Section 4.1, the dominant electric field in the atom will be central, pointing radially from the nucleus, and the potential V is therefore only a function of r. Hence, in the central-field approximation, we can write:

$$\boldsymbol{\mathcal{E}} = -\frac{1}{q}\nabla V = -\frac{1}{e}\frac{\boldsymbol{r}}{r}\frac{\mathrm{d}V}{\mathrm{d}r}\,. \tag{7.28}$$

On making use of this, the spin-orbit energy becomes:

$$\Delta E_{\text{so}} = \frac{g_{\text{s}}\mu_{\text{B}}}{\hbar c^2 e m_{\text{e}}}\left(\frac{1}{r}\frac{\mathrm{d}V}{\mathrm{d}r}\right)\boldsymbol{s}\cdot(\boldsymbol{r} \times \boldsymbol{p})\,, \tag{7.29}$$

where we have substituted $\boldsymbol{v} = \boldsymbol{p}/m_{\text{e}}$. On recalling that the angular momentum $\boldsymbol{l}$ is defined as $\boldsymbol{r} \times \boldsymbol{p}$, and that $\mu_{\text{B}} = e\hbar/2m_{\text{e}}$, we then obtain:

$$\Delta E_{\text{so}} = \frac{g_{\text{s}}}{2c^2 m_{\text{e}}^2}\left(\frac{1}{r}\frac{\mathrm{d}V}{\mathrm{d}r}\right)\boldsymbol{l}\cdot\boldsymbol{s}\,. \tag{7.30}$$

[1] The key result in Jackson (1998) is Eq. (11.149): $\boldsymbol{B}' = -\gamma\boldsymbol{\beta} \times \boldsymbol{\mathcal{E}}$, where $\boldsymbol{\beta} = \boldsymbol{v}/c$ and $\gamma = (1 - v^2/c^2)^{-1/2}$. This shows that a static electric field in one frame of reference looks like a magnetic field in a Lorentz-transformed frame. In the limit of nonrelativistic velocities (i.e., $\gamma \to 1$), we then get $\boldsymbol{B}' = \boldsymbol{\mathcal{E}} \times \boldsymbol{v}/c$. The extra factor of $1/c$ in Eq. (7.24) comes from switching between the Gaussian unit system used in Jackson's text to the S.I. units used throughout this book. In the S.I. unit system, the force on a charged particle is $\boldsymbol{F} = q(\boldsymbol{\mathcal{E}} + \boldsymbol{v} \times \boldsymbol{B})$, which makes it clear that the dimensions of $\boldsymbol{\mathcal{E}}$ are the same as $\boldsymbol{v} \times \boldsymbol{B}$, and hence that a factor such as $1/c^2$ must appear in Eq. (7.24).

Note that the switching of the order of $\boldsymbol{l}$ and $\boldsymbol{s}$ is valid here because spin does not enter into the gross-structure Hamiltonian and therefore the operators commute, so that $\boldsymbol{s}\cdot\boldsymbol{l}=\boldsymbol{l}\cdot\boldsymbol{s}$.

The calculation of ΔE_{so} in Eq. (7.30) does not take proper account of relativistic effects. In particular, we moved the origin from the nucleus to the electron, which is not really valid because the electron is accelerating all the time and is therefore not in an inertial frame. The translation to a rotating frame gives rise to an extra effect called **Thomas precession**,[2] which reduces the energy by a factor of 2. On taking the Thomas precession into account, we obtain the final result:

$$\Delta E_{\mathrm{so}}=\frac{g_{\mathrm{s}}}{2}\frac{1}{2c^2m_{\mathrm{e}}^2}\left(\frac{1}{r}\frac{\mathrm{d}V}{\mathrm{d}r}\right)\boldsymbol{l}\cdot\boldsymbol{s}\,. \tag{7.31}$$

This is the same as the result derived from the Dirac equation, except that g_{s} is exactly equal to 2 in Dirac's theory. Equation (7.31) shows that the spin and orbital angular momenta are coupled together.

The magnitude of the spin-orbit energy can be calculated from Eq. (7.31) as:

$$\Delta E_{\mathrm{so}}=\frac{1}{2c^2m_{\mathrm{e}}^2}\left\langle\frac{1}{r}\frac{\mathrm{d}V}{\mathrm{d}r}\right\rangle\langle\boldsymbol{l}\cdot\boldsymbol{s}\rangle\,, \tag{7.32}$$

where we have taken $g_{\mathrm{s}}=2$, and the $\langle\cdots\rangle$ notation indicates that we take expectation values:

$$\left\langle\frac{1}{r}\frac{\mathrm{d}V}{\mathrm{d}r}\right\rangle=\iiint\psi_{nlm}^{*}\left(\frac{1}{r}\frac{\mathrm{d}V}{\mathrm{d}r}\right)\psi_{nlm}\,r^2\sin\theta\,\mathrm{d}r\,\mathrm{d}\theta\,\mathrm{d}\phi\,. \tag{7.33}$$

The function $(\mathrm{d}V/\mathrm{d}r)/r$ depends only on r, and so we are left to calculate an integral over r only:

$$\left\langle\frac{1}{r}\frac{\mathrm{d}V}{\mathrm{d}r}\right\rangle=\int_0^{\infty}|R_{nl}(r)|^2\left(\frac{1}{r}\frac{\mathrm{d}V}{\mathrm{d}r}\right)r^2\,\mathrm{d}r\,, \tag{7.34}$$

where $R_{nl}(r)$ is the radial wave function. The evaluation of the spin-orbit energy thus requires knowledge of the radial wave functions and the detailed form of $V(r)$. This can be done exactly for hydrogenic atoms, as we shall see in Section 7.4. For other atoms, we need to use numerical computation methods, as the potential (and hence the wave functions) differ from the Coulombic $1/r$ dependence due to the screening of the nuclear field by the other electrons.

One general point can be made immediately for all atoms. It is apparent from Eq. (7.31) that the spin-orbit interaction is zero if $\boldsymbol{l}=0$. Hence electrons

[2] A derivation of the Thomas precession factor may be found, for example, in Jackson (1998), §11.8.

in s-shells with $l = 0$ have no spin-orbit coupling. We therefore only need to work out the spin-orbit coupling for electron states with $l \geq 0$.

7.3.3 Scaling of Spin-Orbit Coupling with Z

The Bohr-model picture of spin-orbit coupling showed that the spin-orbit interaction is proportional to Z^4. (See Eq. [7.22].) This Z^4 scaling will be confirmed by the full quantum-mechanical calculation for hydrogenic atoms in Section 7.4.

The second form of ΔE_{so} in Eq. (7.22) has $\Delta E_{\mathrm{so}} \propto Z^2|E_n|$, which is again confirmed by the full quantum-mechanical analysis. (See Eq. [7.43].) This, of course, is identical to a Z^4 scaling for hydrogenic atoms, as $|E_n| \propto Z^2$. (See Eq. [2.53].) However, the binding energy of the valence electrons of other atoms does not scale as Z^2, due to the screening effect of electrons in inner shells on the nuclear field. In fact, to a very rough approximation, the binding energies are fairly similar, being of order a few eV. This means that the actual scaling of ΔE_{so} can be weaker than a Z^4 dependence. Empirically, we can write:

$$\Delta E_{\mathrm{so}} \propto Z^{\xi}\,, \quad 2 \lesssim \xi \leq 4\,, \tag{7.35}$$

where $\xi = 4$ for an unscreened field, and $\xi \sim 2$ for a screened field. One way or another, spin-orbit effects are expected to be much stronger in heavy atoms with large values of Z, which is indeed the case.

7.4 Evaluation of the Spin-Orbit Energy for Hydrogen

Hydrogenic atoms have pure Coulombic fields with $V(r) = -Ze^2/4\pi\epsilon_0 r$, and hence $(\mathrm{d}V/\mathrm{d}r)/r = Ze^2/4\pi\epsilon_0 r^3$. We thus have to find $\langle r^{-3}\rangle$ from the known radial wave functions. (See Table 2.3.) The result for $l \geq 1$ is:

$$\left\langle \frac{1}{r^3} \right\rangle = \frac{Z^3}{a_0^3 n^3 l(l+\frac{1}{2})(l+1)}\,. \tag{7.36}$$

This shows that we can rewrite Eq. (7.32) in the form:

$$\Delta E_{\mathrm{so}} = C_{nl}\,\langle \boldsymbol{l}\cdot\boldsymbol{s}\rangle\,, \tag{7.37}$$

where C_{nl} depends only on n and l:

$$C_{nl} = \frac{1}{2c^2m_{\mathrm{e}}^2}\left\langle \frac{1}{r}\frac{\mathrm{d}V}{\mathrm{d}r}\right\rangle = \frac{1}{2c^2m_{\mathrm{e}}^2}\,\frac{Ze^2}{4\pi\epsilon_0}\,\frac{Z^3}{a_0^3 n^3 l(l+\frac{1}{2})(l+1)}\,. \tag{7.38}$$

As discussed in Section 5.5, the spin-orbit interaction couples $\boldsymbol{l}$ and $\boldsymbol{s}$ together to form the resultant total angular momentum $\boldsymbol{j}$:

$$\boldsymbol{j} = \boldsymbol{l} + \boldsymbol{s} \, . \tag{7.39}$$

This means that we can evaluate $\langle \boldsymbol{l} \cdot \boldsymbol{s} \rangle$ as follows;

$$\boldsymbol{j}^2 = (\boldsymbol{l} + \boldsymbol{s}) \cdot (\boldsymbol{l} + \boldsymbol{s}) = \boldsymbol{l}^2 + \boldsymbol{s}^2 + 2\boldsymbol{l} \cdot \boldsymbol{s} \, . \tag{7.40}$$

This implies that:

$$\langle \boldsymbol{l} \cdot \boldsymbol{s} \rangle = \frac{1}{2} \left\langle (\boldsymbol{j}^2 - \boldsymbol{l}^2 - \boldsymbol{s}^2) \right\rangle = \frac{\hbar^2}{2} [j(j+1) - l(l+1) - s(s+1)] \, . \tag{7.41}$$

We therefore find:

$$\Delta E_{\text{so}} = C'_{nl} \, [j(j+1) - l(l+1) - s(s+1)] \, , \tag{7.42}$$

where $C'_{nl} = C_{nl}\hbar^2/2$. On putting this all together, we obtain the final result for states with $l \geq 1$:

$$\Delta E_{\text{so}} = -\frac{\alpha^2 Z^2}{2n^2} E_n \frac{n}{l(l+\frac{1}{2})(l+1)} [j(j+1) - l(l+1) - s(s+1)] \, , \tag{7.43}$$

where $\alpha \approx 1/137$ is the fine-structure constant, and $E_n = -R_{\text{H}} Z^2/n^2$ is equal to the gross energy. As mentioned above, Eq. (7.32) shows that states with $l = 0$ have $\Delta E_{\text{so}} = 0$.

The fact that $j = l \oplus s = l \oplus 1/2$ means that j can take values of $l + 1/2$ and $l - 1/2$ for $l \geq 1$. Equation (7.43) then shows that the spin-orbit interaction splits the two j-states with the same value of l. We thus expect the electronic states of hydrogenic atoms with $l \geq 1$ to split into doublets. However, the actual fine structure of hydrogen is more complicated for two reasons:

(i) States with the same n but different l are degenerate.
(ii) The spin-orbit interaction is small.

The first point is a general property of pure one-electron systems, and the second follows from the scaling of ΔE_{so} with Z, as discussed in Section 7.3.3. A consequence of the second point is that other relativistic effects that have been neglected up until now are of a similar magnitude to the spin-orbit coupling. In atoms with higher values of Z, the spin-orbit coupling is the dominant relativistic correction, and we can neglect the other effects.

The fine structure of the $n = 2$ level in hydrogen is illustrated in Figure 7.3. The fully relativistic Dirac theory predicts that states with the same j are degenerate. The degeneracy of the two $j = 1/2$ states is ultimately lifted by a quantum electrodynamic (QED) effect called the Lamb shift.

The complications of the fine structure of hydrogen due to other relativistic and QED effects means that hydrogen is not the paradigm for understanding spin-orbit effects. The alkali metals considered below are in fact simpler to understand.

7.5 Spin-Orbit Coupling in Alkali Atoms

Alkali atoms have a single valence electron outside close shells. Closed shells have no angular momentum, and so the angular momentum state $|L, S, J\rangle$ of the atom is determined entirely by the valence electron. By analogy with the results for hydrogen given in Eq. (7.37) and Eq. (7.42), we can write the spin-orbit interaction term as:

$$\Delta E_{SO} \propto \langle \boldsymbol{L} \cdot \boldsymbol{S} \rangle \propto [J(J+1) - L(L+1) - S(S+1)], \tag{7.44}$$

where

$$\boldsymbol{J} = \boldsymbol{L} + \boldsymbol{S}. \tag{7.45}$$

Note that we are using capital letters here, following the convention mentioned in Section 5.7. In the case of an alkali atom, this makes no practical difference, as we only have one valence electron outside filled shells. In other atoms, however, we will need to be careful to distinguish between individual electrons and resultants for the whole atom.

It follows immediately from Eq. (7.44) that the spin-orbit energy is zero when the valence electron is in an s-shell, since $\boldsymbol{L} \cdot \boldsymbol{S} = 0$ when $L = 0$. We thus focus on the case with $l \neq 0$, where $\boldsymbol{L} \cdot \boldsymbol{S} \neq 0$. J has two possible values, namely $J = L \oplus S = L \oplus 1/2 = L \pm 1/2$. On writing Eq. (7.44) in the form:

$$\Delta E_{SO} = C\,[J(J+1) - L(L+1) - S(S+1)], \tag{7.46}$$

the spin-orbit energy of the $J = (L + 1/2)$ state is given by:

$$\Delta E_{\text{so}} = C\left[(L+\frac{1}{2})(L+\frac{3}{2}) - L(L+1) - \frac{1}{2}\cdot\frac{3}{2}\right] = +CL,$$

while for the $J = (L - 1/2)$ level we have:

$$\Delta E_{\text{so}} = C\left[(L-\frac{1}{2})(L+\frac{1}{2}) - L(L+1) - \frac{1}{2}\cdot\frac{3}{2}\right] = -C(L+1).$$

Hence the term defined by the quantum numbers n and l is split by the spin-orbit coupling into two new states, as illustrated in Figure 7.4(a). This gives rise to doublets in the atomic spectra. The magnitude of the splitting is smaller

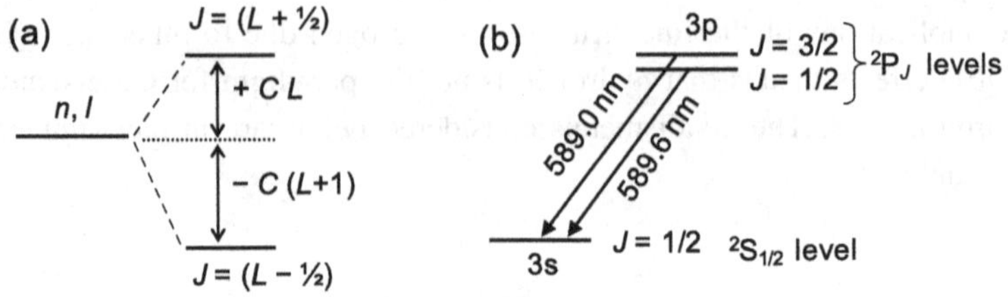

Figure 7.4 Spin-orbit interactions in alkali atoms. (a) The spin-orbit interaction splits the nl states into a doublet if $l \neq 0$. (b) Fine structure in the sodium D-lines.

than the gross energy by a factor $\sim\alpha^2 = 1/137^2$. (See Eq. [7.43].) This is why these effects are called "fine structure," and α is called the "fine-structure constant."

As an example, let us consider the D-line transition of sodium. Sodium has a ground-state configuration of [Ne] $3s^1$, with one valence electron outside filled 1s, 2s, and 2p shells. In the first excited state, the valence electron is promoted to the 3p shell, which has a higher energy due to its smaller quantum defect. (See Section 4.5.) The D-line corresponds to the 3p $\rightarrow$ 3s transition, which occurs in the orange/yellow spectral region. It is well-known that the D-line is a doublet, as shown in Figure 7.4(b). The doublet arises from the spin-orbit coupling. The ground state is a $^2S_{1/2}$ level with zero spin-orbit splitting. The excited state is split into the two levels derived from the different J values for $L = 1$ and $S = 1/2$, namely the $^2P_{3/2}$ and $^2P_{1/2}$ levels. The two transitions in the doublet are therefore:

$$^2P_{3/2} \rightarrow {}^2S_{1/2},$$

and:

$$^2P_{1/2} \rightarrow {}^2S_{1/2}.$$

The energy difference of 17 cm^{-1} between them arises from the spin-orbit splitting of the two J states of the ^2P term.

Similar arguments can be applied to the D-lines of the other alkali elements. The spin–orbit energy splittings are tabulated in Table 7.1. Note that the splitting increases with Z, and that the splitting energy is roughly proportional to Z^2, as shown in Figure 7.5. This is an example of the Z^2 spin-orbit scaling for a screened nuclear field, as discussed in Section 7.3.3.

It should be pointed out that the ordering of the levels shown in Figure 7.4(a) assumes that the constant C in Eq. (7.46) is positive, so that the level with $J = L + 1/2$ lies above the one with $J = L - 1/2$. This is true in most cases, but there are some exceptions. For example, C is negative for the 3d states of

Table 7.1 *Spin-orbit splitting of the D-lines of the alkali elements. The energy splitting ΔE is equal to the difference of the energies of the $J = 3/2$ and $J = 1/2$ levels of the first excited state.*

Element	Z	Ground state	1st excited state	Transition	ΔE (cm^{-1})
Lithium	3	[He] 2s	2p	2p $\rightarrow$2s	0.33
Sodium	11	[Ne] 3s	3p	3p $\rightarrow$3s	17
Potassium	19	[Ar] 4s	4p	4p $\rightarrow$4s	58
Rubidium	37	[Kr] 5s	5p	5p $\rightarrow$5s	238
Caesium	55	[Xe] 6s	6p	6p $\rightarrow$6s	554

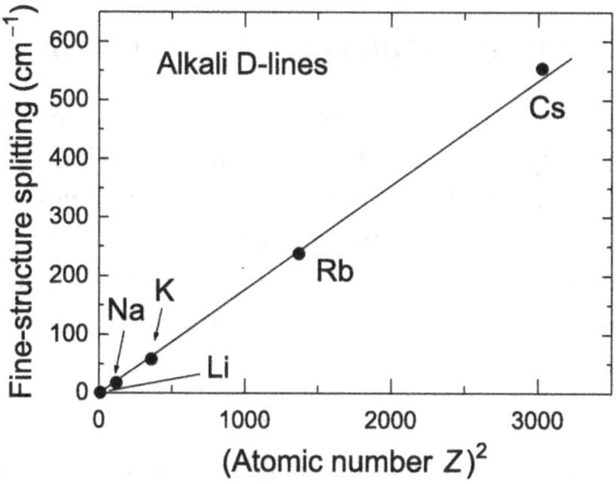

Figure 7.5 Spin-orbit splitting of the first excited state of the alkali atoms versus Z^2, as determined by the fine structure splitting of the D-lines. (See Table 7.1.)

sodium, so that the $^2D_{5/2}$ level lies *below* $^2D_{3/2}$. The 4d term of potassium is also inverted. There is no simple reason why this should be so. It depends on complicated exchange effects.

Example 7.1 The alkali element francium has $Z = 87$ and a ground configuration of [Rn] 7s. The long wavelength component of the D-line doublet has a wavelength of 817 nm. What wavelength do you expect for the other D-line?

Solution: The D-lines correspond to the 7p $\rightarrow$ 7s transition, which are split by the spin-orbit coupling of the 7p shell. The long-wavelength line is the 7p $^2P_{1/2}$ $\rightarrow$ 7s $^2S_{1/2}$ transition, which has an energy of $1/(817 \times 10^{-7}) = 1.22 \times 10^4$ cm^{-1}. The energy of the short-wavelength line is therefore given by:

$$h\nu = 1.22 \times 10^4 \text{ cm}^{-1} + \Delta E_{\text{so}}(\text{Fr, 7p}) .$$

The spin-orbit coupling in alkalis scales as $\sim Z^2$, and so we can estimate ΔE from:

$$\Delta E_{\rm so}(\text{Fr, 7p}) = \frac{87^2}{55^2}\,\Delta E_{\rm so}(\text{Cs, 6p})\,.$$

We read a value of $554\,\text{cm}^{-1}$ for $\Delta E_{\rm so}(\text{Cs 6p})$ from Table 7.1, and hence estimate $1390\,\text{cm}^{-1}$ for the 7p splitting in francium. Hence the transition energy is $1.36 \times 10^4\,\text{cm}^{-1}$, implying a wavelength of 736 nm. The experimental value is 718 nm, which shows that there is some departure from exact Z^2 scaling.

7.6 Spin-Orbit Coupling in Many-Electron Atoms

We have seen in Chapter 5 that atoms with more than one valence electron can have different types of angular momentum coupling. We restrict our attention here to atoms with LS coupling, which is the most common type. In LS coupling, the residual electrostatic interaction couples the orbital and spin angular momenta together according to Eqs. (5.31) and (5.32). (See Section 5.7.) The resultants are then coupled together to give the total angular momentum $\boldsymbol{J}$:

$$\boldsymbol{J} = \boldsymbol{L} + \boldsymbol{S}\,. \tag{7.47}$$

The rules for coupling of angular momenta produce several J-states for each LS-term, with J running from $L+S$ down to $|L-S|$ in integer steps. These J-states experience different spin-orbit interactions, and so are shifted in energy from each other. Hence the spin-orbit coupling splits the J-states of a particular LS-term into fine-structure multiplets.

The splitting of the J-states can be evaluated as follows: The spin-orbit interaction takes the form:

$$\Delta E_{\rm so} = -\boldsymbol{\mu}_{\rm spin} \cdot \boldsymbol{B}_{\rm orbital} \propto \langle \boldsymbol{L} \cdot \boldsymbol{S} \rangle\,, \tag{7.48}$$

which implies (see Eqs. [7.37] – [7.42]):

$$\Delta E_{SO} = C_{LS}\,[J(J+1) - L(L+1) - S(S+1)]\,. \tag{7.49}$$

It follows from Eq. (7.49) that the spin-orbit splitting between adjacent levels within a multiplet of an LS-term is given by:

$$\Delta E_{SO}^{LS}(J) - \Delta E_{SO}^{LS}(J-1) = 2C_{LS}\,J\,. \tag{7.50}$$

This result, which is called the **interval rule**, shows that the level splittings are in the ratio of the J-values of the upper level. Figure 5.3 shows an example of the interval rule for the ^{3}P term of the (3s,3p) configuration of magnesium.

	m_l				
m_s	-2	-1	0	+1	+2
+1/2				↑	↑
-1/2					

(a)

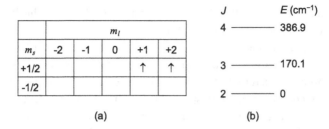

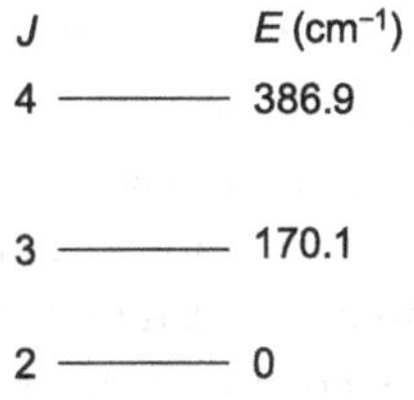

(b)

Figure 7.6 Fine-structure levels of the ground-state of titanium with configuration $3d^2$. (a) Arrangement of the electrons for the lowest level following Hund's rules. (b) Fine-structure levels with their J-values assigned.

Example 7.2 The ground state of titanium (configuration $[\mathrm{Ar}]\,4s^2\,3d^2$) consists of a triplet of levels with energies of 0, 170.1, and $386.9\,\mathrm{cm}^{-1}$.
(a) Assign angular momentum quantum numbers to these levels.
(b) Account for the relative magnitude of the splitting between the levels.

Solution (a) We find the ground-state level by using Hund's rules, as in Example 5.4. We have two d electrons, and these are arranged within the ten $\{m_s, m_l\}$ states of the 3d shell as shown in Figure 7.6(a). The ground state has $S = 1/2 + 1/2 = 1$ and $L = 2 + 1 = 3$, and is therefore a $^3\mathrm{F}$ term. The possible values of J are then $3 \oplus 1 = 4, 3, 2$. The shell is less than half full, so Hund's third rule says that the $J = 2$ level is the ground state. The other two are excited states, split by the spin-orbit energy. The three levels, in order of increasing energy, are thus: $^3\mathrm{F}_2$, $^3\mathrm{F}_3$, and $^3\mathrm{F}_4$, as shown in Figure 7.6(b).
(b) The splitting of the levels should follow the interval rule given in Eq. (7.50), which implies:

$$\frac{E(^3\mathrm{F}_4) - E(^3\mathrm{F}_3)}{E(^3\mathrm{F}_3) - E(^3\mathrm{F}_2)} = \frac{4}{3}\,.$$

The experimental ratio of $216.8/170.1 = 1.27$ is close to the interval rule prediction of 4/3.

7.7 Fine Structure in X-Ray Spectra

It was mentioned in the discussion of X-ray spectra in Section 4.4.3 that the K-shell with $n = 1$ is a unique level, but higher shells (e.g., L and M corresponding to $n = 2$ and 3, respectively) show substructure. We can now explain this properly, using what we know about spin-orbit coupling.

Consider first the K-shell, which corresponds to the 1s shell. This has $l = 0$, and therefore has no spin-orbit coupling. The K-shell is thus a unique level. Now consider the L-shell, which has $n = 2$, and therefore can have $l = 0$ or 1, corresponding to the 2s and 2p sub-shells. The 2s sub-shell, like the K-shell, has no fine structure, as the spin-orbit coupling is zero for $l = 0$. The 2p sub-shell, by contrast, is split by spin-orbit coupling into two j-levels, namely $j = 3/2$ and $j = 1/2$. The three L sub-shells thus correspond to the 2s, $2p_{1/2}$, and $2p_{3/2}$ levels. These are labelled L_1, L_2, and L_3 in X-ray notation.

The M-shell with $n = 3$ shows even more substructure. For $n = 3$, l can take values of 0, 1, or 2, corresponding to the 3s, 3p, and 3d sub-shells. The 3s sub-shell has no spin-orbit coupling, but the 3p and 3d sub-shells are both spin-orbit doublets. For the 3p sub-shell, we have $j = 1 \oplus 1/2$, giving $j = 1/2$ and 3/2, while for 3d we have $j = 2 \oplus 1/2$, giving $j = 3/2$ and 5/2. This gives a total of five levels, namely 3s, $3p_{1/2}$, $3p_{3/2}$, $3d_{3/2}$, and $3d_{5/2}$, which are labeled $M_1 \ldots M_5$ in X-ray notation.

The splitting of the higher edges is clearly visible in the absorption spectrum of gold ($Z = 79$) in Figure 4.6(b). The L_1, L_2, and L_3 edges occur at 14.35, 13.74, and 11.92 keV, respectively, with the L_2–L_3 splitting giving the spin-orbit splitting of the 2p shell as 1.82 keV. It is interesting to note that this splitting fits to the value predicted by the theory developed in Section 7.4 for a hydrogenic atom with $Z = 79$ to within about 3%. (See Exercise 7.3.) This shows that the inner shells of heavy atoms are very hydrogenic in character.

7.8 Nuclear Effects in Atoms

For most of the time in atomic physics we just take the nucleus to be a heavy charged particle sitting at the center of the atom. However, careful analysis of the spectral lines can reveal small effects that give us direct information about the nucleus. The main effects that can be observed generally fall into two categories, namely **isotope shifts** and **hyperfine structure**.

7.8.1 Isotope Shifts

There are two main processes that give rise to isotope shifts:

Mass effects The mass m that enters the Schrödinger equation is the *reduced mass*, not the bare electron mass m_e (see Eq. [2.5]). Changes in the nuclear mass therefore make small changes to m and hence to the atomic energies.

Field effects Electrons in s-shells have a small, but finite, probability of penetrating the nucleus (see Exercise 2.8), and are therefore sensitive to its charge distribution.

Both effects cause small shifts in the wavelengths of the spectral lines from different isotopes of the same element. The mass shifts are most noticeable for light elements. As a case in point, the heavy isotope of hydrogen, namely deuterium, was discovered through its mass effect. (See Example 2.2.)

7.8.2 Hyperfine Structure

High-resolution spectroscopy reveals small shifts and splittings in spectral lines due to **hyperfine interactions**. These arise from the interaction between the magnetic dipole due to the nuclear spin and the magnetic field produced at the nucleus by the electrons:

$$\Delta E_{\text{hyperfine}} = -\boldsymbol{\mu}_{\text{nucleus}} \cdot \boldsymbol{B}_{\text{electron}}. \tag{7.51}$$

Most nuclei possess spin, $\boldsymbol{I}$, which is quantized with:

$$|\boldsymbol{I}|^2 = I(I+1)\hbar^2 , \tag{7.52}$$

where the nuclear spin quantum number I can take integer or half-integer values. Just as for electrons, the magnetic dipole moment of the nucleus is directly proportional to its spin:

$$\boldsymbol{\mu}_{\text{nucleus}} = \gamma_I \boldsymbol{I} = g_I \frac{\mu_{\text{N}}}{\hbar} \boldsymbol{I} , \tag{7.53}$$

where γ_I is the nuclear gyromagnetic ratio, g_I is the nuclear g-factor, and $\mu_{\text{N}} \equiv e\hbar/2m_{\text{P}}$ is the nuclear magneton, with m_{P} being the proton mass. The value of μ_{N} in S.I. units is 5.050783×10^{-27} A m^2. There are two interesting points that can be made here in comparison to the equivalent result for electrons:

- The nuclear gyromagnetic ratio is about 2000 times smaller than the electron gyromagnetic ratio on account of the heavier proton mass.
- The presence of the nuclear g-factor in Eq. (7.53) highlights the quantum-mechanical origin of nuclear spin. The g-factors of protons and neutrons are +5.5857 and −3.8261, respectively. These non-integer values point to the fact that protons and neutrons are actually composite rather than elementary particles. The negative, non-zero value for the neutron is particularly striking, given that the neutron is uncharged.

The magnetic field generated by the electrons at the nucleus points in the same direction as the total electronic angular momentum $\boldsymbol{J}$. The hyperfine interaction in Eq. (7.51) is therefore of the form:

$$\Delta E_{\text{hyperfine}} \propto \langle \boldsymbol{I} \cdot \boldsymbol{J} \rangle \,. \tag{7.54}$$

This shows that hyperfine interactions cause a coupling between the nuclear spin ($\boldsymbol{I}$) and total electron angular momentum ($\boldsymbol{J}$). The magnitudes of hyperfine interactions are about three orders of magnitude smaller than fine-structure interactions, due to the small nuclear gyromagnetic ratio: hence the name "hyperfine."

Hyperfine states are labeled by the total angular momentum $\boldsymbol{F}$ of the whole atom (i.e., nucleus plus electrons), where:

$$\begin{aligned} \boldsymbol{F} &= \boldsymbol{I} + \boldsymbol{J} \,, \\ |\boldsymbol{F}|^2 &= F(F+1)\,\hbar^2 \,. \end{aligned} \tag{7.55}$$

The quantum number F can take integer or half-integer values. Just as for electrons (see Eq. [7.41]), the hyperfine interaction in Eq. (7.54) can be evaluated from:

$$\begin{aligned} \boldsymbol{F} \cdot \boldsymbol{F} &= (\boldsymbol{I} + \boldsymbol{J}) \cdot (\boldsymbol{I} + \boldsymbol{J}) \,, \\ &= \boldsymbol{I} \cdot \boldsymbol{I} + \boldsymbol{J} \cdot \boldsymbol{J} + 2\boldsymbol{I} \cdot \boldsymbol{J} \,. \end{aligned}$$

We therefore obtain:

$$\Delta E_{\text{hyperfine}} = \frac{A}{2} \left(F(F+1) - I(I+1) - J(J+1) \right) \,, \tag{7.56}$$

where A is the hyperfine constant.

The selection rule for optical transitions between hyperfine levels can be deduced by applying conservation of angular momentum to the initial and final states. The photon carries one unit of angular momentum, and so we must have:

$$\Delta F = 0, \pm 1 \,, \tag{7.57}$$

with the exception that $F = 0 \to 0$ transitions are forbidden. Let us consider two examples of how this works.

The 21 cm hydrogen line

Consider the ground state of hydrogen. The nucleus consists of just a single proton, and we therefore have $I = 1/2$. The hydrogen ground state is the $1s\ {}^2S_{1/2}$ term, which has $J = 1/2$. The hyperfine quantum number F is then found from $F = I \oplus J = 1/2 \oplus 1/2 = 1$ or 0. These two hyperfine states correspond to the cases in which the spins of the electron and the nucleus are

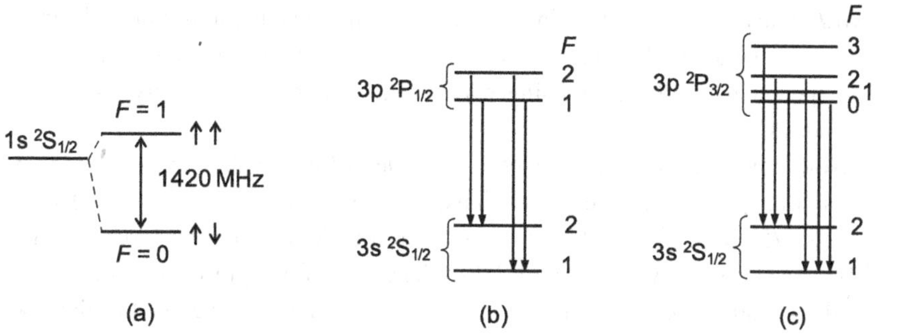

Figure 7.7 (a) Hyperfine structure of the 1s ground state of hydrogen. The arrows indicate the relative directions of the electron and nuclear spin. (b) Hyperfine transitions for the sodium D_1 line. (c) Hyperfine transitions for the sodium D_2 line. Note that the hyperfine splittings are not drawn to scale. The splittings of the sodium levels are as follows: $^2S_{1/2}$, 1772 MHz; $^2P_{1/2}$, 190 MHz; $^2P_{3/2}$ ($3 \rightarrow 2$), 59 MHz; $^2P_{3/2}$ ($2 \rightarrow 1$), 34 MHz; $^2P_{3/2}$ ($1 \rightarrow 0$), 16 MHz.

aligned parallel ($F = 1$) or antiparallel ($F = 0$). The two F states are split by the hyperfine interaction by 0.0475 cm^{-1} (5.9×10^{-6} eV). (See Figure 7.7[a].) Transitions between these levels occur at 1420 MHz ($\lambda = 21$ cm), and are very important in radio astronomy. (See Section 12.4.2.) Radio frequency transitions such as these are also routinely exploited in nuclear magnetic resonance (NMR) spectroscopy. (See Section 8.3.2.)

Hyperfine Structure of the Sodium D-Lines

The sodium D-lines originate from 3p $\rightarrow$ 3s transitions. As discussed in Section 7.5, there are two lines with energies split by the spin-orbit coupling, as indicated in Figure 7.4(b).

Consider first the lower energy D_1 line, which is the $^2P_{1/2} \rightarrow {}^2S_{1/2}$ transition. The nuclear spin of sodium is 3/2, and so we have $F = I \oplus J = 3/2 \oplus 1/2 = 2$ or 1 for both the upper and lower levels of the transition, as shown in Figure 7.7(b). Note that the hyperfine splittings are not drawn to scale in Figure 7.7(b): the splitting of the $^2S_{1/2}$ level is 1772 MHz, which is much larger than that of the $^2P_{1/2}$ level, namely 190 MHz. This is a consequence of the fact that s-electrons have higher probability densities at the nucleus, and hence experience stronger hyperfine interactions. All four transitions are allowed by the selection rules, and so we observe four lines. Since the splitting of the upper and lower levels are so different, we obtain two doublets with relative frequencies of (0, 190) MHz and (1772, 1962) MHz. These splittings should be compared to the much larger ($\sim 5 \times 10^{11}$ Hz) splitting between the two J-states caused by the spin-orbit interaction. Since the hyperfine splittings are much smaller, they are not routinely observed in optical spectroscopy, and specialized techniques using narrow band lasers are typically employed nowadays.

Now consider the higher energy D_2 line, which is the ${}^2P_{3/2} \rightarrow {}^2S_{1/2}$ transition. In the upper level, we have $J = 3/2$, and hence $F = I \oplus J = 3/2 \oplus 3/2 = 3, 2, 1$, or 0. There are therefore four hyperfine levels for the ${}^2P_{3/2}$ level, as shown in Figure 7.7(c). The hyperfine splittings of the ${}^2P_{3/2}$ level are again much smaller than that of the ${}^2S_{1/2}$ level, on account of the low probability density of p-electrons near the nucleus. Six transitions are allowed, with the $F = 3 \rightarrow 1$ and $F = 0 \rightarrow 2$ transitions being forbidden by the $|\Delta F| \leq 1$ selection rule. We thus have six hyperfine lines, which split into two triplets at relative frequencies of (0, 34, 59) MHz and (1756, 1772, 1806) MHz.

Exercises

7.1 Consider an electron moving with velocity $\boldsymbol{v}$ through the electric field of a nucleus with charge $+Ze$. In the rest-frame of the electron, the nucleus has a velocity $-\boldsymbol{v}$, as shown in Fig. 7.2. The magnetic field experienced by the electron can be calculated by treating the orbiting nucleus as a current loop and applying Biot–Savart's law:

$$\boldsymbol{B} = \frac{\mu_0}{4\pi} \oint_{\text{loop}} i \, \frac{\mathrm{d}\boldsymbol{u} \times \boldsymbol{r}}{r^3} \,,$$

where i is the current and $\mathrm{d}\boldsymbol{u}$ is an orbital path element.

(a) Show that the magnetic field reduces to the following form for a circular orbit with radius r:

$$\boldsymbol{B} = \frac{\mu_0}{4\pi} \frac{Ze}{r^3} \boldsymbol{r} \times \boldsymbol{v} \,.$$

(b) Show that for the case of a pure Coulomb field, $\boldsymbol{B}$ can be written:

$$\boldsymbol{B} = \mu_0 \epsilon_0 \, \boldsymbol{\mathcal{E}} \times \boldsymbol{v} \,.$$

(c) Show that $\boldsymbol{B}$ satisfies Eq. (7.24).

7.2 Verify Eq. (7.36) for a 2p electron of a hydrogenic atom using the radial wave function given in Table 2.3.

7.3 Show that the spin-orbit coupling of a 2p shell is equal to $\alpha^2 Z^4 R_H/16$ for a hydrogenic atom. Evaluate this energy for (a) hydrogen ($Z = 1$), and (b) Au^{78+} ($Z = 79$).

7.4 The 3p → 3s transition of the Mg^+ ion ($Z = 12$) consists of a fine-structure doublet with wavelengths of 279.55 and 280.27 nm. Find the value of the constant C defined in Eq. (7.46) for the 3p state of Mg^+.

7.5 The alkali metal caesium has a ground-state electronic configuration of [Xe] 6s. The 6p → 6s D-line transition is a doublet with wavelengths of 852.11 nm and 894.35 nm. The 5d level is split into a doublet with energies of 14499 and 14597 cm^{-1} relative to the 6s ground state. Find the wavelengths of the 5d → 6p transitions.

7.6 The Ca^+ ion is isoelectronic to potassium, with 19 electrons in a configuration of [Ar] $4s^1$. The long wavelength line of the 4p → 4s doublet occurs at 396.8 nm. Use the fact that the energies of the 4p levels in Ca^+ and K are −8.75 eV and −2.73 eV respectively, together with the $Z^2|E_n|$ scaling of the spin-orbit energy (see Eq. [7.43]) and the data in Table 7.1, to estimate the wavelength of the other line in the doublet.

7.7 The wavelengths of the 3s4s → 3s3p triplet lines of magnesium ($Z = 12$) are 516.73, 517.27, and 518.36 nm.

(a) Explain the origin of the triplet of lines.
(b) What is the value of the fine-structure constant C defined in Eq. (7.49) for the 3s3p configuration?
(c) Account for the relative size of the splitting between the lines.

7.8 The ground state configuration of praseodymium ($Z = 59$) is [Xe] $4f^3 6s^2$.

(a) Use Hund's rules to deduce the ground state level.
(b) What other J-levels are possible for the ground state LS-term?
(c) The splitting between the ground state and the highest J-level of the ground state LS-term is 4381.1 cm^{-1}. Use the interval rule to estimate the energies of the other levels relative to the ground state.

7.9 Use the data in Exercise 6.8 to work out the fine-structure splitting of the 5s5p ^3P term of cadmium. Are the the splittings consistent with the interval rule?

7.10 The K absorption edge of tungsten ($Z = 74$) occurs at 69.52 keV, while the L_1, L_2, and L_3 edges occur at 12.10, 11.54, and 10.20 keV respectively. Find the wavelengths of the L → K emission lines that would be generated in a tungsten X-ray tube.

7.11 A cadmium discharge tube contains an equal mixture of the isotopes ^{112}Cd and ^{114}Cd, causing a splitting of the 441.6 nm line by 1.5 GHz.

(a) Show that the mass shift between two isotopes with mass numbers A and A' is expected to cause a change in the emission frequency by a factor $(m_e/m_p)(A^{-1} - A'^{-1})$.

(b) Can the mass shift account for the splitting observed in the Cd 441.6 nm line? If not, what might?

(c) Estimate the maximum temperature at which it would be possible to resolve the separate emission lines from the two isotopes.

7.12 Caesium is an alkali metal with electronic configuration [Xe] 6s and nuclear spin 7/2. Deduce the possible hyperfine transitions for both components of the caesium D-line doublet.

8

External Fields: The Zeeman and Stark Effects

In the previous chapter, we considered the effects of the internal magnetic fields within atoms. We now wish to consider the effects of external fields. Table 8.1 defines the nomenclature of the effects that we shall be considering.

8.1 Magnetic Fields

The first person to study the effects of magnetic fields on the optical spectra of atoms was Zeeman in 1896. Later work showed that the interaction between an atom and a magnetic field can be classified into two regimes:

- Weak fields: the **Zeeman effect**, either **normal** or **anomalous**;
- Strong fields: the **Paschen–Back effect**.

The "normal" Zeeman effect is so called because it agrees with the classical theory developed by Lorentz. The "anomalous" Zeeman effect is caused by electron spin, and is therefore a completely quantum result. The criterion for deciding whether a particular field is "weak" or "strong" will be discussed in Section 8.1.3. In practice, we usually work in the weak-field (i.e., Zeeman) limit.

8.1.1 The Normal Zeeman Effect

The normal Zeeman effect is observed in atoms with no net electronic spin. It is therefore only observed in a relatively small subset of cases. The total spin of an N-electron atom is given by:

$$S = \sum_{i=1}^{N} s_i \,. \tag{8.1}$$

Table 8.1 *Names of the effects of external fields in atomic physics.*

Applied field	Field strength	Effect
Magnetic	Weak	Zeeman
	Strong	Paschen–Back
Electric	All	Stark

Filled shells have no net spin, and so we only need to consider the valence electrons here. Since all the individual electrons have spin $1/2$, it will not be possible to obtain $S = 0$ from atoms with an odd number of valence electrons. On the other hand, spin-singlet $S = 0$ states are possible if there is an even number of valence electrons. For example, if we have two valence electrons, then the total spin quantum number $S = 1/2 \oplus 1/2$ can be either 0 or 1. In fact, the ground states of divalent atoms from group II of the periodic table (electronic configuration ns^2) always have $S = 0$ because the two electrons align with their spins antiparallel.

The magnetic moment of an atom with $S = 0$ will originate entirely from its orbital motion:

$$\boldsymbol{\mu} = -\frac{\mu_\mathrm{B}}{\hbar}\boldsymbol{L}, \tag{8.2}$$

where $\mu_\mathrm{B}/\hbar = e/2m_\mathrm{e}$ is the gyromagnetic ratio. (See Eq. [7.8].) The interaction energy between a magnetic dipole $\boldsymbol{\mu}$ and a uniform magnetic field $\boldsymbol{B}$ is given by:

$$\Delta E = -\boldsymbol{\mu} \cdot \boldsymbol{B}. \tag{8.3}$$

We set up the axes of our spherically symmetric atom so that the z-axis coincides with the direction of the field. In this case we have $\boldsymbol{B} = B_z\,\hat{\mathbf{z}}$, and the interaction energy of the atom is therefore:

$$\Delta E = -\mu_z B_z = \mu_\mathrm{B} B_z M_L, \tag{8.4}$$

where M_L is the orbital magnetic quantum number. Equation (8.4) shows us that the application of an external $\boldsymbol{B}$-field splits the degenerate M_L states evenly. This is illustrated for the case where $L = 2$ in Figure 8.1(a). The fact that M_L states split in magnetic fields explains why m_l is called the magnetic quantum number. Note that capital letters are being used here for the angular momentum states, as they refer to resultants for the whole atom. (See discussion in Section 5.7.) Note also that $J = L$ when $S = 0$, and hence $M_J = M_L$, so that we could have also written Eq. (8.4) as $\Delta E = \mu_\mathrm{B} B_z M_J$.

The effect of the magnetic field on the spectral lines can be worked out from the splitting of the levels. Consider the transitions between two Zeeman-split

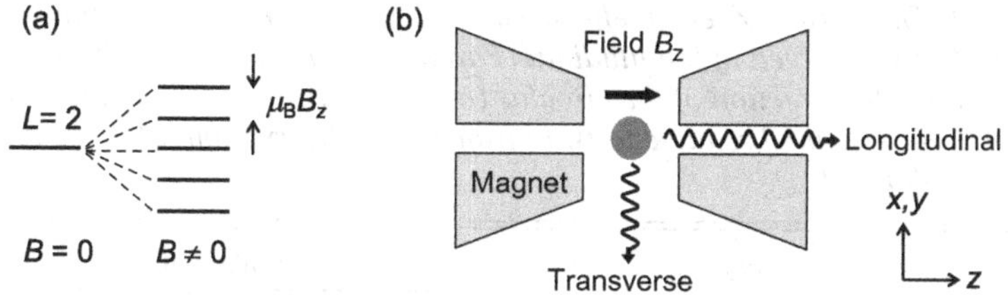

Figure 8.1 The normal Zeeman effect. (a) Splitting of the degenerate M_L states of an atomic level with $L = 2$ by a magnetic field. (b) Definition of longitudinal (Faraday) and transverse (Voigt) observations. The direction of the field defines the z-axis.

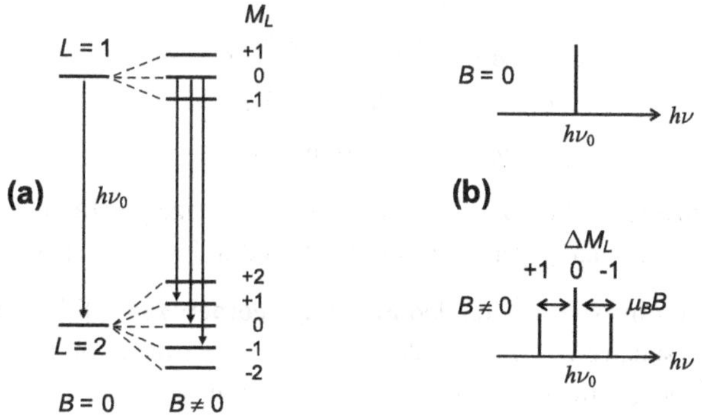

Figure 8.2 The normal Zeeman effect for a P → D transition. (a) The field splits the degenerate M_L levels equally. Optical transitions can occur if $\Delta M_L = 0, \pm 1$. (Only the transitions originating from the $M_L = 0$ level of the P-state are identified here for the sake of clarity.) (b) The spectral line splits into a triplet when observed transversely to the field. The $\Delta M_L = 0$ transition is unshifted, but the $\Delta M_L = \pm 1$ transitions occur at $(h\nu_0 \mp \mu_B B_z)$.

states as shown in Figure 8.2. The selection rules listed in Table 3.1 indicate that we can have transitions with $\Delta M_L = 0$ or ± 1. This gives rise to three transitions whose frequencies are given by:

$$
\begin{aligned}
h\nu &= h\nu_0 + \mu_B B_z \qquad & \Delta M_L &= -1\,, \\
h\nu &= h\nu_0 & \Delta M_L &= 0\,, \\
h\nu &= h\nu_0 - \mu_B B_z & \Delta M_L &= +1\,.
\end{aligned}
\tag{8.5}
$$

This is the same result as that derived by classical theory. In calculating the Zeeman shifts, the following energy equivalents for μ_B are useful:

Table 8.2 *The normal Zeeman effect. The last two columns refer to the polarizations observed in longitudinal (Faraday) and transverse (Voigt) geometries. The direction of the circular ($\sigma^{\pm}$) polarization in longitudinal observation is defined relative to **B**. In transverse observation, all lines are linearly polarized.*

ΔM_L	Energy	Polarization	
		Longitudinal observation	Transverse observation
+1	$h\nu_0 - \mu_B B$	σ^+	$\boldsymbol{\mathcal{E}} \perp \boldsymbol{B}$
0	$h\nu_0$	Not observed	$\boldsymbol{\mathcal{E}} \parallel \boldsymbol{B}$
−1	$h\nu_0 + \mu_B B$	σ^-	$\boldsymbol{\mathcal{E}} \perp \boldsymbol{B}$

$$\begin{aligned}\mu_B &= 5.7884 \times 10^{-5}\,\mathrm{eV\,T^{-1}}\,,\\ \mu_B/h &= 13.996 \times 10^{9}\,\mathrm{Hz\,T^{-1}}\,.\\ \mu_B/hc &= 0.46686\,\mathrm{cm^{-1}\,T^{-1}}\,.\end{aligned} \tag{8.6}$$

The polarizations of the Zeeman lines are summarized in Table 8.2. The two different observational geometries are illustrated in Figure 8.1(b).

Longitudinal observation (called **Faraday geometry** in solid-state physics): The spectrum is viewed along the field, with the photons propagating in the z direction. Light waves are transverse, and so only the x and y polarizations are possible. The z-polarized $\Delta M_L = 0$ line is therefore absent, and we just observe the σ^+ and σ^- circularly polarized $\Delta M_L = \pm 1$ transitions.

Transverse observation (called **Voigt geometry** in solid-state physics): The spectrum is viewed at right angles to the field, making the z-polarized transition observable. The $\Delta M_L = 0$ transition is linearly polarized parallel to the field, while the $\Delta M_L = \pm 1$ transitions are linearly polarized at right angles to the field.

Example 8.1 What emission lines would be observed for $1s3s\,^1S_0 \rightarrow 1s2p\,^1P_1$ transition of helium at 728.13507 nm when observing (a) transversely and (b) longitudinally to a magnetic field of 1.5 T?

Solution: This is a singlet transition with $S = 0$, so that the normal Zeeman effect will be observed. The transition will be split into three lines with energies given by Eq. (8.5). From Eq. (8.6) we calculate that the $\Delta M_L = \pm 1$ transitions will be shifted by $\mp 0.467 \times 1.5 = \mp 0.700\,\mathrm{cm^{-1}}$. The transition energies are thus:

$$h\nu = \left(\frac{10^7}{728.13507} \pm 0.700\right)\mathrm{cm^{-1}} = (13733.716 \pm 0.700)\,\mathrm{cm^{-1}}\,.$$

The shifted wavelengths can then be worked out as 728.09796 and 728.17218 nm using Eq. (1.4).

(a) All lines are observed when viewing transversely. Hence there will be three lines at 728.09796, 728.13507, and 728.17218 nm.
(b) The $\Delta M_L = 0$ line is not observed when viewed longitudinally. There will therefore be just two lines at 728.09796 and 728.17218 nm.

Note that the wavelength shifts induced by the Zeeman effect are quite small ($\sim$ 0.005%), and so it is necessary to have a good spectrometer to observe them.

8.1.2 The Anomalous Zeeman Effect

The anomalous Zeeman effect is observed in atoms with non-zero spin. This will include all atoms with an odd number of electrons. It will also include transitions between $S \neq 0$ states of atoms with an even number of electrons, e.g., triplet transitions in divalent atoms such as helium or group II elements. It is therefore the general case, and is more commonly observed than the normal Zeeman effect. As we shall see, the normal Zeeman effect is just the limit of the anomalous effect for the special case where $S = 0$.

In the LS-coupling regime, the spin-orbit interaction couples the spin and orbital angular momenta together to give the resultant total angular momentum $\boldsymbol{J}$ according to:

$$\boldsymbol{J} = \boldsymbol{L} + \boldsymbol{S}\,. \tag{8.7}$$

The orbiting electrons in the atom are equivalent to a classical magnetic gyroscope. The torque applied by the field causes the atomic magnetic dipole to precess around $\boldsymbol{B}$, an effect called **Larmor precession**. The external magnetic field therefore causes $\boldsymbol{J}$ to precess slowly about $\boldsymbol{B}$. Meanwhile, $\boldsymbol{L}$ and $\boldsymbol{S}$ precess more rapidly about $\boldsymbol{J}$ due to the spin-orbit interaction. This situation is illustrated in Figure 8.3(a). The speed of the precession about $\boldsymbol{B}$ is proportional to the field strength. If we turn up the field, the Larmor precession frequency will eventually be faster than the spin-orbit precession of $\boldsymbol{L}$ and $\boldsymbol{S}$ around $\boldsymbol{J}$. This is the point where the behavior ceases to be Zeeman-like, and we are in the strong-field regime of the Paschen–Back effect discussed in Section 8.1.3.

The interaction energy of the atom is equal to the sum of the interactions of the spin and orbital magnetic moments with the field:

$$\Delta E = -\mu_z B_z = -(\mu_z^{\text{orbital}} + \mu_z^{\text{spin}})B_z = \langle \boldsymbol{L}_z + g_s \boldsymbol{S}_z \rangle \frac{\mu_{\text{B}}}{\hbar} B_z\,, \tag{8.8}$$

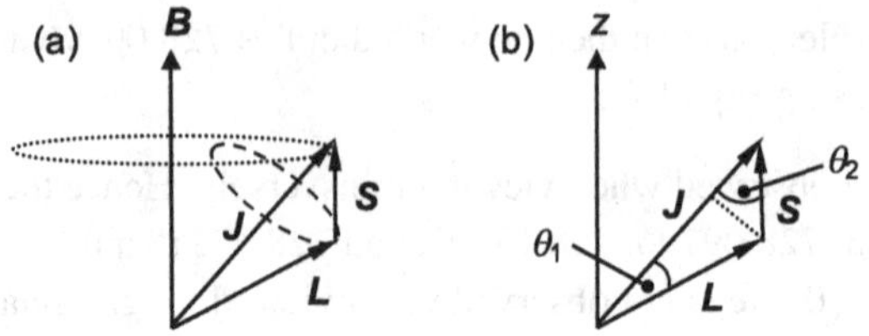

Figure 8.3 (a) Slow precession of $\boldsymbol{J}$ around $\boldsymbol{B}$ in the anomalous Zeeman effect. The spin-orbit interaction causes $\boldsymbol{L}$ and $\boldsymbol{S}$ to precess much more rapidly around $\boldsymbol{J}$. (b) Definition of the projection angles θ_1 and θ_2 used in the calculation of the Landé g-factor, given in Eq. (8.15).

where $g_s = 2$, and the symbol $\langle\cdots\rangle$ implies, as usual, that we take expectation values. The normal Zeeman effect is obtained by setting $S_z = 0$ and $L_z = M_L\hbar$ in this formula. In the case of the precessing atomic magnet shown in Figure 8.3(a), neither S_z nor L_z are constant. Only $J_z = M_J\hbar$ is well defined. We must therefore first project $\boldsymbol{L}$ and $\boldsymbol{S}$ onto $\boldsymbol{J}$, and then reproject this component onto the z-axis. The effective dipole moment of the atom is therefore given by:

$$\boldsymbol{\mu} = -\left\langle |\boldsymbol{L}|\cos\theta_1 \frac{\boldsymbol{J}}{|\boldsymbol{J}|} + 2|\boldsymbol{S}|\cos\theta_2 \frac{\boldsymbol{J}}{|\boldsymbol{J}|}\right\rangle \frac{\mu_\mathrm{B}}{\hbar}, \tag{8.9}$$

where the factor of 2 in the second term comes from the fact that $g_s = 2$. The angles θ_1 and θ_2 that appear here are defined in Figure 8.3(b), and can be calculated from the scalar products of the respective vectors:

$$\begin{aligned} \boldsymbol{L}\cdot\boldsymbol{J} &= |\boldsymbol{L}|\,|\boldsymbol{J}|\cos\theta_1\,, \\ \boldsymbol{S}\cdot\boldsymbol{J} &= |\boldsymbol{S}|\,|\boldsymbol{J}|\cos\theta_2\,, \end{aligned} \tag{8.10}$$

which implies that:

$$\boldsymbol{\mu} = -\left\langle \frac{\boldsymbol{L}\cdot\boldsymbol{J}}{|\boldsymbol{J}|^2} + 2\frac{\boldsymbol{S}\cdot\boldsymbol{J}}{|\boldsymbol{J}|^2}\right\rangle \frac{\mu_\mathrm{B}}{\hbar}\boldsymbol{J}. \tag{8.11}$$

Now Eq. (8.7) implies that $\boldsymbol{S} = \boldsymbol{J} - \boldsymbol{L}$, and hence that:

$$\boldsymbol{S}\cdot\boldsymbol{S} = (\boldsymbol{J}-\boldsymbol{L})\cdot(\boldsymbol{J}-\boldsymbol{L}) = \boldsymbol{J}\cdot\boldsymbol{J} + \boldsymbol{L}\cdot\boldsymbol{L} - 2\boldsymbol{L}\cdot\boldsymbol{J}.$$

We therefore find that:

$$\boldsymbol{L}\cdot\boldsymbol{J} = (\boldsymbol{J}\cdot\boldsymbol{J} + \boldsymbol{L}\cdot\boldsymbol{L} - \boldsymbol{S}\cdot\boldsymbol{S})/2,$$

so that:

$$\begin{aligned} \left\langle \frac{\boldsymbol{L}\cdot\boldsymbol{J}}{|\boldsymbol{J}|^2}\right\rangle &= \frac{[J(J+1) + L(L+1) - S(S+1)]\hbar^2/2}{J(J+1)\hbar^2}, \\ &= \frac{[J(J+1) + L(L+1) - S(S+1)]}{2J(J+1)}. \end{aligned} \tag{8.12}$$

Similarly:

$$\boldsymbol{S}\cdot\boldsymbol{J} = (\boldsymbol{J}\cdot\boldsymbol{J} + \boldsymbol{S}\cdot\boldsymbol{S} - \boldsymbol{L}\cdot\boldsymbol{L})/2\,,$$

and so:

$$\begin{aligned}\left\langle\frac{\boldsymbol{S}\cdot\boldsymbol{J}}{|\boldsymbol{J}|^2}\right\rangle &= \frac{[J(J+1)+S(S+1)-L(L+1)]\hbar^2/2}{J(J+1)\hbar^2}\,,\\ &= \frac{[J(J+1)+S(S+1)-L(L+1)]}{2J(J+1)}\,.\end{aligned} \tag{8.13}$$

We therefore conclude that:

$$\boldsymbol{\mu} = -g_J\frac{\mu_{\rm B}}{\hbar}\boldsymbol{J}\,, \tag{8.14}$$

where g_J is the **Landé *g*-factor** given by:

$$\begin{aligned}g_J &= \frac{[J(J+1)+L(L+1)-S(S+1)]}{2J(J+1)} + 2\frac{[J(J+1)+S(S+1)-L(L+1)]}{2J(J+1)}\\ &= 1+\frac{J(J+1)+S(S+1)-L(L+1)}{2J(J+1)}\,.\end{aligned} \tag{8.15}$$

This implies that

$$\mu_z = -g_J\mu_{\rm B}M_J\,, \tag{8.16}$$

and hence that the interaction energy with the field is:

$$\Delta E = -\mu_z B_z = g_J\mu_{\rm B}B_z M_J\,. \tag{8.17}$$

This is the final result for the energy shift of an atomic state in the anomalous Zeeman effect. Note that we just obtain $g_J = 1$ if $S = 0$, as we would expect for an atom with only orbital angular momentum. Similarly, if $L = 0$ so that the atom only has spin angular momentum, we find $g_J = 2$. Classical theories always predict $g_J = 1$. The departure of g_J from unity is caused by the spin part of the magnetic moment, and is a purely quantum effect.

The spectra can be understood by applying the following selection rules on J and M_J:

$$\begin{aligned}\Delta J &= 0, \pm 1\,,\ \ J = 0 \to 0 \ \text{ forbidden}\,;\\ \Delta M_J &= 0, \pm 1\,,\ \ M_J = 0 \to 0 \text{ forbidden if } \Delta J = 0\,.\end{aligned} \tag{8.18}$$

These rules have to be applied in addition to the $\Delta l = \pm 1$ and $\Delta S = 0$ rules. (See discussion in Section 5.8.) There are no selection rules on M_L and M_S here because L_z and S_z are not constants of the motion when L and S are coupled by the spin-orbit interaction.

Table 8.3 *Landé g-factors evaluated from Eq. (8.15) for the levels involved in the sodium D-lines.*

Level	J	L	S	g_J
$^2P_{3/2}$	3/2	1	1/2	4/3
$^2P_{1/2}$	1/2	1	1/2	2/3
$^2S_{1/2}$	1/2	0	1/2	2

The transition energy shift that follows from Eq. (8.17) is given by :

$$\begin{aligned} h\Delta\nu &= (h\nu - h\nu_0)\,, \\ &= \left(g_J^{\text{upper}} M_J^{\text{upper}} - g_J^{\text{lower}} M_J^{\text{lower}}\right) \mu_B B_z\,, \end{aligned} \tag{8.19}$$

where $h\nu_0$ is the transition energy at $B_z = 0$ and the superscripts refer to the upper and lower states respectively. This reduces to the normal Zeeman effect if $S = 0$ so that $g_J^{\text{upper}} = g_J^{\text{lower}} = 1$.

The polarizations of the transitions follow the same patterns as for the normal Zeeman effect:

- With *longitudinal* observation, the $\Delta M_J = 0$ transitions are absent and the $\Delta M_J = \pm 1$ transitions are $\sigma^\pm$ circularly polarized.
- With *transverse* observation, the $\Delta M_J = 0$ transitions are linearly polarized along the z-axis (i.e., parallel to $\boldsymbol{B}$) and the $\Delta M_J = \pm 1$ transitions are linearly polarized in the x-y plane (i.e., perpendicular to $\boldsymbol{B}$).

We can see how this works by considering the Zeeman effect on the sodium 3p → 3s D-lines. As shown in Figure 7.4, this is split into a doublet at $\boldsymbol{B} = 0$ by the spin-orbit interaction of the 3p term. The Landé g-factors of the three relevant states calculated using Eq. (8.15) are given in Table 8.3. Note that $g_J = 2$ for the $^2S_{1/2}$ level as it only has spin angular momentum. The splitting of the levels in the field is shown schematically in Figure 8.4. The $^2P_{3/2}$ level splits into four M_J states, while the $^2P_{1/2}$ and $^2S_{1/2}$ levels both split into two states. The splittings are different for each level because of the different Landé factors.

The transition energies can be calculated by using Eq. (8.19). Consider first the $^2P_{1/2} \rightarrow {}^2S_{1/2}$ D_1 line. All four combinations of levels are allowed by the selection rules, giving rise to four lines with frequency shifts given in Table 8.4. Now consider the $^2P_{3/2} \rightarrow {}^2S_{1/2}$ D_2 line. In principle, there could be eight combinations of upper and lower levels. However, the $M_J = \pm 3/2 \rightarrow \mp 1/2$ transitions are forbidden as these involve $\Delta M_J = \mp 2$. There are therefore six

Table 8.4 *Anomalous Zeeman effect for the sodium D-lines. The transition energy shifts are worked out from Eq. (8.19) and are quoted in units of* $\mu_B B_z$.

M_J^{upper}	M_J^{lower}	ΔM_J	Transition Energy shift	
			D_1 line	D_2 line
+3/2	+1/2	−1		+1
+1/2	+1/2	0	−2/3	−1/3
+1/2	−1/2	−1	+4/3	+5/3
−1/2	+1/2	+1	−4/3	−5/3
−1/2	−1/2	0	+2/3	+1/3
−3/2	−1/2	+1		−1

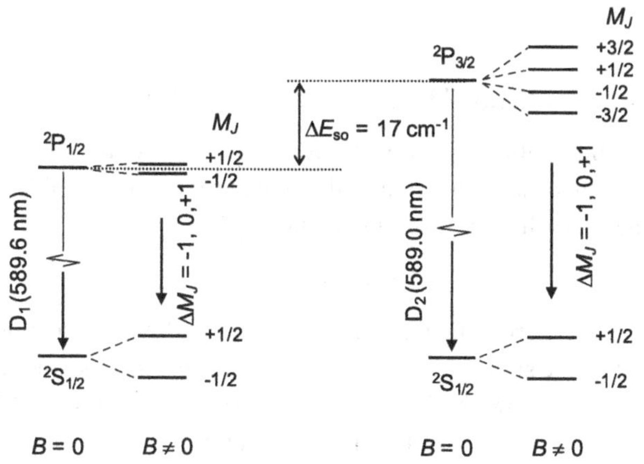

Figure 8.4 Splitting of the sodium D-line levels by a weak magnetic field. Note that the Zeeman splittings are exaggerated in the diagram to show them clearly. In the weak-field limit, they must be much smaller than the spin-orbit splitting.

transitions, with energy shifts given in Table 8.4. The splitting of the lines is shown schematically in the middle panel of Figure 8.6.

The results tabulated in Table 8.4 can be compared to those predicted by the normal Zeeman effect. In the normal Zeeman effect we observe three lines with an energy spacing equal to $\mu_B B$. In the anomalous effect, there can be more than three lines, and the spacing is different to the classical value. In the case of the alkali D-lines, we get four and six transitions, but this is not a general rule, and each case has to be worked out from first principles. See Exercises 8.4–8.6.

Example 8.2 What emission lines would be observed for the sodium D_1 line at 589.5924 nm when observed (a) transversely and (b) longitudinally to a magnetic field of 1.8 T?

Solution: The energy shifts of the four Zeeman lines are given in Table 8.4, i.e., $\pm(2/3)\mu_B B$ and $\pm(4/3)\mu_B B$. On using Eq. (8.6), we work this out to be $\pm 0.560\,\text{cm}^{-1}$ and $\pm 1.120\,\text{cm}^{-1}$. At $B = 0$ we have $h\nu_0 = 16960.87\,\text{cm}^{-1}$, and so the wave numbers of the Zeeman lines are 16959.75, 16960.31, 16961.43, and 16961.99 cm^{-1}. These correspond to wavelengths of 589.6314, 589.6119, 589.5729, and 589.5535 nm.

(a) All four lines are observed in transverse observation.
(b) The $\Delta M_J = 0$ lines are are not possible in longitudinal geometry, and so only the lines with $h\Delta\nu = \pm(4/3)\mu_B B$ will be observed, i.e., the outer two lines at 589.6314 and 589.5535 nm.

8.1.3 The Paschen–Back Effect

The Paschen–Back effect is observed at very strong fields. The criterion for observing it is that the interaction with the external magnetic field should be much stronger than the spin-orbit interaction:

$$\mu_B B_z \gg \Delta E_{so} \,. \tag{8.20}$$

If we satisfy this criterion, then the precession around the external field will be much faster than the spin-orbit precession. This means that the interaction with the external field is now the largest perturbation, and so it should be treated first, before the perturbation of the spin-orbit interaction.

Another way to think of the strong-field limit is that it occurs when the external field is much stronger than the internal field of the atom arising from the orbital motion. We saw in Section 7.3 that the internal fields in most atoms are large. For example, the Bohr model predicts an internal field of 12 T for the $n = 1$ shell of hydrogen. (See Eq. [7.19].) This is a very strong field that can only be obtained in the laboratory by using powerful superconducting magnets. It will therefore usually be the case that the field required to observe the Paschen–Back effect is so large that we never go beyond the Zeeman regime in the laboratory.[1] For example, the field strength equivalent to the spin-orbit splitting of the 3p state in sodium is given by:

$$B_z = \frac{\Delta E_{so}}{\mu_B} = \frac{17\,\text{cm}^{-1}}{9.27 \times 10^{-24}\,\text{JT}^{-1}} = 36\,\text{T},$$

[1] There can be extremely large magnetic fields present in some astrophysical environments. See Section 12.3.

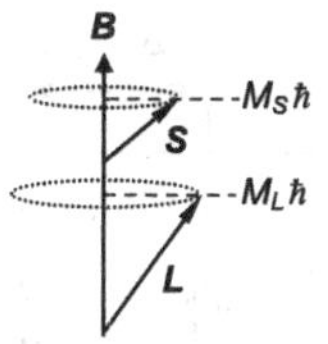

Figure 8.5 Precession of $\boldsymbol{L}$ and $\boldsymbol{S}$ around $\boldsymbol{B}$ in the Paschen–Back effect.

which is not achievable in normal laboratory conditions. On the other hand, since the spin-orbit interaction decreases with decreasing atomic number Z (see Section 7.3.3), the splitting of the 2p state in lithium with $Z = 3$ is only $0.3\,\text{cm}^{-1}$. This means that we can reach the strong-field regime for fields $\gg 0.6\,\text{T}$. This is readily achievable, and allows the Paschen–Back effect to be observed. Another important example is the $n = 2$ shell of hydrogen. The spin-orbit coupling is small ($0.365\,\text{cm}^{-1}$; see Figure 7.3), and the strong-field limit requires $B \gg 0.8\,\text{T}$.

In the Paschen–Back effect, the spin-orbit interaction is assumed to be negligibly small, and $\boldsymbol{L}$ and $\boldsymbol{S}$ are therefore no longer coupled together. Each precesses independently around $\boldsymbol{B}$, as sketched in Figure 8.5. The precession rates for $\boldsymbol{L}$ and $\boldsymbol{S}$ are different because of the different g-factors. Hence the magnitude of $(\boldsymbol{L}+\boldsymbol{S})$ varies with time: $\boldsymbol{J}$ is no longer a constant of the motion.

The interaction energy is calculated by adding the separate contributions of the spin and orbital energies:

$$\Delta E = -\mu_z B_z = -(\mu_z^{\text{orbital}} + \mu_z^{\text{spin}})B_z = (M_L + g_s M_S)\mu_B B_z\,. \tag{8.21}$$

The shift of the spectral lines is given by:

$$h\Delta\nu = (\Delta M_L + g_s \Delta M_S)\mu_B B_z\,. \tag{8.22}$$

We have noted before that optical transitions do not affect the spin, and so we must have $\Delta M_S = 0$. The frequency shift is thus given by:

$$h\Delta\nu = \mu_B B_z\,\Delta M_L\,, \tag{8.23}$$

where $\Delta M_L = 0$ or ± 1. In other words, we revert to the normal Zeeman effect.

Figure 8.6 illustrates the change of the spectra as we increase B from zero for the p $\rightarrow$ s transitions of an alkali atom. At $B = 0$, the lines are split by the spin-orbit interaction. At weak fields we observe the anomalous Zeeman effect, while at strong fields we change to the Paschen–Back effect.

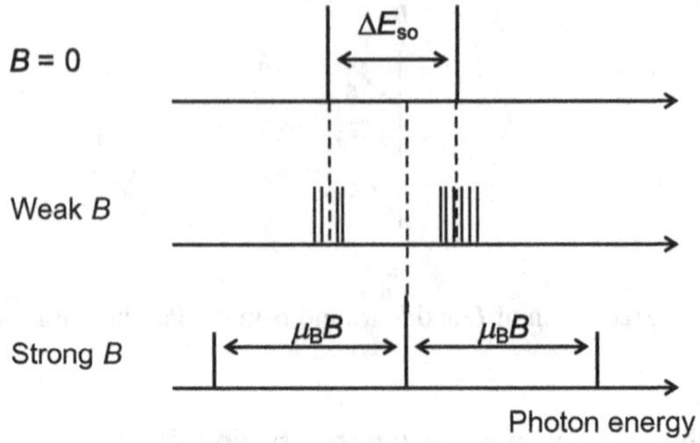

Figure 8.6 Schematic progression of the optical spectra for the p → s transitions of an alkali atom with increasing field.

8.2 The Concept of "Good" Quantum Numbers

It is customary to refer to quantum numbers that relate to constants of the motion as "good" quantum numbers. In this discussion of the effects of magnetic fields, we have used six different quantum numbers to describe the angular momentum state of the atom: J, M_J, L, M_L, S, M_S. However, we cannot know all of these at the same time. In fact, we can only know four: (L, S, J, M_J) in the weak-field limit, or (L, S, M_L, M_S) in the strong-field limit. In the weak-field limit, $\boldsymbol{J}$ and J_z are constant, and so J and M_J are good quantum numbers. On the other hand, L_z and S_z are not constant, which implies that M_L and M_S are not good quantum numbers. Similarly, in the strong-field limit, the coupling between $\boldsymbol{L}$ and $\boldsymbol{S}$ is broken, and so $\boldsymbol{J}$ and J_z are not constants of the motion; M_L and M_S are good quantum numbers, but J and M_J are not.

A similar type of argument applies to the two angular momentum coupling schemes discussed in Section 5.6, namely LS coupling and jj coupling. As an example, consider the total angular momentum state of a two-electron atom. In the LS-coupling scheme, we specify (L, S, J, M_J), whereas in the jj-coupling scheme we have (j_1, j_2, J, M_J). In both cases, we have four "good" quantum numbers, which tell us the quantities that are constant within the validity of the approximations that underpin the coupling schemes. The other quantum numbers are unknown because the physical quantities they represent are not constant. In LS coupling we cannot know the j values of the individual electrons because the residual electrostatic potential overpowers the spin-orbit effect whereas in the jj-coupling scheme we cannot know L and S. Note, however, that J and M_J are good quantum numbers in both coupling limits, as a consequence of the conservation of angular momentum in the absence of external perturbations. (See Section 5.1.). This means that we can always describe the Zeeman energy of the atom by Eq. (8.17) – although in the case of jj coupling, the formula for the g_J factor given in Eq. (8.15) will not be valid because L and S are not good quantum numbers.

8.3 Nuclear Effects

8.3.1 Magnetic Field Effects for Hyperfine Levels

The Zeeman effect on hyperfine levels can be calculated by a method analogous to Section 8.1.2. As discussed in Section 7.8.2, the hyperfine interaction couples $\boldsymbol{J}$ to the nuclear spin $\boldsymbol{I}$ to produce the resultant:

$$\boldsymbol{F} = \boldsymbol{I} + \boldsymbol{J}\,. \tag{8.24}$$

The Zeeman shift is then given by:[2]

$$\begin{aligned} \Delta E &= g_F \mu_B B M_F\,, \\ g_F &\approx g_J \frac{F(F+1) + J(J+1) - I(I+1)}{2F(F+1)}\,. \end{aligned} \tag{8.25}$$

These effects can only be observed when the Zeeman shift is smaller than the hyperfine splitting at $B = 0$. As discussed in Section 7.8.2, these hyperfine splittings are much smaller than fine-structure effects due to the small gyromagnetic ratio of the nucleus. This implies that the change from the weak- to strong-field limit occurs at much smaller fields than for the electrons. In the strong field limit, $\boldsymbol{J}$ is decoupled from $\boldsymbol{I}$, and we just revert to the Zeeman effect of the electron states. The strong-field limit for hyperfine structure thus corresponds to the weak-field limit for the electrons.

8.3.2 Nuclear Magnetic Resonance

The hyperfine splitting of the ground-state of hydrogen is 1420 MHz. (See Figure 7.7[a].) The hyperfine strong-field regime therefore applies when $\mu_B B/h \gg 1.42 \times 10^9$ Hz, i.e. $B \gg 0.1$ T (see Eq. [8.6]). Hence the nuclear spin $\boldsymbol{I}$ is decoupled from $\boldsymbol{J}$ at very modest field strengths, and it is then possible to observe field-split states entirely related to the nucleus. Transitions between these states lie at the basis of nuclear magnetic resonance (NMR) techniques.

We have seen in Section 7.8.2 that a nucleus with spin has a magnetic dipole moment given by (see Eq. [7.53]):

[2] See, for example, Woodgate (1980), § 9.6.

$$\boldsymbol{\mu}_{\text{nucleus}} = g_I \frac{\mu_{\text{N}}}{\hbar} \boldsymbol{I}, \tag{8.26}$$

where g_I is the nuclear g-factor, and $\mu_{\text{N}} = e\hbar/2m_{\text{p}}$ is the nuclear magneton. If an external magnetic field is applied along the z-direction, the energy of the nucleus will shift by:

$$\Delta E = -\boldsymbol{\mu}_{\text{nucleus}} \cdot \boldsymbol{B} = -\mu_z^{\text{nucleus}} B_z\,. \tag{8.27}$$

On substituting from Eq. (8.26), the Zeeman energy of the nucleus becomes:

$$\Delta E = -g_I \frac{\mu_{\text{N}}}{\hbar} I_z B_z = -g_I \mu_{\text{N}} B_z M_I\,, \tag{8.28}$$

where $I_z = M_I\hbar$, and M_I runs in integer steps from $-I$ to $+I$. In an NMR experiment, an electromagnetic pulse is applied, which induces magnetic-dipole (M1) transitions between the Zeeman-split levels. The selection rules for M1 transitions will be discussed in Section 12.2.2. At this stage, the key point is that the angular momentum of the nucleus changes by one unit when the photon is absorbed, so that the selection rule is $\Delta M_I = \pm 1$. The energy of the photon required to induce this transition is thus given by:

$$h\nu = g_I \mu_{\text{N}} B_z\,. \tag{8.29}$$

The frequency typically lies in the radio-frequency (RF) range and can be measured either by scanning ν at fixed B_z, or by scanning B_z at fixed ν. In working out NMR frequencies, it is convenient to use MHz and Tesla units, with $\mu_{\text{N}}/h = 7.6226\,\text{MHz}\,\text{T}^{-1}$.

In the magnetic resonance systems used in medical imaging, the RF photons are brought to resonance with the hydrogen atoms or ions in the body. The g-factor of the proton is 5.586, which implies that $\nu = 42.6\,\text{MHz}$ at a field of 1 T. Magnetic resonance can also be observed from other nuclei in a variety of liquid- and solid-state environments, and this gives rise to a host of techniques used in chemistry and biology to obtain information about the structure and bonding of molecules.

8.4 Electric Fields

The shift of the electronic levels in an electric field is called the **Stark effect**. The effect is named after Johannes Stark, who was the first to measure the splitting of the hydrogen Balmer lines in an electric field in 1913. In most atoms, we observe the **quadratic Stark effect** and we therefore consider this effect first. We then move on to consider the **linear Stark effect**, which is observed for the excited states of hydrogen, and in other atoms at very strong fields. The Stark shift of an atom is harder to observe than the Zeeman shift, which explains why magnetic effects are more widely studied in atomic physics. However, large Stark effects are readily observable in solid-state physics, as will be discussed in Section 11.4.1.

8.4.1 The Quadratic Stark Effect

The quadratic Stark effect causes a small *red shift* (i.e., a shift to lower energy) which is proportional to the *square* of the electric field. It can be explained by considering the energy shift of an atom in an electric field $\boldsymbol{\mathcal{E}}$:

$$E = -\boldsymbol{p} \cdot \boldsymbol{\mathcal{E}}\,, \tag{8.30}$$

where $\boldsymbol{p}$ is the electric dipole of the atom. The negatively charged electron clouds of an atom are spherically symmetric about the positively charged nucleus in the absence of applied fields, as shown in Figure 8.7(a). A charged sphere acts like a point charge at its center, and it is thus apparent that atoms do not normally possess a dipole moment. When a field is applied, the electron cloud and the nucleus experience opposite forces, which results in a net displacement of the electron cloud with respect to the nucleus, as shown in Figure 8.7(b). This creates a dipole $\boldsymbol{p}$ which is parallel to $\mathcal{E}$ and whose magnitude is proportional to $|\boldsymbol{\mathcal{E}}|$. This can be expressed mathematically by writing:

$$\boldsymbol{p} = \alpha \boldsymbol{\mathcal{E}}\,, \tag{8.31}$$

where α is the **polarizability** of the atom. The energy shift of the atom is found by calculating the energy change on increasing the field strength from zero:

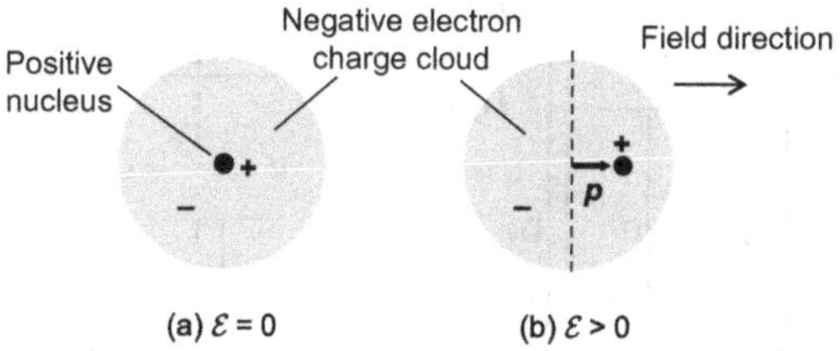

Figure 8.7 Effect of an electric field $\mathcal{E}$ on the electron cloud of an atom. (a) When $\mathcal{E} = 0$, the negatively charged electron cloud is arranged symmetrically about the nucleus, and there is no electric dipole. (b) When the electric field is applied, the electron cloud is displaced, and a net dipole parallel to the field is induced. Note that the magnitude of the dipole has been greatly exaggerated to illustrate the effect more clearly.

$$\Delta E = -\int_0^{\mathcal{E}} \boldsymbol{p}\cdot \mathrm{d}\boldsymbol{\mathcal{E}}' = -\int_0^{\mathcal{E}} \alpha \mathcal{E}' \mathrm{d}\mathcal{E}' = -\frac{1}{2}\alpha\mathcal{E}^2, \tag{8.32}$$

which predicts a quadratic red shift, as required. The magnitude of the red shift is generally rather small. This is because the electron clouds are tightly bound to the nucleus, and it therefore requires very strong electric fields to induce a significant dipole.

A more detailed description of the quadratic Stark effect based on second-order perturbation theory is given in Section D.1 of Appendix D. It is shown there that the energy shift of the ith state is given by:

$$\Delta E_i = \sum_{j\neq i} \frac{|\langle \psi_i | H' | \psi_j \rangle|^2}{E_i - E_j}, \tag{8.33}$$

where the summation runs over all the other states of the system, and E_i and E_j are the unperturbed energies of the states. Explicit evaluation of the matrix elements for sodium indicates that the Stark shift at a given field strength depends on M_J^2. This means that electric fields do not completely break the degeneracy of the M_J sub-levels of a particular $|J\rangle$ level, which contrasts with the Zeeman effect, where the degeneracy in M_J is fully lifted.

The quadratic Stark shift of the sodium D-lines is shown schematically in Figure 8.8. All states are shifted to lower energy, with those of the same M_J values being shifted equally for a given level, as indicated in Figure 8.8(a). As discussed in Section D.1, the shifts of the upper 3p levels are larger than that of the lower 3s $^2S_{1/2}$ term, and both spectral lines therefore show a net shift to lower energy, as indicated in Figure 8.8(b). Owing to the degeneracy of the

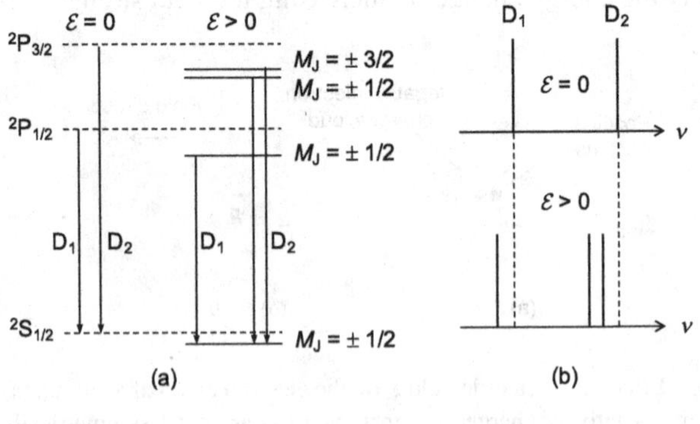

Figure 8.8 (a) Shift of the $^2S_{1/2}$, $^2P_{1/2}$, and $^2P_{3/2}$ levels of an alkali atom in an electric field. Note that the red shifts of the upper levels are larger than that of the lower level. (b) Red shift of the D_1 ($^2P_{1/2} \rightarrow {}^2S_{1/2}$) and the D_2 ($^2P_{3/2} \rightarrow {}^2S_{1/2}$) lines in the field.

sub-levels with the same $|M_J|$, the D_1 ($^2P_{1/2} \rightarrow {}^2S_{1/2}$) line does not split, while the D_2 ($^2P_{3/2} \rightarrow {}^2S_{1/2}$) line splits into a doublet.

An interesting consequence of the perturbation caused by the electric field is that the unperturbed atomic states get mixed with other states of the opposite parity. For example, the 3s state has even parity at $\mathcal{E} = 0$, but acquires a small admixture of the odd parity 3p state as the field is increased. This means that parity-forbidden transitions (eg s→s, p→p, d→s, etc.) become weakly allowed as the field is increased. Since we are dealing with a second-order perturbation, the intensity of these forbidden transitions increases in proportion to $\mathcal{E}^2$.

Example 8.3 The polarizabilities of the 4p $^2P_{1/2}$ and 4s $^2S_{1/2}$ levels of potassium ($Z = 19$) are $+5.0 \times 10^{-16}$ and $+2.4 \times 10^{-16}\,\text{cm}^{-1}\,(\text{V/m})^{-2}$, respectively. What is the wavelength shift of the D_1 line at 769.89645 nm in a field of 250 kV/cm?

Solution: Both the upper and lower levels shift to lower energy, as shown in Figure 8.8(a). The energy shift of the levels is given by Eq. (8.32), and the red shift of the spectral line is equal to their difference:

$$\Delta E = -\frac{1}{2}\left(\alpha^{\text{upper}} - \alpha^{\text{lower}}\right)\mathcal{E}^2 = -\frac{1}{2}\left(\alpha^{4p} - \alpha^{4s}\right)\mathcal{E}^2 .$$

The field is equivalent to 2.5×10^7 V/m, and so the shift in cm^{-1} is:

$$\Delta E = -\frac{1}{2}\left(5.0 \times 10^{-16} - 2.4 \times 10^{-16}\right) \times (2.5 \times 10^7)^2 = -0.081\,\text{cm}^{-1} .$$

The unperturbed transition occurs at $12989\,\text{cm}^{-1}$, and so we have a fractional red shift of $(-0.081/12989) = -6.3 \times 10^{-6}$. The wavelength shift is thus $-(-6.3 \times 10^{-6}) \times \lambda = +4.8 \times 10^{-3}\,\text{nm} = +4.8\,\text{pm}$. This is a small shift, even though the field is very large, and explains why the quadratic Stark effect is hard to observe.

8.4.2 The Linear Stark Effect

Stark's original experiment of 1913 was performed on the visible-frequency Balmer lines of hydrogen, which correspond to transitions that terminate on the $n = 2$ level. In contrast to what has been discussed in the previous subsection, the shift was quite large and varied linearly with the field. The explanation of the linear Stark effect in hydrogen by degenerate perturbation theory is given in Section D.2 of Appendix D. It is shown there that the linear Stark effect is observed when an atom possesses degenerate states of opposite parities. The classic example is the 2s and 2p states of hydrogen, which are degenerate in the absence of fine-structure effects. Perturbation theory predicts that the $n = 2$ shell splits into a triplet, with energies of $-3ea_0\mathcal{E}$, 0, and $+3ea_0\mathcal{E}$ with respect to the unperturbed level, where a_0 is the Bohr radius of hydrogen. The splitting is linear in the field and has a much larger magnitude than that calculated for the quadratic Stark effect. For example, at $\mathcal{E} = 250\,\text{kV/cm}$, we find shifts of $\pm 4 \times 10^{-3}$ eV (32 cm^{-1}), which are more than two orders of magnitude larger than the shifts of the levels in alkalis at the same field strength. (See Example 8.3.) This, of course, explains why the linear Stark effect in hydrogen was the first electric-field-induced phenomenon to be discovered. Note that the Stark shift is also two orders of magnitude larger than the fine-structure splitting of the $n = 2$ shell, (see Figure 7.3), which justifies the use of degenerate perturbation theory for the 2s and 2p sub-shells.

The second-order perturbation analysis discussed in Section D.1 is expected to break down at large field strengths when the field-induced perturbation becomes comparable to the splittings of the unperturbed levels. The field required to reach this limit for the 3s ground-state level of sodium is shown in Section D.1 to be extremely large ($\sim 10^{10}$ V/m). However, the fields required for excited states can be significantly smaller, because some atoms have different parity excited states that are relatively close to each other. We would then expect the behavior to change as the field is increased. At low fields we would observe the quadratic Stark effect, but when the field is sufficiently large that the perturbation is comparable to the energy splitting, we would effectively have degenerate levels with different parities, giving rise to a *linear* shift. This change from the quadratic to linear Stark effect at high fields was first studied for the excited states of helium by Foster in 1927.

Exercises

8.1 (a) Compare the Zeeman splitting of helium (atomic weight 4.00) in Example 8.1 to the Doppler linewidth at room temperature.
(b) Work out the minimum field for the Zeeman splitting of the D_1 line of sodium (atomic weight 23.0) to exceed the Doppler linewidth at room temperature.
(c) Explain why laboratory demonstrations of the Zeeman effect frequently use the 546.1 nm line of a low-pressure mercury lamp. (The atomic weight of mercury is 200.6.)

8.2 Consider a spectral line with wavelength λ measured in nm. Show that the wavelength shift in nm corresponding to a Zeeman energy shift of ΔE measured in cm^{-1} is given by: $\delta\lambda(\text{nm}) = -[\lambda(\text{nm})]^2 \Delta E(\text{cm}^{-1}) \times 10^{-7}$. Verify that this is correct for Example 8.2

8.3 Repeat Example 8.2 for the sodium D_2 line at 588.9950 nm.

8.4 A Zeeman experiment is carried out on helium in the weak-field limit. Sketch the spectrum that would be observed when viewing transversely:

(a) for the 1s3p $^1P_1 \rightarrow$ 1s2s $^1S_0$ transition;
(b) for the 1s3p $^3P_1 \rightarrow$ 1s2s $^3S_1$ transition.

8.5 A Zeeman experiment with $B = 0.8\,\text{T}$ is carried out on the 3s4s $^3S_1 \rightarrow$ 3s3p $^3P_0$ transition of magnesium at 516.7322 nm. Sketch the spectrum that would be observed when viewing transversely to the field, stating the wavelengths of any lines that occur.

8.6 The 6s7s $^3S_1 \rightarrow$ 6s6p $^3P_2$ transition of mercury occurs at 546.0735 nm.

(a) Deduce the energy shifts of the Zeeman lines in units of $\mu_B B$ for transverse observation.
(b) What would be the wavelength shifts in nm of the Zeeman lines observed in longitudinal observation for $B = 1\,\text{T}$?

8.7 Calculate the ratio of g_F to g_J for the 3p hyperfine levels of sodium ($I = 3/2$) shown in Figure 7.7(b) and (c).

8.8 The ^{12}C isotope of carbon has $I = 0$, and so NMR experiments on organic materials focus on ^{13}C, which has $I = 1/2$, and a relative abundance of 1.1%. Given that $g_I = 1.4048$ for ^{13}C, calculate the magnetic field required to observe an NMR resonance at 90 MHz, and compare it to the value for ^{1}H.

8.9 The deuteron (^{2}H) nucleus has $I = 1$ and $g_I = 0.8574$.

(a) What values would be possible for a measurement of μ_z on a deuteron in a uniform magnetic field?
(b) What would be the ratio of the resonant RF frequencies for the two isotopes of hydrogen in an NMR experiment on water?

8.10 A Stark effect experiment is performed on the 5p $^2P_{1/2} \rightarrow$ 5s $^2S_{1/2}$ D_1 line of rubidium at 780 nm. The polarizabilities of the $^2P_{1/2}$ and $^2S_{1/2}$ levels are 6.5×10^{-16} and $2.6 \times 10^{-16}\,\text{cm}^{-1}\,(\text{V/m})^{-2}$, respectively.

(a) Calculate the shift of the transition energy in wave-number units for an electric field strength of 400 kV/cm.
(b) Deduce the field strength that would have to be applied to shift the wavelength by 0.001 nm.

8.11 What electric field would give the same level splitting in the $n = 2$ shell of hydrogen as a magnetic field of 2 T?

Part II

Applications of Atomic Physics

9

Stimulated Emission and Lasers

Lasers are an icon of the modern technological world, performing mundane tasks such as bar-code reading in supermarkets, and imagination-catching applications such as Star Wars weaponry. The word *LASER* is an acronym standing for "Light Amplification by Stimulated Emission of Radiation." In this chapter, we develop the theory of stimulated emission, which is central to understanding how lasers work, and then we discuss how it is used in practical laser systems.

9.1 Stimulated Emission

The spontaneous tendency for atoms in excited states to drop to lower levels by emitting radiation was considered in Chapter 3. We now consider the optical transitions that occur when the atom is subjected to electromagnetic radiation with its frequency resonant with the energy difference of the two levels. We follow Einstein's treatment from 1917.

Consider an atom with two levels 1 and 2, with energy difference $E_2 - E_1$. As discussed in Section 1.4 and illustrated in Figure 9.1(a) and (b), there are two obvious ways that the atom can interact with photons. The first process is called **spontaneous emission**. Here, an atom in the upper level has a spontaneous tendency to drop to the lower level by emitting a photon of energy $h\nu$, where:

$$h\nu = E_2 - E_1 \,. \tag{9.1}$$

If the atom starts in the lower level, it can be induced to make a transition to the upper level by absorbing a photon of the same energy. This process is called **absorption**. It is these two processes that we have been considering up to now whenever we have discussed radiative transitions. (See, for example, Chapter 3.)

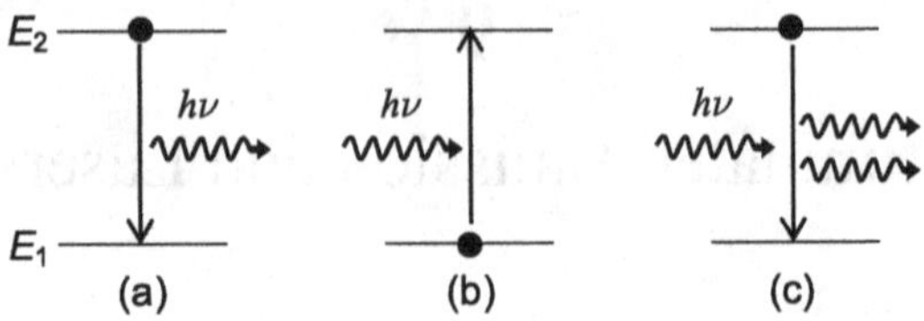

Figure 9.1 Different types of optical transitions between two levels of an atom. (a) Spontaneous emission. (b) Absorption. (c) Stimulated emission.

Einstein considered the interaction between a gas of atoms and black-body radiation in thermal equilibrium with each other at temperature T. He realized that he could not reach a consistent analysis of the problem without introducing a third process called **stimulated emission**. This process is the reverse of absorption: an atom in level 2 drops to level 1 stimulated by photons with energy $h\nu = (E_2 - E_1)$, as illustrated in Figure 9.1(c).

Consider a gas of atoms irradiated by black-body radiation, with N_2 atoms in level 2 and N_1 atoms in level 1. Let W_{ij} represent the transition rate from level $i \to j$. The part of black-body spectrum at frequency ν that satisfies Eq. (9.1) can induce absorption and stimulated emission transitions. The rates for the three types of transition shown in Figure 9.1 can then be written:

- Spontaneous emission ($2 \to 1$):

$$W_{21}^{\text{spon}} = \frac{\mathrm{d}N_2}{\mathrm{d}t} = -\frac{\mathrm{d}N_1}{\mathrm{d}t} = -A_{21}N_2\,. \tag{9.2}$$

- Stimulated emission ($2 \to 1$):

$$W_{21}^{\text{stim}} = \frac{\mathrm{d}N_2}{\mathrm{d}t} = -\frac{\mathrm{d}N_1}{\mathrm{d}t} = -B_{21}N_2u(\nu)\,. \tag{9.3}$$

- Absorption ($1 \to 2$):

$$W_{12} = \frac{\mathrm{d}N_1}{\mathrm{d}t} = -\frac{\mathrm{d}N_2}{\mathrm{d}t} = -B_{12}N_1u(\nu)\,. \tag{9.4}$$

These three equations are effectively the definitions of the Einstein A and B coefficients. The A coefficient was previously introduced in Section 3.6. Equation 9.2 is the same as Eq. (3.24), and says that the spontaneous emission rate is proportional to the number of atoms in the upper level and the transition probability – as given by the A coefficient, which has units s^{-1}. The B coefficients give the probablities for the stimulated processes. Equations (9.3) and (9.4) state that the rate for stimulated transitions is proportional to the number of atoms in the starting level, the spectral energy density of the radiation at frequency ν, namely $u(\nu)$ (units $\mathrm{J\,m^{-3}\,Hz^{-1}}$), and the appropriate

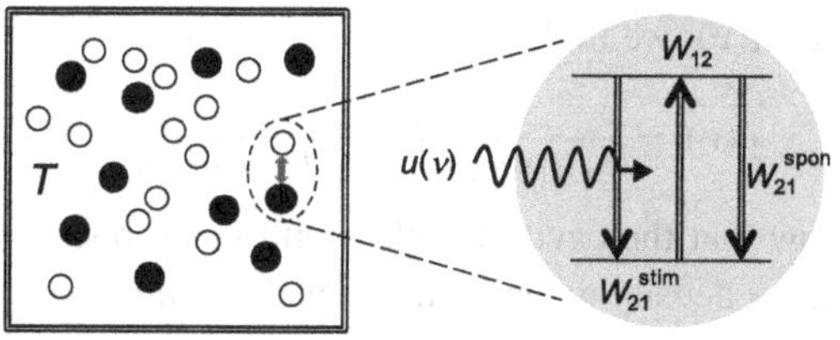

Figure 9.2 Atoms (•) in thermal equilibrium with black-body radiation at temperature T. The photons are represented by open circles (○). The radiation at frequency ν induces stimulated emission and absorption transitions, as shown on the right. Spontaneous emission also occurs.

B coefficient. The units of the B coefficients can be worked out from Eqs. (9.3) and (9.4) to be $\mathrm{m^3\,J^{-1}\,s^{-2}}$.

We might be inclined to think that the three Einstein coefficients are independent parameters, but this is not, in fact, the case. To see this, we imagine a gas of atoms inside a box at temperature T with black walls, as shown in Figure 9.2. The atoms will interact with the black-body radiation that fills the cavity, and will be absorbing and emitting photons all the time. If we leave the system for long enough, the atoms and radiation will come to equilibrium. In the steady-state conditions that occur when a system is in equilibrium, the rate of upward transitions must exactly balance the rate of downward transitions:

$$\begin{aligned} W_{12} &= W_{21} \\ &= W_{21}^{\text{spon}} + W_{21}^{\text{stim}} \\ \therefore \quad B_{12}N_1u(\nu) &= A_{21}N_2 + B_{21}N_2u(\nu)\,, \end{aligned} \tag{9.5}$$

which implies that:

$$\frac{N_2}{N_1} = \frac{B_{12}u(\nu)}{A_{21} + B_{21}u(\nu)}\,. \tag{9.6}$$

Furthermore, since the atoms are in thermal equilibrium at temperature T, the ratio of N_2 to N_1 must satisfy Boltzmann's law:

$$\frac{N_2}{N_1} = \frac{g_2}{g_1}\exp\left(-\frac{h\nu}{k_BT}\right)\,, \tag{9.7}$$

where g_2 and g_1 are the degeneracies of levels 2 and 1 respectively, and $h\nu$ is the energy difference given by Eq. (9.1). Equations (9.6) and (9.7) together imply that:

$$\frac{B_{12}u(\nu)}{A_{21} + B_{21}u(\nu)} = \frac{g_2}{g_1}\exp\left(-\frac{h\nu}{k_BT}\right)\,. \tag{9.8}$$

On solving this for $u(\nu)$, we find:

$$u(\nu) = \frac{g_2 A_{21}}{g_1 B_{12} \exp(h\nu/k_B T) - g_2 B_{21}} . \tag{9.9}$$

However, we know that the cavity is filled with black-body radiation, which has a spectral energy density given by the Planck formula:

$$u(\nu) = \frac{8\pi h\nu^3}{(c/n)^3} \frac{1}{\exp(h\nu/k_B T) - 1} , \tag{9.10}$$

where c/n is the speed of light in a medium with refractive index n. The only way to make Eq. (9.9) and (9.10) consistent with each other at all temperatures and frequencies is if:

$$\begin{aligned} g_1 B_{12} &= g_2 B_{21} , \\ A_{21} &= \frac{8\pi n^3 h\nu^3}{c^3} B_{21} . \end{aligned} \tag{9.11}$$

A moment's thought will convince us that it is not possible to get consistency between the equations without the stimulated emission term. This is what led Einstein to introduce the concept.

The relationships between the Einstein coefficients in Eq. (9.11) have been derived for the special case of an atom in thermal equilibrium with black-body radiation. However, once we have derived the interrelationships, they will apply in all other cases as well. This is very useful, since we only need to know one of the coefficients to work out the other two. For example, we can measure the radiative lifetime to determine A_{21} from (see Eq. [3.26]),

$$A_{21} = \frac{1}{\tau} , \tag{9.12}$$

and then work out the B coefficients from Eq. (9.11).

Equation (9.11) shows that the probabilities for absorption and stimulated emission are the same apart from the degeneracy factors, and that the ratio of the probability for spontaneous emission to stimulated emission increases in proportion to ν^3. In a laser we want to encourage stimulated emission and suppress spontaneous emission. Hence it gets progressively more difficult to make lasers work as the frequency increases, all other things being equal.

9.2 Population Inversion

It is apparent from Figure 9.1(c) that the process of stimulated emission increases the numbers of photons in a beam of light. We now wish to study how we can use stimulated emission to make an amplifier for light. Let us first consider a system with nondegenerate levels. In this case, we have

$g_1 = g_2 = 1$ and hence, from Eq. (9.11), $B_{12} = B_{21}$. In a gas of atoms in thermal equilibrium, the population of the lower level will always be greater than the population of the upper level. (See Eq. [9.7].) Therefore, if a light beam is incident on the medium, the rate of upward transitions due to absorption – namely W_{12} given in Eq. (9.4) – will always exceed the rate of stimulated downward transitions – namely W_{21}^{stim} given in Eq. (9.3). Hence there will be net absorption, and the intensity of the beam will diminish on progressing through the medium. Amplification requires that $N_2 > N_1$. This nonequilibrium situation is called **population inversion**.

The more general case of population inversion with nondegenerate levels can be considered by noting that amplification requires that $W_{21} > W_{12}$. The light intensity is very high in an operating laser, so that $W_{21}^{\text{stim}} \gg W_{21}^{\text{spon}}$. The condition for amplification then simplifies to $W_{21}^{\text{stim}} > W_{12}$, which can written using Eqs. (9.3) and (9.4) as:

$$B_{21}N_2u(\nu) > B_{12}N_1u(\nu)\,. \tag{9.13}$$

Substituting from Eq. (9.11) leads to the conclusion:

$$N_2 > \frac{g_2}{g_1}N_1\,, \tag{9.14}$$

which reduces to:

$$N_2 > N_1\,, \tag{9.15}$$

for nondegenerate levels. On comparing Eq. (9.14) and Eq. (9.7), it is apparent that population inversion corresponds to *negative* temperatures. This is not as ridiculous as it sounds, because the atoms with population inversion are not in thermal equilibrium.

Once we have population inversion, we have a mechanism for generating amplification in a laser medium. The art of making a laser is to work out how to get population inversion for the relevant transition.

Example 9.1 The degeneracies of the upper and lower levels of the 488.0 nm line of the argon ion laser are 6 and 4 respectively. Deduce the effective temperature of the laser levels when the population of the upper level is twice that of the lower level.

Soluton: The effective temperature can be deduced from Eq. (9.7) with $g_2 = 6$, $g_1 = 4$, $h\nu = hc/(488 \times 10^{-7}) = 2.54\,\text{eV}$, and $N_2/N_1 = 2$:

$$\frac{N_2}{N_1} = 2 = \frac{6}{4}\exp\left(-\frac{2.54\text{eV}}{k_\text{B}T}\right)\,.$$

On solving, we find $T = -1.02 \times 10^5$ K. The temperature is negative because the population is inverted, with $N_2 > (g_2/g_1)N_1$.

9.3 Optical Amplification

Ordinary optical materials do not amplify light. Instead, they tend to absorb or scatter light, so that the light intensity decreases as it propagates through the medium. To get amplification, energy must be pumped into the medium, putting it in a nonequilibrium state. The amplification is quantified by the **gain coefficient** γ, which is defined by the following equation:

$$I(x + \mathrm{d}x) = I(x) + \gamma I(x)\,\mathrm{d}x \equiv I(x) + \mathrm{d}I\,, \tag{9.16}$$

where $I(x)$ represents the intensity (i.e., power per unit area) at a point x within the gain medium, as shown in Figure 9.3(a). The differential equation can be solved as follows:

$$\begin{aligned} \mathrm{d}I &= \gamma I \mathrm{d}x \\ \therefore \frac{\mathrm{d}I}{\mathrm{d}x} &= \gamma I \\ \therefore I(x) &= I(0)\,\mathrm{e}^{\gamma x}\,. \end{aligned} \tag{9.17}$$

Thus the intensity grows exponentially within the gain medium.

In the previous section, we have seen that population inversion gives rise to amplification of light. We now want to work out a relationship between the gain coefficient and the population inversion density. Before we can proceed, we must first refine our analysis of the absorption and stimulated emission rates. Einstein's analysis considered the interaction between an atom and the continuous spectrum of black-body radiation. In practice, we are more interested in the interaction between the spectral line of an atom with a narrow band of light that will eventually become the laser mode.

The energy density $u(\nu)$ that appears in Eq. (9.3)–(9.10) is the *spectral* energy density (units: $\mathrm{J\,m^{-3}\,Hz^{-1}} \equiv \mathrm{J\,s\,m^{-3}}$). We now consider the interaction

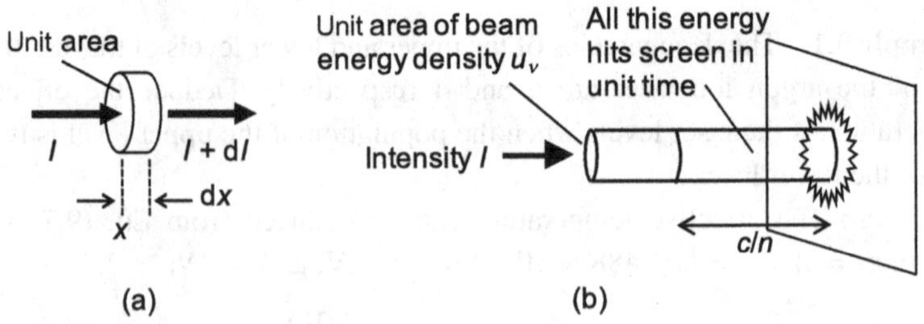

Figure 9.3 (a) Incremental intensity increase in a gain medium. (b) Relationship between the intensity I and energy density u_ν of a light beam.

between an atom with a normalized line shape function $g(\nu)$ as defined in Section 3.7 (units: Hz^{-1}) and a beam of light whose emission spectrum is much narrower than the spectral linewidth of the atomic transition. This situation is considered in Appendix E.1, where it is shown that the rates of absorption and stimulated emission per unit volume can be written respectively as:

$$\begin{aligned} W_{12} &= B_{12} N_1 u_\nu g(\nu)\,, \\ W_{21}^{\mathrm{stim}} &= B_{21} N_2 u_\nu g(\nu)\,. \end{aligned} \tag{9.18}$$

(Compare with Eqs. [E.4] and [E.5] with the subscript on the laser frequency omitted.)

The light source is considered to have a Dirac delta function spectrum at frequency ν with energy density u_ν per unit volume (units $\mathrm{J\,m^{-3}}$). Energy density u_ν is related to the intensity I of the optical beam by (see Figure 9.3[b]):

$$I = u_\nu \frac{c}{n}\,, \tag{9.19}$$

where n is the refractive index of the medium. On making use of the relationship between B_{12} and B_{21}, given in Eq. (9.11), we can then write the net stimulated rate downwards from level 2 to level 1 as:

$$W_{21}^{\mathrm{net}} \equiv W_{21} - W_{12} = \Delta N B_{21} g(\nu) \frac{n}{c} I\,, \tag{9.20}$$

where

$$\Delta N = \left(N_2 - \frac{g_2}{g_1} N_1 \right)\,, \tag{9.21}$$

is the population inversion density (see Eq. [9.14]), which reduces to $\Delta N = N_2 - N_1$ for nondegenerate levels. Note that, again, we are ignoring spontaneous emission in this argument, as we are assuming $W_{21}^{\mathrm{stim}} \gg W_{21}^{\mathrm{spon}}$ in a working laser.

For each net transition, a photon of energy $h\nu$ is added to the beam. The energy added to a unit volume of beam per unit time is thus $W_{21}^{\mathrm{net}} h\nu$. Consider a small increment of the light beam inside the gain medium with length $\mathrm{d}x$, as shown in Figure 9.3(a). The energy added to this increment of beam per unit time is $W_{21}^{\mathrm{net}} h\nu \times \mathbb{A}\,\mathrm{d}x$, where $\mathbb{A}$ is the beam area. Because the intensity equals the energy per unit time per unit area, we can write:

$$\begin{aligned} \mathrm{d}I &= W_{21}^{\mathrm{net}} h\nu\, \mathbb{A} \mathrm{d}x / \mathbb{A}\,, \\ &= W_{21}^{\mathrm{net}} h\nu\, \mathrm{d}x\,, \\ &= \Delta N B_{21} g(\nu) \frac{n}{c} I h\nu\, \mathrm{d}x\,. \end{aligned} \tag{9.22}$$

On comparing this to Eq. (9.17), we see that the gain coefficient γ is given by:

$$\gamma(\nu) = \Delta N B_{21} g(\nu) \frac{n}{c} h\nu \,. \tag{9.23}$$

This result shows that the gain is directly proportional to the population inversion density, and also follows the spectrum of the emission line. By using Eq. (9.11) to express B_{21} in terms of A_{21}, we can rewrite the gain coefficient in terms of the natural radiative lifetime τ using Eq. (9.12) to obtain:

$$\gamma(\nu) = \Delta N \frac{\lambda^2}{8\pi n^2 \tau} g(\nu) \,, \tag{9.24}$$

where λ is the vacuum wavelength of the emission line. This is the required result. Equation (9.24) tells us how to relate the gain in the medium to the population inversion density using experimentally measurable parameters: λ, τ, n, and $g(\nu)$.

Example 9.2 The 694.3 nm laser transition of ruby has a radiative lifetime 3 ms, and a full width at half-maximum of 2×10^{11} Hz. The refractive index of the crystal is 1.78. Estimate the gain coefficient for a population inversion density of $1.5 \times 10^{23}\ \mathrm{m}^{-3}$.

Solution: The gain coefficient can be worked out from Eq. (9.24) if we know $g(\nu)$. The laser will operate at line center where $g(\nu)$ is maximized, and thus we need to work out the value of $g(\nu_0)$. The spectral line-shape function has an area of unity (see Eq. [3.28]), and so it must be the case that $g(\nu_0) = C/\Delta\nu$, where $\Delta\nu$ is the FWHM, and C is of order unity. On putting $C \approx 1$, we then substitute to find:

$$\gamma \approx 1.5 \times 10^{23} \frac{(694.3 \times 10^{-9})^2}{8\pi\,(1.78)^2\,(3 \times 10^{-3})} \frac{1}{2 \times 10^{11}} \approx 1.5\,\mathrm{m}^{-1} \,.$$

The exact formulas for Lorentzian or Gaussian line shapes in Eqs. (3.31) and (3.42) give $C = 2/\pi = 0.637$ and $\sqrt{4 \ln 2/\pi} = 0.939$, which change the value of γ to $0.96\,\mathrm{m}^{-1}$ and $1.4\,\mathrm{m}^{-1}$, respectively. Since we do not know what type of line shape we have, the use of $C \approx 1$ is good enough to get a rough estimate.

9.4 Principles of Laser Oscillation

The word LASER is an acronym that stands for "light amplification by stimulated emission of radiation." We have seen that light amplification is achieved by stimulated emission in a medium with population inversion. However, there

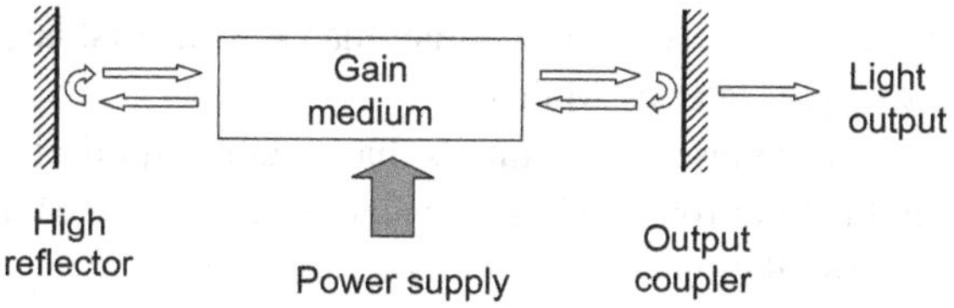

Figure 9.4 Schematic diagram of a laser.

is more to a laser than just light amplification. A laser is actually an oscillator, and a more accurate acronym might therefore have been LOSER. (It is easy to understand why that name never caught on.) The difference is that an oscillator has positive feedback in addition to amplification. The key ingredients of a laser may thus be summarized as:

LASER = light amplification + positive optical feedback

Figure 9.4 shows a schematic diagram of a laser. The gain medium (i.e., the light amplifier) lies at the heart of the system. The population of the laser levels must be inverted for gain to be present, and this highly nonequilibrium situation is only achieved by "pumping" energy from some form of power supply. (See Section 9.5–9.7.) The gain medium is surrounded by the optical cavity, which provides the positive optical feedback. Light inside the cavity passes through the gain medium and is amplified. It then bounces off the end mirrors and passes through the gain medium again, getting amplified further. The process of stimulated emission is a coherent quantum-mechanical effect. The photons emitted by stimulated emission are therefore in phase with the photons that induce the transition. This means that repeated amplification by stimulated emission as the light bounces around the cavity leads to the build-up of an intense optical field within the cavity.

Einstein introduced the concept of stimulated emission in 1917, but it was not until the early 1950s that the first practical devices that exploited the phenomenon, namely masers, were developed. A maser is a microwave oscillator that relies on stimulated emission. The jump from microwaves to optical frequencies was not straightforward, as microwave cavities are usually designed with dimensions that are comparable to the wavelength of the radiation, which might typically be around 10 cm. Such designs cannot be scaled easily to optical wavelengths, where $\lambda \sim 1\ \mu\text{m}$, and it required some lateral thinking by Schawlow and Townes in 1958 to come up with the idea of using the end mirrors of a cavity much larger than λ.[1] The first laser, ruby, was demonstrated two years later by Maimen.

The condition for the laser to operate is that a stable equilibrium condition is reached when the total round-trip gain balances all the losses in the cavity. The condition for oscillation is thus:

$$\boxed{\text{round-trip gain} = \text{round-trip loss}} \tag{9.25}$$

The losses in the cavity fall into two categories: useful and useless.

- Useful loss comes from the output coupling. One of the mirrors (called the **output coupler**) has reflectivity less than unity, and allows some of the light oscillating around the cavity to be transmitted as the output of the laser.
- Useless losses arise from absorption in the optical components (including the laser medium), scattering, and the imperfect reflectivity of the other mirror (the **high reflector**).

The value of the transmission of the output coupler is chosen to maximize the output power. If the transmission is too low, very little of the light inside the cavity can escape, and thus we get very little output power. On the other hand, if the transmission is too high, there may not be enough gain to sustain oscillation, and there would be no output power. The optimum value is somewhere between these two extremes.

By taking into account the fact that the light passes twice through the gain medium during a round trip, the condition for oscillation in a laser can be written:

$$\mathrm{e}^{2\gamma l} R_{\mathrm{OC}} R_{\mathrm{HR}} (1 - L) = 1 \,, \tag{9.26}$$

where l is the length of the gain medium, R_{OC} is the reflectivity of the output coupler, R_{HR} is the reflectivity of the high reflector, and L is the fractional round-trip loss due to absorption and scattering. If the total round-trip losses are small ($\lesssim 10\%$), then the gain required to sustain lasing will also be small, and Eq. (9.26) simplifies to:

$$2\gamma l = (1 - R_{\mathrm{OC}}) + (1 - R_{\mathrm{HR}}) + L \,. \tag{9.27}$$

This shows more clearly how the gain in the laser medium must exactly balance the losses in the cavity, as prescribed in Eq. (9.25).

In general we expect the gain to increase as we pump more energy into the laser medium. At low pump powers, the gain will be insufficient to reach the

[1] See Schawlow and Townes (1958). It is only relatively recently that it has been possible to make "microcavity lasers" and "nanolasers" that have physical dimensions comparable to the wavelength of light.

oscillation condition. The laser will not start to oscillate until there is enough gain to overcome all of the losses. This implies that the laser will have a *threshold* in terms of the pump power, as will be discussed in Sections 9.5 and 9.7.

Example 9.3 A laser contains a ruby rod of length 5 cm at the center of a cavity that has an output coupler with 10% transmissivity. The high reflector has a reflectivity of 99%, and the total round-trip loss aside from the mirrors is 3%. What is gain coefficient in the laser crystal?

Solution: The 10% transmissivity of the output coupler implies $R_{OC} = 1 - 0.1 = 90\%$. The gain is then found by using Eq. (9.26) with $l = 0.05$ m, $R_{OC} = 0.9$, $R_{HR} = 0.99$, and $L = 0.03$:

$$\exp(2\gamma \times 0.05) \times 0.9 \times 0.99 \times 0.97 = 1\,.$$

This gives $\gamma = 1.46\,\text{m}^{-1}$. If we had used Eq. (9.27) instead, we would have concluded:

$$2 \times \gamma \times 0.05 = 0.1 + 0.01 + 0.03\,,$$

which gives $\gamma = 1.4\,\text{m}^{-1}$. The small-loss approximation is thus good in this case, as the total round-trip loss is only 14%.

Example 9.4 A semiconductor laser of length 0.5 mm has end-mirrors with reflectivities of 99% and 30%. Calculate the minimum value of the gain coefficient to make the laser oscillate.

Solution The minimum gain for lasing will occur when $L = 0$. We thus substitute into Eq. (9.26) with $l = 5 \times 10^{-4}$ m, $R_{OC} = 0.3$, $R_{HR} = 0.99$, and $L = 0$. This gives $\gamma = 1.2 \times 10^3\,\text{m}^{-1}$. If we had used Eq. (9.27) instead, we would have incorrectly found $\gamma = (0.7 + 0.01)/(2 \times 0.0005) = 710\,\text{m}^{-1}$, which highlights that Eq. (9.27) is not valid when the round-trip losses are $\gtrsim 0.1$.

9.5 Four-Level Lasers

It is not possible to achieve population inversion in a system with just two energy levels. As we pump atoms from the lower level to the upper one, the increasing rate of stimulated emission reduces the net absorption, and we would get stuck at the point with equal populations, where there is neither net absorption nor net stimulated emission. Laser systems must therefore have more than two levels. In practice, we only need to consider two scenarios, namely **three-level** or **four-level** systems. We consider four-level lasers first.

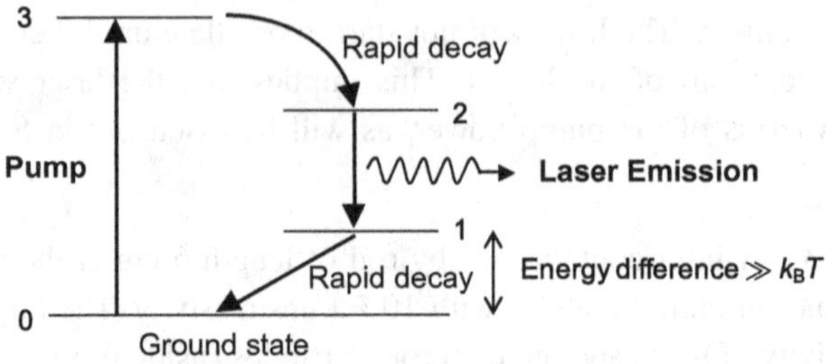

Figure 9.5 Level scheme for a four-level laser.

The level scheme for an ideal four-level laser is shown in Figure 9.5. The four levels are: the ground state (0), the two lasing levels (1 and 2), and a fourth level (3) that is used as part of the pumping process. The feature that differentiates it from a three-level system is that the lower laser level is at an energy $\gg k_B T$ above the ground state. This means that the thermal population of level 1 is negligible, and so level 1 is empty before we turn on the pumping mechanism.

We assume that the atoms are inside a cavity and being pumped into the upper laser level (level 2) at a constant rate of $\mathbb{R}_2$ per unit volume. This is done by exciting atoms to level 3 (for example, with a bright flash lamp or by an electrical discharge), from where they rapidly decay to level 2. It is shown in Appendix E (see Section E.2, Eq. [E.11]) that the population inversion density is given by:

$$\Delta N \equiv N_2 - \frac{g_2}{g_1} N_1 = \frac{\mathbb{R}}{W + 1/\tau_2}, \tag{9.28}$$

where:

$$\mathbb{R} = \mathbb{R}_2 \left(1 - \frac{g_2 \tau_1}{g_1 \tau_2}\right),$$

$$W = \frac{W_{21}^{\text{net}}}{\Delta N} = B_{21} g(\nu) \frac{n}{c} I. \tag{9.29}$$

Here, τ_2 and τ_1 are the lifetimes of the upper and lower laser levels respectively, and I is the intensity inside the laser cavity. $\mathbb{R}$ is the net pumping rate per unit volume, after allowing for accumulation of atoms in level 1. Equation (9.28) shows that the population inversion is directly proportional to the net pumping rate, and Eq. (9.29) shows that it is not possible to achieve $\Delta N > 0$ unless $\tau_2 > (g_2/g_1)\tau_1$, which reduces to $\tau_2 > \tau_1$ for nondegenerate levels. The latter conclusion is a consequence of the fact that unless the lower laser level empties quickly, atoms will pile up in the lower laser level; this will destroy the population inversion.

As explained in the previous section, the laser will not oscillate unless the gain medium provides sufficient amplification to overcome the cavity losses. This implies that the laser has a **threshold**. All lasers have a threshold: they will not oscillate unless they are pumped hard enough. If the laser is below the threshold, there will be very few photons in the cavity. Therefore, W will be very small because I is very small (see Eq. [9.29].) The population inversion is simply $\mathbb{R}\tau_2$, which increases linearly with the pumping rate. Equation (9.24) implies that the gain coefficient similarly increases linearly with the pumping rate below threshold.

Eventually, we reach the threshold condition where there is enough gain to balance the round-trip losses. The value of the gain coefficient required to do this, $\gamma^{\rm th}$, is worked out from Eq. (9.26) or (9.27). Equation (9.24) then allows us to work out the population inversion density at threshold:

$$\Delta N^{\rm th} = \frac{8\pi n^2 \tau_2}{\lambda^2 g(\nu)} \gamma^{\rm th} . \tag{9.30}$$

Since the light intensity in the cavity is effectively zero, we can put $W = 0$ into Eq. (9.28) and then find the threshold pumping rate, namely $\mathbb{R}^{\rm th} = \Delta N^{\rm th}/\tau_2$.

In a practical laser, we want the pumping rate to be well above the threshold value so that there is a large power output. With the steady-state oscillation conditions governed by Eq. (9.26), the gain cannot increase once the threshold is passed. This implies that the population inversion is clamped at the value given by Eq. (9.30) even when $\mathbb{R}$ exceeds $\mathbb{R}^{\rm th}$. This is shown in Figure 9.6(a).

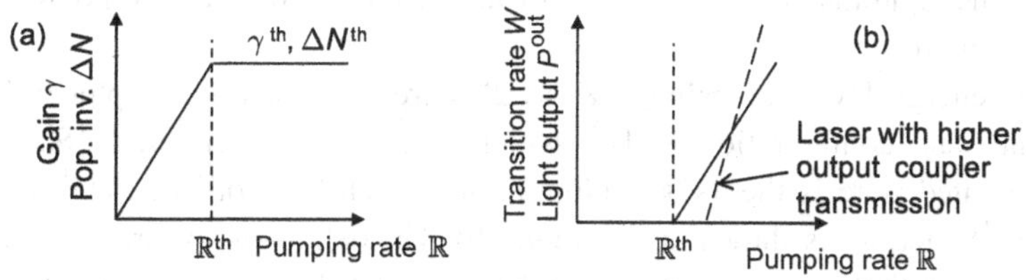

Figure 9.6 (a) Variation of the gain coefficient and population inversion in a laser with the pumping rate. (b) Comparison of the threshold and light outputs for two different values of the transmission of the output coupler. Note that these curves only apply to four-level laser systems.

The output power above threshold can be deduced by rearranging Eq. (9.28) with ΔN clamped at the value set by Eq. (9.30):

$$W = \frac{\mathbb{R}}{\Delta N^{\text{th}}} - \frac{1}{\tau_2} = \frac{1}{\tau_2}\left(\frac{\mathbb{R}}{\mathbb{R}^{\text{th}}} - 1\right) \quad \ldots \text{ for } \mathbb{R} > \mathbb{R}^{\text{th}}. \tag{9.31}$$

Now W is proportional to the intensity I inside the cavity (see Eq. [9.29]), which in turn is proportional to the output power P^{out} emitted by the laser. Thus P^{out} is proportional to W, and we may write:

$$P^{\text{out}} \propto \left(\frac{\mathbb{R}}{\mathbb{R}^{\text{th}}} - 1\right) \quad \ldots \text{ for } \mathbb{R} > \mathbb{R}^{\text{th}}. \tag{9.32}$$

This shows that the output power increases linearly with the pumping rate once the threshold has been achieved, as shown in Figure 9.6(b).

The choice of the reflectivity of the output coupler affects the threshold because it determines the oscillation conditions. (See Eq. [9.26] or [9.27].) If the output coupler transmission $(1 - R_{\text{OC}})$ is small, the laser will have a low threshold, but the output coupling efficiency will be low. As the transmission increases, the threshold increases, but the power is coupled out more efficiently. This point is illustrated in Figure 9.6(b). The final choice for R_{OC} depends on how much pump energy is available, which will govern the optimal choice to get the maximum output power.

9.6 The Helium–Neon Laser

The helium-neon laser is a four-level system, and it is a good example of how the principles developed in the previous section are put into practice. A typical He:Ne laser consists of a discharge tube inserted between highly reflecting mirrors, as shown schematically in Figure 9.7(a). The tube contains a mixture of helium and neon gas in the approximate ratio of 5:1. The light is emitted by the neon atoms, and the purpose of the helium is to assist the population inversion process.

The energy levels of helium ($Z = 2$) were discussed in Chapter 6. The ground-state configuration is $1s^2$, which is a spin singlet term: ${}^1S_0$. The first excited state is the 1s2s configuration, which has both singlet ${}^1S_0$ and triplet ${}^3S_1$ terms, as shown in Figure 9.7(b). The helium atoms are excited by collisions with the electrons in the discharge tube and cascade down the levels. When they get to the 1s2s configuration, however, the cascade process slows right down, as the $2s1s \rightarrow 1s^2$ transitions are forbidden. Both the singlet and triplet transitions violate the $\Delta l = \pm 1$ selection rule. The triplet ${}^3S_1 \rightarrow {}^1S_0$ transition violates the $\Delta S = 0$ selection rule, while the singlet ${}^1S_0 \rightarrow {}^1S_0$

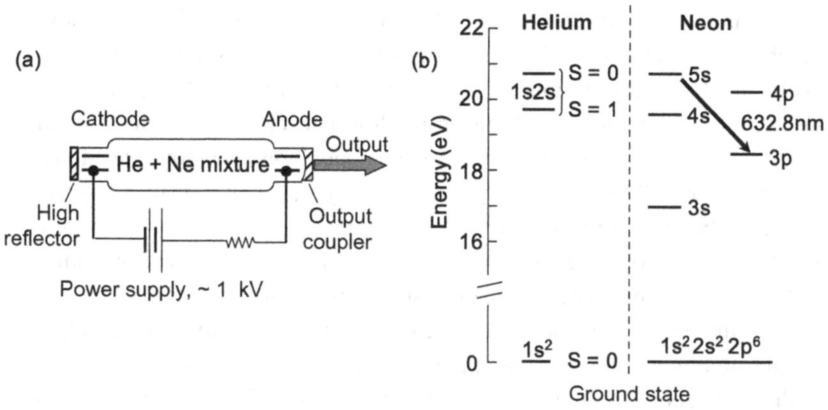

Figure 9.7 (a) Schematic diagram of a helium–neon laser. (b) Level scheme for the He:Ne laser. The neon excited states have one of the 2p electrons in a higher shell and are labeled accordingly.

transition is $J = 0 \rightarrow 0$. The net result is that all transitions from the 1s2s levels are strongly forbidden. The 1s2s level therefore has a very long lifetime, and is called **metastable**.[2]

Neon has ten electrons, with ground-state configuration $1s^2 2s^2 2p^6$. The excited states correspond to the promotion of one of the 2p electrons to higher levels. This gives the level scheme shown on the right of Figure 9.7(b). The symbols of the excited states refer to the level of this single excited electron. By good luck, the 5s and 4s levels of the neon atoms are almost degenerate with the metastable $S = 0$ and $S = 1$ terms of the 1s2s configuration of helium. The helium atoms can then easily deexcite by collisions with neon atoms in the ground state according to the following scheme:

$$\mathrm{He}^* + \mathrm{Ne} \Rightarrow \mathrm{He} + \mathrm{Ne}^* . \tag{9.33}$$

The star indicates that the atom is in an excited state. Any small differences in the energy between the excited states of the two atoms are taken up, for example, as kinetic energy. This scheme leads to a large population of neon atoms in the 5s and 4s excited states, generating population inversion with respect to the 3p and 4p levels. It would not be easy to get this population inversion without the helium because collisions between the neon atoms and the electrons in the tube would tend to excite all the neon levels equally. This is why there is more helium than neon in the tube.

[2] Optical transitions from these metastable states will be discussed in Section 12.5 in the context of astrophysics.

The main laser transition at 632.8 nm occurs between the 5s and 3p levels, which have lifetimes of 170 ns and 10 ns, respectively. This transition therefore easily satisfies the criterion $\tau_2 > \tau_1$. (See discussion of Eq. [9.29].) This ensures that atoms do not pile up in the lower level once they have emitted the laser photons, as this would destroy the population inversion. The atoms in the 3p level rapidly relax to the ground state by radiative transitions to the 3s level and then by collisional deexcitation to the original 2p level. Lasing can also be obtained on other transitions: for example, 5s $\rightarrow$ 4p at 3391 nm and 4s $\rightarrow$ 3p at 1152 nm. These are not as strong as the main 632.8 nm line.

The gain in a He:Ne tube tends to be rather low because of the relatively low density of atoms in the gas (compared to a solid). This is partly compensated by the fairly short lifetime of 170 ns. (See Eq. [9.24].) The round-trip gain may only be a few percent, and so very highly reflecting mirrors are needed. With relatively small gain, the output powers are not very high – only a few mW. However, the ease of manufacture of He:Ne lasers makes them very popular for low-power applications: bar-code readers, laser alignment tools, classroom demonstrations, etc. Nowadays they are gradually being replaced by visible semiconductor laser diodes, such as the type used in laser pointers.

9.7 Three-Level Lasers

The key difference between a three- and four-level laser is that the lower laser level is the ground state, as shown in Figure 9.8(a). On comparing Figures 9.5 and 9.8, it is apparent that the lower laser level of the four-level system has merged with the ground state in the three-level system. This makes it much more difficult to obtain population inversion because the lower laser level initially has a very large population.

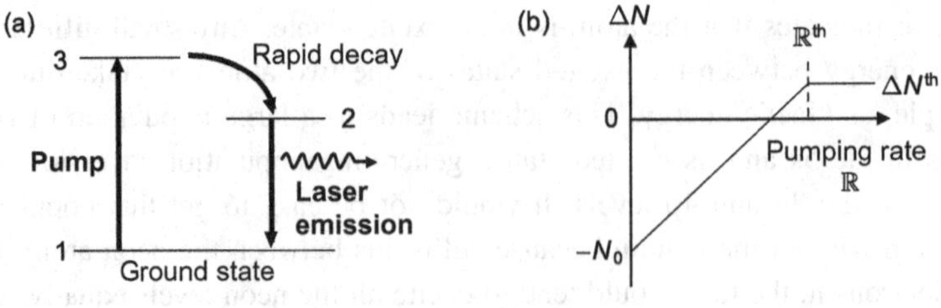

Figure 9.8 (a) Level scheme for a three-level laser, for example: ruby. (b) Variation of the population inversion density ΔN with pumping rate $\mathbb{R}$ in a three-level laser.

Consider a system with N_0 atoms. With the pump turned off, all of the atoms will be in the lower laser level, so that $N_1 = N_0$, $N_2 = 0$, and $N_2 - N_1 = -N_0$. By turning on the pump, we excite dN atoms to level 3, which then decay to level 2. The population of level 2 is thus dN, while the population of level 1 is $(N_0 - \mathrm{d}N)$. For population inversion we require $N_2 > (g_2/g_1)N_1$, where g_2 and g_1 are the level degeneracies. (See Eq. [9.14].) We therefore need:

$$\mathrm{d}N > \frac{g_2}{g_1}(N_0 - \mathrm{d}N),$$

$$\therefore \mathrm{d}N > \frac{g_2}{g_1 + g_2} N_0 . \tag{9.34}$$

For $g_1 = g_2$, this becomes d$N > N_0/2$. Therefore, in order to obtain population inversion we have to pump more than half of the atoms out of the ground state into the upper laser level. This obviously requires a very large amount of energy, which contrasts with four-level lasers, where level 1 is empty before pumping starts, and much less energy is required to reach threshold.

The variation of the inversion density ΔN, with pumping rate $\mathbb{R}$ for a three-level laser is shown schematically in Figure 9.8(b). As explained earlier, the ΔN is equal to $-N_0$ at $\mathbb{R} = 0$, and only becomes positive when a very significant fraction of the atoms (more than half if $g_1 = g_2$) have been pumped to the upper level. Once ΔN is positive, amplification occurs, and the lasing threshold will be reached when ΔN is sufficiently large to provide enough gain to overcome the cavity losses. As with the four-level laser, the gain (and hence ΔN) above threshold are fixed at the level set by the oscillation condition in Eq. (9.26), which is first reached at the threshold pumping rate $\mathbb{R}^{\mathrm{th}}$.

The standard example of a three-level laser is ruby, which was the first laser ever produced, and therefore has historical significance. Ruby crystals have Cr^{3+} ions doped into Al_2O_3 at a low concentration (typically $\lesssim 0.1\%$). The Al_2O_3 host crystal is colorless, and the light is emitted by transitions of the Cr^{3+} ions. The level scheme follows Figure 9.8 – with the exception that level 3 corresponds to broad absorption bands in the blue and green spectral regions, rather than a specific level. (See Section 11.5.1.) Electrons are excited to the bands (level 3) from the ground state (level 1) by a powerful pulsed flashlamp, and then relax rapidly to the upper laser level (level 2) by fast nonradiative transitions. With a sufficiently powerful flashlamp, it is possible to generate population inversion. Lasing can then occur on the $2 \to 1$ transtion at 694.3 nm, if a suitable cavity is provided.

The upper level in a ruby laser has a very long lifetime (3 ms), as the $2 \to 1$ transition involves an E1-forbidden jump between 3d levels split by the crystal field. (See Section 11.5.1.) This long lifetime makes it easier to build up a large population in level 2 and enables the crystal to store a lot of energy, leading to pulsed emission with energies as high as 100 J per pulse. However, ruby lasers have declined in importance, and modern high-power solid-state lasers tend to be based on four-level systems such as Nd:YAG. A more contemporary example of a three-level system is the semiconductor laser (see Section 11.2.4), which has very high technological importance in the modern world.

9.8 Classification of Lasers

We have seen that lasers can be classified as either three- or four-level systems. Several other general classifications are also useful.

(i) **Wavelength**: The letter L in laser stands for "light," but this does not mean that lasers have to be visible. Light is understood here to mean electromagnetic radiation with a frequency of $\sim 10^{14}$–10^{15} Hz, which therefore includes infrared and ultraviolet frequencies. Most lasers have fixed wavelengths, but some can be tuneable. In the latter case, the tuning range is determined by the spectral width of the laser transition.

(ii) **Gain medium**: The gain medium can be solid, liquid, or gas. The first laser, ruby, was a solid-state laser. However, many of the other early lasers (e.g., He:Ne; see Section 9.6) were gas lasers. At the time of writing, gas lasers are becoming obsolete, and modern laser technology is based predominantly on solid-state systems.

(iii) **CW/pulsed**: A laser can operate either in continuous wave (CW) or pulsed mode. Pulsed operation is quite common for solid-state lasers (e.g., ruby) because they tend to have long upper state lifetimes, which allows the storage of a large amount of energy in the crystal. It is seldom used in gas lasers because the lifetimes are usually shorter, making it difficult to store energy in the gain medium.

(iv) **Mode structure**: The mode structure of the cavity determines the spatial properties of the beam that is emitted, and the spectral properties of the light. The former is characterized by the *transverse* modes of the cavity, and the latter by its *longitudinal* modes (i.e., resonant frequencies). In a **mode-locked laser**, the phases of the longitudinal modes are locked together. A single pulse bounces around the cavity, generating a train of pulses separated in time by $2nL/c$, where L is the cavity length and n is its average refractive index. Fourier analysis governs that the minimum pulse duration is inversely related to the spectral width, $\Delta\nu^{\text{gain}}$, of the laser transition:

Table 9.1 *Common lasers. Some of the lasers can operate on more than one line, in which case the most common wavelength(s) is (are) listed. There are many different types of semiconductor lasers available. The wavelengths listed are for the lasers used in Blu-ray and DVD technology, laser pointers, and fiber-optic systems.*

Laser	Gain medium	Wavelength(s)
He:Ne	gas	632.8 nm
Argon ion	gas	514 nm, 488 nm
Carbon dioxide	gas	10.6 μm
Nd:YAG	solid-state	1064 nm
Nd:glass	solid-state	1054 nm
Ruby	solid-state	694.3 nm
Ti:sapphire	solid-state	690–1100nm, tuneable
Semiconductor	solid-state	405 nm, 635 nm, 780 nm 850 nm, 1300 nm, 1550 nm

$$\Delta t_{\text{min}} \geq \frac{C}{\Delta\nu^{\text{gain}}}, \tag{9.35}$$

where C is a numerical constant of order unity (e.g., $C = 0.441$ for Gaussian pulses). Some lasers (e.g., Ti:sapphire) have very broad emission bands, and hence can be used to generate extremely short pulses with durations in the femtosecond range. These are extensively used for studying fast processes in physics, chemistry, and biology, and for transmitting high-speed data down optical fibres.

An incomplete list of lasers commonly used in science and technology is given in Table 9.1. The reader is referred to specialized texts for a more detailed discussion of lasers, for example Silfvast (2004) or Hooker and Webb (2010). Supplementary notes on laser cavities, nonlinear frequency conversion, and solid-state lasers are available online.

Exercises

9.1 A gas contains a mixture of atoms with nondegenerate levels. The gas is in equilibrium with black-body radiation at temperature T. Show that $W_{21}^{\text{stim}} > W_{21}^{\text{spon}}$ in those atoms that have $\nu < k_{\text{B}}T \ln 2/h$. Evaluate ν for $T = 300\text{K}$.

9.2 The degeneracies of the upper and lower levels of the 694.3 nm laser transition in ruby are 2 and 4 respectively. What is the effective temperature of the laser levels when the populations of the upper and lower levels are equal?

9.3 The effective temperature of the laser levels in an Nd:YAG laser operating on the $^4F_{3/2} \rightarrow {}^4I_{11/2}$ transition at 1064 nm is $-15,000$ K. What is the ratio of N_2/N_1? Is the population inverted?

9.4 A laser crystal of length 10 cm has a gain coefficient of $5\,\text{m}^{-1}$. What is the percentage gain for a single pass through the crystal?

9.5 A helium-neon laser of length 0.6 m operating at 632.8 nm has end mirrors of reflectivity 99.9% and 99%. The laser transition has a width of 1.5 GHz and an Einstein A coefficient of $3.4 \times 10^6\,\text{s}^{-1}$.

(a) Calculate the gain coefficient.
(b) Estimate the population inversion density in the gain medium.

9.6 The Nd:YAG laser is a four-level system. The $^4F_{3/2} \rightarrow {}^4I_{11/2}$ 1064 nm laser transition has a radiative lifetime of 0.23 ms and a spectral width of 0.45 nm. The refractive index of the laser crystal is 1.82. A rod of length 10 cm is placed inside a cavity with a high reflector of reflectivity 99.9% and an output coupler of reflectivity 95%.

(a) Account for the long radiative lifetime of the transition. What can you deduce about the lifetime of $^4I_{11/2}$ level?
(b) What is the value of B_{12} for the laser transition?
(c) Calculate the gain coefficient in the laser crystal.
(d) Estimate the population inversion density in the laser rod.

9.7 A semiconductor laser has end mirrors with reflectivities of 99% and 35%. Calculate the minimum chip length that can be used if the maximum gain coefficient that can be obtained is $2 \times 10^4\,\text{m}^{-1}$.

9.8 A laser contains a Nd:YAG rod of length 5 cm. Find the lowest value of the reflectivity of the output coupler that will still sustain lasing if the maximum value of the gain coefficient in the laser rod is $0.5\,\text{m}^{-1}$.

9.9 A He:Ne laser of length 0.3 m has a high reflector mirror with a reflectivity of 99.9%. The maximum gain coefficient that can be achieved in the laser tube is 0.02m^{-1}. Three output coupler mirrors are available with reflectivities of 99.5%, 99.0%, and 98.5% respectively. State, with reasons, which one would be expected to give the largest output power.

9.10 A ruby laser (694.3 nm) contains a laser rod of volume $10^{-6}\,\text{m}^3$ with a Cr^{3+} doping density of $2 \times 10^{24}\,\text{m}^{-3}$. A flash lamp pumps 40% of the atoms from the ground state to the upper laser level, and then emits a short laser pulse. Calculate the maximum energy of the pulse, bearing in mind that ruby is a three-level system, with $g_2 = 2$ and $g_1 = 4$.

9.11 A semiconductor laser has a chip with length $l = 1$ mm and uncoated edge facets with $R_{HR} = R_{OC} = 30\%$. The laser medium has internal losses that can be characterized by a distributed loss coefficient $\alpha = 200\,\text{m}^{-1}$, such that $(1 - L) = \text{e}^{-2\alpha l}$.

(a) Find the gain coefficient in the laser.

(b) A highly reflective coating is applied to the rear facet, increasing R_{HR} to 99%. Given that the original laser had a threshold current of 100 mA, deduce the threshold current of the coated laser. (Assume that α is unaffected by the coating of the rear facet.)

(c) How would the power output of the coated laser compare to the uncoated one when both are driven at 200 mA?

10
Cold Atoms

10.1 Introduction

The resonant force between atoms and light was first observed in 1933, when Otto Frisch measured the deflection of a sodium beam by a sodium lamp. The invention of lasers opened up new possibilities, leading to the development of the **laser-cooling** techniques that are the subject of this chapter.

There are two aspects of laser cooling that make it particularly remarkable:

(i) It is highly surprising that the technique works at all. We would normally expect a powerful laser to cause heating rather than cooling. This makes us realize that the technique will only work when special conditions are satisfied.

(ii) The very low temperatures achieved by laser cooling are extremely impressive, but this in itself is not the main point, as techniques for achieving very low temperatures have been used for decades by condensed-matter physicists. For example, commercial dilution refrigerators routinely achieve temperatures in the milli-Kelvin range, and as early as the 1950s, Nicholas Kurti and coworkers at Oxford University used adiabatic demagnetisation to achieve nuclear spin temperatures in the micro-Kelvin range. The novelty of laser cooling is that it produces an ultracold *gas* of atoms, in contrast to the condensed-matter techniques that work on all liquids or solids. These ultracold atoms only interact weakly with each other, which makes it possible to study them with unsurpassed precision.

The ability to cool a gas of atoms to very low temperatures has given rise to a whole host of related benefits. Atomic clocks have been made with greater accuracy, and a whole range of new quantum phenomena have been discovered. The most spectacular of these is **Bose–Einstein condensation**, which was first observed in 1995 and is discussed in Section 10.7.

The description of laser cooling and Bose–Einstein condensation in this chapter focuses on the basic principles. The reader is referred to specialized texts or articles for a more detailed discussion. See, for example, Foot (2004), Metcalf and van der Straten (1999), or Phillips (1998).

10.2 Gas Temperatures

In order to understand how laser cooling works, we first need to clarify how the temperature of a gas of atoms is measured. The key point is the link between the thermal motion of the atoms and the temperature. Starting from the Maxwell–Boltzmann distribution (see Eq. [3.39]), it is possible to define a number of different characteristic velocities for the gas. (See Exercises 10.1 and 10.2.) The simplest of these is the root-mean-square (rms) velocity, which can be evaluated through the principle of **equipartition of energy**. This states that the average thermal energy per degree of freedom is equal to $\frac{1}{2}k_{\mathrm{B}}T$. For an atom of mass *m*, each component of the velocity must therefore satisfy:

$$\frac{1}{2}m\overline{v_i^2} = \frac{1}{2}k_{\mathrm{B}}T\,, \tag{10.1}$$

which implies that the rms velocity is given by:

$$\frac{1}{2}m(v^{\mathrm{rms}})^2 = \frac{3}{2}k_{\mathrm{B}}T\,. \tag{10.2}$$

We therefore conclude that:

$$v_x^{\mathrm{rms}} = \sqrt{\frac{k_{\mathrm{B}}T}{m}}\,, \tag{10.3}$$

$$v^{\mathrm{rms}} = \sqrt{\frac{3k_{\mathrm{B}}T}{m}}\,. \tag{10.4}$$

These simple relationships allow us to work out, for example, that the atoms in a typical gas at room temperature jostle about in a random way with thermal velocities of around 1000 km/hour. This random thermal motion is the cause of the Doppler broadening of spectral lines considered in Section 3.10.

The link between temperature and the velocity distribution tells us that we can cool the gas if we can slow the atoms down, which is the strategy adopted in laser-cooling experiments. Furthermore, the temperature of the gas can be inferred from a measurement of the velocity distribution of the atoms. This is the method that is used to determine the temperature of an ultra-cold gas cooled by a laser.

10.3 Doppler Cooling

10.3.1 The Laser-Cooling Process

Consider an atom absorbing and emitting at ν_0, moving in the $+x$ direction with velocity v_x toward a laser beam of frequency ν_L, as shown in Figure 10.1. The laser is tuned so that its frequency is below the absorption line by a small amount δ:

$$\nu_L = \nu_0 - \delta\,. \tag{10.5}$$

The Doppler-shifted frequency ν_L^{observed} of the laser in the atom's frame of reference is given by:

$$\nu_L^{\text{observed}} = \nu_L\left(1+\frac{v_x}{c}\right) = (\nu_0-\delta)\left(1+\frac{v_x}{c}\right) = \nu_0-\delta+\frac{v_x}{c}\nu_0-\frac{v_x}{c}\delta\,. \tag{10.6}$$

The last term is small because $\delta \ll \nu_0$ and $v_x \ll c$. Hence, if we choose:

$$\delta = \nu_0\frac{v_x}{c} \equiv \frac{v_x}{\lambda}\,, \tag{10.7}$$

we find $\nu_L^{\text{observed}} = \nu_0$. This situation is depicted in Figure 10.2(a). The laser is in resonance with atoms moving in the $+x$ direction, but not with those moving away or obliquely. This means that only those atoms moving toward the laser absorb photons from the laser beam.

Now consider what happens after the atom has absorbed a photon from the laser beam. The atom goes into an excited state and then emits another photon by spontaneous emission. This occurs on average after a time τ (the radiative lifetime), and the direction of the emitted photon is random. The absorption-emission cycle is illustrated schematically in Figure 10.2(b).

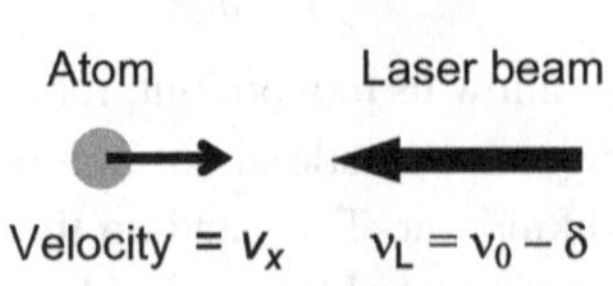

Figure 10.1 In Doppler cooling, the laser frequency is tuned below the atomic resonance by δ. The frequency seen by an atom moving toward the laser is Doppler shifted up by $\nu_0(v_x/c)$.

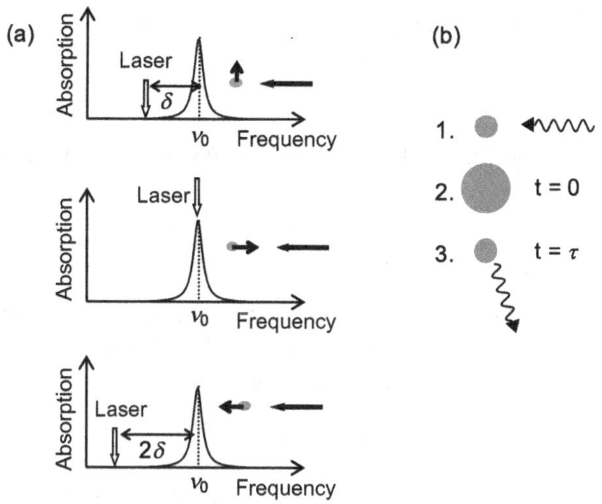

Figure 10.2 Doppler cooling. (a) Doppler-shifted laser frequency in the rest-frame of the atom. A laser with frequency $\nu_0 - \delta$ is in resonance with the atoms when they are moving toward the laser and $\delta = \nu_0(v_x/c)$, but not if they are moving sideways or away. (b) An absorption-emission cycle. (1) A laser photon impinges on the atom. (2) The atom absorbs the photon and goes into an excited state. (3) The atom re-emits a photon in a random direction by spontaneous emission after a time τ.

Repeated absorption-emission cycles generate a net force in the same direction as the laser beam, that is, the $-x$ direction. This happens because each photon of wavelength λ has a momentum of h/λ. Conservation of momentum demands that every time a photon is absorbed from the laser beam, the momentum of the atom changes by $(-h/\lambda)$. On the other hand, the change of momentum due to the recoil of the atom after spontaneous emission averages to zero, because the photons are emitted in random directions. Hence the net change of momentum per absorption-emission cycle is given by:

$$\Delta p_x = -\frac{h}{\lambda}. \tag{10.8}$$

If the laser intensity is large, the probability for absorption will also be large, leading to a fast time to absorb the laser photon. Hence the time to complete the absorption-emission cycle is determined by the radiative lifetime τ. The maximum force exerted on the atom is thus given by:

$$F_x = \frac{\mathrm{d}p}{\mathrm{d}t} = \frac{\Delta p_x}{\tau} = -\frac{h}{\lambda\tau}, \tag{10.9}$$

and the deceleration is given by

$$\dot{v}_x = \frac{F_x}{m} = -\frac{h}{m\lambda\tau} \,. \tag{10.10}$$

The number of absorption-emission cycles required to stop the atom is given by:

$$\mathsf{N}_{\text{stop}} = \frac{mu_x}{\Delta p_x} = \frac{mu_x\lambda}{h} \equiv \frac{p_{\text{initial}}}{p_{\text{photon}}} \,, \tag{10.11}$$

where u_x is the initial velocity of the atom; p_{initial} and p_{photon} are the initial momentum and photon momentum, respectively. This sets a minimum time for the laser beam to slow the atoms to a near halt:

$$t_{\min} = \mathsf{N}_{\text{stop}} \times \tau = \frac{mu_x\lambda\tau}{h} \,. \tag{10.12}$$

In this time, the atoms move a minimum distance $d_{\min}$ given by:

$$0 - u_x^2 = 2\,\dot{v}_x\,d_{\min} \,, \tag{10.13}$$

where $\dot{v}_x$ is the deceleration given by Eq. (10.10), and we have assumed that the final velocity of the atom is very small. This gives:

$$d_{\min} = -\frac{u_x^2}{2\dot{v}_x} = \frac{m\lambda\tau u_x^2}{2h} \,. \tag{10.14}$$

The analysis above ignores stimulated emission. The atom in the excited state – step 2 in Figure 10.2(b) – can be triggered to emit a photon by stimulated emission from other impinging laser photons. The stimulated photon will be emitted in the same direction as the incident photon, and the photon recoil exactly cancels the momentum kick given by the absorption process. When stimulated emission is considered, the maximum force is reduced by a factor two. This happens because the population of levels 1 and 2 equalize at a value of $N_0/2$, where N_0 is the total number of atoms. The atom then only spends a maximum of half its time in the excited state, and so the shortest time to absorb and emit a photon is twice the radiative lifetime. The final result is that the time to stop the atoms and the distance traveled in that time are both doubled.

Additional insight into the cooling process can be gained by thinking in terms of energy, rather than momentum. During an absorption-emission cycle, a photon of energy $h\nu_{\text{L}} = h(\nu_0 - \delta)$ is absorbed, and then a photon of energy $h\nu_0$ is emitted in the atom's rest-frame. The lab-frame energy of the emitted photon varies from $h(\nu_0 - \delta)$ to $h(\nu_0 + \delta)$ depending on its direction relative to the moving atom. Hence the average lab-frame emission energy is higher than $h\nu_{\text{L}}$, and this average energy difference of $+h\delta$ must come from the atom's kinetic energy. The interaction with a red-detuned laser therefore causes a reduction in the kinetic energy, and hence slows the atom.

Example 10.1 A beam of sodium atoms with average temperature 700 K is cooled by a laser beam tuned to near resonance with one of the D-lines at 589 nm. The transition has a radiative lifetime of 16 ns, and the atomic weight of sodium is 23.0.

(a) What is the rms velocity of the atoms?
(b) What initial detuning is required to instigate the laser-cooling process?
(c) What is the force imparted to the atom by the laser beam? What deceleration does it produce, in units of g, the acceleration due to gravity?
(d) What is the minimum number of absorption-emission cycles required to cool the atoms to their minimum temperature?
(e) How long would the cooling process take, and how far would the atoms travel in that time?

Solution We first work out the parameters for the problem without considering stimulated emission. We assume that the final temperature is low, so that we can assume that the final velocity is small compared to the initial velocity.

(a) A beam is a one-dimensional problem, and so we work out the rms velocity using Eq. (10.3). This gives:

$$v_x^{\mathrm{rms}} = \sqrt{\frac{k_{\mathrm{B}} \times 700}{23.0 m_{\mathrm{H}}}} = 500\,\mathrm{m/s}\,.$$

(b) The required detuning is worked out from Eq. (10.7):

$$\delta = \frac{500}{589 \times 10^{-9}} = 850\,\mathrm{MHz}\,.$$

(c) The force and deceleration are worked out using Eqs. (10.9) and (10.10):

$$F_x = -\frac{h}{589\,\mathrm{nm} \times 16\,\mathrm{ns}} = -7.0 \times 10^{-20}\,\mathrm{N}\,,$$

$$\dot{v}_x = \frac{F_x}{23.0 m_{\mathrm{H}}} = -1.8 \times 10^{6}\,\mathrm{ms}^{-2} \sim 2 \times 10^{5} g\,.$$

We thus see that the deceleration is very large, even though the force is very small. This is not unreasonable: the force acts on only one atom, which has a very small mass.

(d) The minimum number of absorption-emission cycles is worked out from Eq. (10.11), using the initial velocity from (a):

$$\mathsf{N}_{\mathsf{stop}} = \frac{23.0 m_{\mathrm{H}} \times 500 \times 589\,\mathrm{nm}}{h} = 1.7 \times 10^{4}\,.$$

(e) The time for the cooling process and the distance traveled are given in Eqs. (10.12) and (10.14):

$$t_{\rm min} = (1.7 \times 10^4) \times 16\,{\rm ns} = 0.27\,{\rm ms}\,,$$
$$d_{\rm min} = \frac{23.0 m_{\rm H} \times 589\,{\rm nm} \times 16\,{\rm ns} \times 500^2}{2h} = 0.07\,{\rm m}\,.$$

These calculations ignore stimulated emission, which would halve the force and deceleration, and double $\mathsf{N}_{\rm stop}$, $t_{\rm min}$, and $d_{\rm min}$.

10.3.2 The Doppler-Limit Temperature

At first sight, we might think that we would be able to completely stop the atoms by the Doppler-cooling technique, but this is not, in fact, the case. The rigorous derivation of the minimum temperature is given in the more advanced texts (see Foot [2004]), but it is possible to give a simpler, less thorough argument that arrives at the same conclusion.

The cooling effect only works if we have the right detuning frequency δ for the particular velocity. However, from Eq. (3.32) we see that the radiative lifetime τ of the transition causes broadening. This gives rise to an intrinsic uncertainty in the energy of the atom, and we will therefore never be able to reduce the thermal energy below:

$$E_{\rm min} \sim \frac{1}{2} h \Delta\nu_{\rm lifetime} = \frac{1}{2} h \frac{1}{2\pi\tau} = \frac{\hbar}{2\tau}\,. \tag{10.15}$$

On equating $E_{\rm min}$ with $k_{\rm B} T_{\rm min}$, we then find:

$$T_{\rm min} \sim \frac{\hbar}{2 k_{\rm B} \tau}\,. \tag{10.16}$$

The minimum temperature in Eq. (10.16) is the same as the one derived rigorously, and is called the **Doppler limit**. The equivalent minimum speed is found by setting $E_{\rm min}$ equal to $\frac{1}{2} m v_{\rm min}^2$.

Example 10.2 What is the Doppler-limit temperature and velocity for the sodium D-line cooling experiment considered in Example 10.1?

Solution The Doppler-limit temperature is found by inserting $\tau = 16\,{\rm ns}$ into Eq. (10.16). This gives $T_{\rm min} = 2.4 \times 10^{-4}\,{\rm K} \equiv 240\,\mu{\rm K}$. The minimum speed is found from:

$$\frac{1}{2} m v_{\rm min}^2 = k_{\rm B} T_{\rm min}\,,$$

with $m = 23.0\, m_{\rm H}$. This gives $v_{\rm min} \approx 0.4\,{\rm ms}^{-1}$.

10.4 Optical Molasses and Magneto-Optical Traps

The arrangement with a single laser beam shown in Figure 10.1 is able to stop the atoms moving in the positive direction for one of the components of the velocity (i.e., the $+x$ direction). To stop the atoms in both directions for all three velocity components (i.e., the $\pm x$, $\pm y$, and $\pm z$ directions), we need a six-beam arrangement as shown in Figure 10.3(a). This counter-propagating, six-beam technique was given the name **optical molasses**. Molasses (called "treacle" in British English) is a thick juice produced as a by-product of the sugar-refining process, and it gives a good description of how the Doppler-cooling force acts like a viscous medium for the trapped atoms.

The optical molasses experiment becomes a **magneto-optical trap** when magnetic coils are added above and below the intersection point, as shown in Figure 10.3(b). The current flows in opposite directions through the coils, which produces a quadrupole field where the field at the center of the apparatus cancels. The atoms can be classified as low-field seeking or high-field seeking, depending on the direction of their spin relative to the field.[1] The low-field seeking atoms experience a potential minimum at the center. This has the effect of trapping the atoms close to the origin if their thermal energy is less than the depth of the potential well. The combination of optical molasses and the quadrupole field thus provides a method to cool and trap a gas of atoms at very low temperatures.

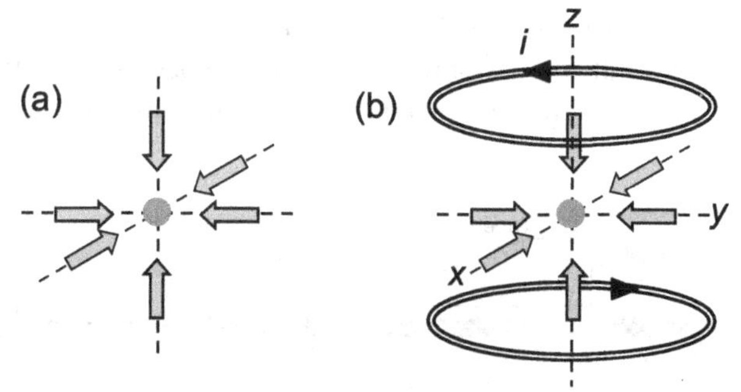

Figure 10.3 (a) Optical molasses. Six laser beams are used to annul the three velocity components of the atom's velocity in both directions. (b) Magneto-optical trap, comprising the optical molasses lasers and a quadrupole magnetic field.

[1] The single s-shell electron of an alkali has $M_J = \pm 1/2$ and $g_J = 2$. The Zeeman energy (see Eq. [8.17]) is therefore $\pm \mu_B B$. Note, however, that the detailed understanding of the mechanism requires consideration of the hyperfine states.

10.5 Experimental Considerations

Efficient cooling requires that the laser should exert the optimal force on the atoms, which only occurs when the laser is detuned by v_x/λ. (See Eq. [10.7].) However, v_x decreases as the atoms cool, which means that the optimal frequency also decreases. Two different strategies have been devised to keep the detuning at the optimal value as the atoms slow down:

(i) Tune the laser frequency.
(ii) Keep the laser frequency fixed and tune the transition frequency using a magnetic field.

These two methods are called chirp cooling and Zeeman slowing, respectively.

The chirp cooling method gets its name from the chirping sound made by birds, in which the frequency changes during the birdsong. In the experiment, the laser frequency needs to be tuned in a programmed way as the atoms slow down. Early experiments on sodium used tunable dye lasers emitting around 589 nm, but more modern experiments on rubidium or caesium use tunable semiconductor lasers emitting around 780 nm or 852 nm, respectively.

The Zeeman slowing method requires a custom-designed tapered solenoid in which the field strength decreases as the atoms pass along its bore, as shown in Figure 10.4. The transition energy is shifted by the Zeeman effect (see Eq. [8.19]), and the laser detuning is set at the value required for the slow atoms emerging from the low-field region at the end of solenoid. The reduction of B along the bore compensates for the reduction in v_x as the atoms slow down. The method is typically used to slow fast atoms to velocities where they can be captured by a magneto-optical trap, typically around 20 m/s. This is especially

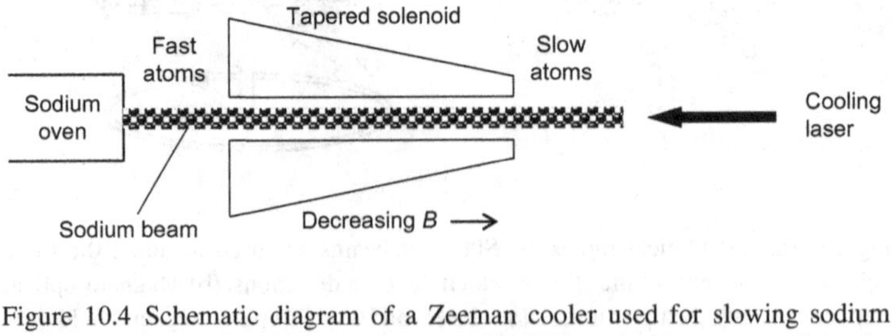

Figure 10.4 Schematic diagram of a Zeeman cooler used for slowing sodium atoms. A tapered solenoid is used to vary the field as the atoms pass along the bore of the magnet. The frequency of the cooling laser is fixed, and the varying Zeeman shift of the transition energy keeps the detuning at the optimal value. The emerging atoms drift to a second region where they can be captured and cooled in a magneto-optical trap. Adapted from Phillips (1998).

important for light atoms (e.g., Li, Na) where the starting velocity from the source might be so high that there is a negligible number of atoms in the initial distribution with velocities within the capture range. The velocity distribution of heavier atoms (e.g., Rb, Cs) peaks at lower values (see Eq. [10.3]), and pre-slowing might not be needed to trap a significant number of atoms.

The temperature of the cold atomic gas can be measured by the "time-of-flight" technique, in which the magnetic field and laser are turned off and the expansion of the atomic gas is recorded as a function of time. This is done by illuminating the gas cloud with a probe laser at a specific time later, and then imaging the bright, fluorescing spot onto a camera to determine its dimensions. The expansion is determined by the velocity distribution of the atoms, which is in turn determined by the temperature. Therefore, by making many measurements of the size of the gas cloud at different expansion times, the velocity distribution can be deduced, and hence the temperature.

10.6 Cooling below the Doppler Limit

Careful measurements led to the rather startling result that the temperature of the laser-cooled atoms in an optical molasses experiment could be substantially less than the Doppler limit given in Eq. (10.16). The explanation of the discrepancy comes from realizing that the single-beam mechanism described in Section 10.3 is too simplistic. The counter-propagating laser beams in an optical molasses experiment form an interference pattern, and this leads to a new cooling mechanism called **Sisyphus cooling**. The mechanism is named after the character in Greek mythology who was condemned to roll a stone up a hill forever, only for it to roll down again every time he got near the top. This is an analogy for the way Sisyphus cooling works: the atoms repeatedly climb to the top of a potential barrier created by the optical Stark effect of the interfering laser beams,[2] and then drop to the bottom of the potential barrier after absorption and emission of a photon. The energy loss in the process is taken from the atom's thermal energy.

The detailed mechanism for Sisyphus cooling is too complicated to describe fully at this level of treatment. The key point is that the minimum temperature that can be achieved is set by the **recoil limit**, rather than the Doppler limit. The atoms are constantly emitting photons of wavelength λ in random directions

[2] The optical Stark effect is a small red-shift in the energy of the atoms caused by the AC electric field of the laser beam. The Stark shift varies with position, following the intensity pattern of the interference fringes.

Table 10.1 *Parameters for laser cooling of sodium and caesium atoms.* T_{min} *and* T_{recoil} *are the minimum temperature set by the Doppler and photon recoil limits given in eqns (10.16) and* (10.18)*, respectively.*

		Sodium	Caesium
Laser		Rhodamine dye	Semiconductor diode
Atomic transition		3p → 3s	6p → 6s
Wavelength	λ	589 nm	852 nm
Atomic mass	m	23.0 m_{H}	132.9 m_{H}
Radiative lifetime	τ	16 ns	31 ns
Doppler limit	T_{min}	240 μK	120 μK
Recoil limit	T_{recoil}	2.4 μK	0.2 μK

by spontaneous emisison. The atom recoils each time with momentum h/λ, so it ends up with a random thermal energy given by:

$$\frac{1}{2}k_{\text{B}}T_{\text{recoil}} = \frac{(h/\lambda)^2}{2m} = \frac{h^2}{2m\lambda^2}\,. \tag{10.17}$$

This gives a minimum temperature of:

$$T_{\text{recoil}} = \frac{h^2}{mk_{\text{B}}\lambda^2}\,. \tag{10.18}$$

In the pioneering experiments in the 1980s on sodium, the temperature in the optical molasses was measured to be around 40 μK, i.e., six times lower than the Doppler limit, and almost within an order of magnitude of the recoil limit.

Table 10.1 compares the key parameters of the sodium and caesium atoms that are frequently used in laser-cooling experiments. Note that caesium offers potentially lower temperatures, on account of its larger mass.

10.7 Bose–Einstein Condensation

Laser-cooling techniques can produce a very cold gas of atoms. Nevertheless, the motion of the atoms at the focus of the laser beams is still *classical* in terms of statistical mechanics. At even lower temperatures, a phase transition to a quantum state proposed by Bose and Einstein in 1924 and 1925 can occur. The theory of Bose–Einstein condensation (BEC) is discussed in statistical mechanics texts. (See Mandl [1988].) In this section we give a brief summary of the key ideas, and then discuss the experimental observation of BEC in atomic gases. Additional notes on the concepts of BEC are given in the online supplement, and further details of the experiments may be found in the specialist literature, for example, Cornell (1996).

10.7.1 Atomic Bosons

The quantized behavior of a gas of identical particles at low temperatures depends on the spin of the particle. Particles with integer spins are called **bosons**, while those with half-integer spins are called **fermions**. Fermions obey the Pauli exclusion principle, making it impossible to put more than one particle into a particular quantum state. (See Section 6.3.) Bosons, by contrast, do not obey the Pauli principle. There is no limit to the number of particles that can be put into a particular level, paving the way for new quantum effects, such as BEC.

Atoms are composite particles, made up of protons, neutrons, and electrons. These are all spin-1/2 particles, but the composite atom can be either a fermion or a boson depending on its total spin, which can be worked out from:

$$S_{\text{atom}} = S_{\text{electrons}} \oplus I\,, \tag{10.19}$$

where I is the nuclear spin. Since the number of electrons and protons in a neutral atom is equal, it is easy to see that the atom will be a boson if the number of neutrons is an even number, and a fermion if it is odd.

The simplest example to consider is hydrogen. ^{1}H has one proton and one electron, and so we find $S_{\text{atom}} = 0$ or 1. ^{1}H atoms are therefore bosons. Deuterium atoms (^{2}H), by contrast, are fermions. Now consider helium, which has two common isotopes: ^{4}He and ^{3}He. The ground state of the ^{4}He nucleus is the α-particle with $I = 0$, and the electron ground state also has $S = 0$. (See Chapter 6.) Thus the spin of the ^{4}He atom in its ground state is zero, which makes it a boson. ^{3}He atoms, by contrast, have $I = 1/2$, making them fermions. For this reason, ^{3}He and ^{4}He behave very differently at low temperatures.

10.7.2 The Condensation Temperature

Statistical mechanics tells us that a gas of *noninteracting* bosons will condense at a critical temperature T_{c}. The word "noninteracting" is very important here, implying that the particles are completely free, with only kinetic energy. The picture that emerges from statistical mechanics is as follows:

(i) Above the critical temperature, the particles are distributed among the energy states of the system according to the **Bose–Einstein distribution**:

$$n_{\mathrm{BE}}(E) = \frac{1}{\exp[(E-\mu)/k_{\mathrm{B}}T]-1}, \tag{10.20}$$

where μ is the chemical potential. Noninteracting particles only have kinetic energy, and so the minimum value of E is zero. The chemical potential must therefore be negative to keep n_{BE} well-behaved for all possible values of E.

(ii) The chemical potential increases with decreasing T, and at T_{c} it reaches its maximum value of zero. In these conditions, there is a singularity in Eq. (10.20) for the zero-velocity state with $E = 0$. A phase transition then occurs in which a macroscopic fraction of the particles condenses into the zero-velocity state. The remainder of the particles continue to be distributed thermally among the finite-velocity states.

(iii) The critical temperature of a gas of nondegenerate bosons of mass m, with N particles per unit volume, is given by (see Mandl [1988]):

$$T_{\mathrm{c}} = 0.0839\frac{h^2}{mk_{\mathrm{B}}}N^{2/3}. \tag{10.21}$$

(iv) The fraction of the particles in the zero-velocity state is given by:

$$N_0(T) = N\left[1-\left(\frac{T}{T_{\mathrm{c}}}\right)^{3/2}\right]. \tag{10.22}$$

This dependence is plotted in Figure 10.5(a). We see that N_0 is zero for $T \geq T_{\mathrm{c}}$ and increases to the maximum value of N at $T = 0$.

The theory of Bose–Einstein condensation was first applied to liquid ^{4}He. Below T_{c} some of the liquid shows superfluid behavior, while the remainder

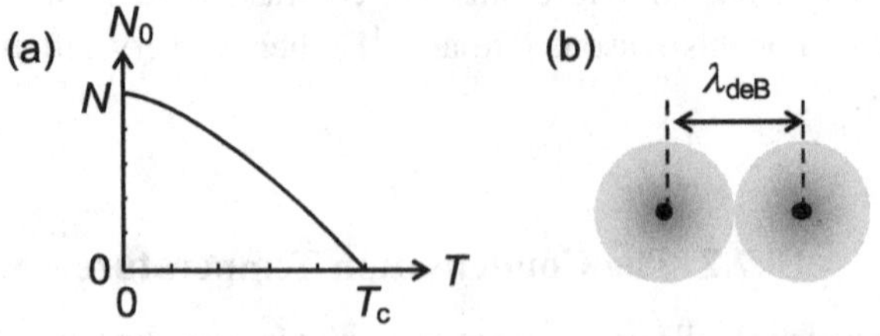

Figure 10.5 (a) Number of particles in the Bose-condensed state versus temperature. T_{c} is the condensation temperature given by Eq. (10.21). (b) Overlapping wave functions of two atoms separated by λ_{deB}.

remains "normal." The value of T_c calculated from Eq. (10.21) is close, but not exactly equal, to the actual superfluid transition temperature of 2.17 K. (See Exercise 10.6.) The discrepancy is a consequence of the fact that the ^{4}He atoms in the liquid phase are not truly noninteracting. To observe BEC in its pure form, we want a low-density system, such as a gas. However, Eq. (10.21) shows that $T_c \propto N^{2/3}$, and so low-density systems, are expected to have very low transition temperatures. It is only with the advent of laser-cooling techniques that it has been possible to get close to the temperatures that are required.

A more intuitive notion of the condensation process can be given by considering the **de Broglie wavelength** of the particles. If the particles are noninteracting, they only have kinetic energy with no forces between them. In these circumstances, λ_{deB} is determined by the free thermal motion:

$$\frac{p^2}{2m} = \frac{1}{2m}\left(\frac{h}{\lambda_{\text{deB}}}\right)^2 = \frac{3}{2}k_B T . \tag{10.23}$$

This implies that

$$\lambda_{\text{deB}} = \frac{h}{\sqrt{3mk_B T}} . \tag{10.24}$$

The thermal de Broglie wavelength thus increases as T decreases.

The quantum-mechanical wave function of a free atom extends over a distance of $\sim \lambda_{\text{deB}}$. As λ_{deB} increases with decreasing T, a temperature will eventually be reached when the wave functions of neighboring atoms begin to overlap. This situation is depicted in Figure 10.5(b). The atoms will interact with each other and coalesce to form a "super atom" with a common wave function. This is the Bose–Einstein condensed state. The condition for this to occur is that the reciprocal of the effective particle volume determined by λ_{deB} should be equal to the particle density:

$$N \sim \frac{1}{\lambda_{\text{deB}}^3} . \tag{10.25}$$

On solving for T using λ_{deB} from Eq. (10.24), we find:

$$T_c \sim \frac{1}{3}\frac{h^2}{mk_B}N^{2/3} . \tag{10.26}$$

This is the same as Eq. (10.21) apart from the numerical factor.

Example 10.3 ^{87}Rb ($Z = 37$) has a ground-state electronic configuration of [Kr] $5s^1$, and a nuclear spin of 3/2. Confirm that ^{87}Rb is a boson, and find T_c for a gas with an atomic density of $1.0 \times 10^{19}\ \text{m}^{-3}$.

Solution The $5s^1$ ground-state has a single electron outside a filled shell, and therefore has $S = 1/2$. The total spin of the whole atom is thus $S_{\text{atom}} = 1/2 \oplus 3/2 = 2$ or 1. Hence ^{87}Rb is a boson. Note that it has $(87 - 37) = 50$ neutrons, an even number. T_c is found from Eq. (10.21):

$$T_c = 0.0839 \frac{h^2}{87 m_H \cdot k_B} (1.0 \times 10^{19})^{2/3} = 8.5 \times 10^{-8}\,\text{K} \equiv 85\,\text{nK}.$$

10.7.3 Experimental Techniques for Atomic BEC

The conditions required to achieve BEC in a gas impose severe technical challenges. If we want to observe pure BEC without the complication of other effects such as liquefaction, we have to keep the atoms well apart from each other. This means that the particle density must be small, which in turn implies that the transition temperature is very low.

We have seen in Section 10.6 that laser cooling can produce temperatures in the μK range. This is not quite cold enough, as temperatures well below 1 μK are typically needed. (See Example 10.3.)[3] We therefore have to use new techniques. The general procedure usually follows three steps:

(i) Trap a gas of atoms and cool them toward the recoil-limit temperature using laser-cooling techniques.
(ii) Turn the cooling laser off to permit cooling below the recoil limit.
(iii) Cool the gas again by evaporative cooling until condensation occurs.

The first step has been discussed previously in Section 10.6. Once the gas has been trapped, the cooling lasers have to be turned off since the temperature will not fall below the recoil limit given in Eq. (10.18) while the lasers are on. The final step is called **evaporative cooling**, in analogy to the cooling of a liquid by evaporation. In this technique, the magnetic field strength is slowly ramped down in order to reduce the depth of the magnetic potential as shown in Figure 10.6(b). The fastest-moving atoms now have enough kinetic energy to escape, leaving the slower ones behind. This causes an overall reduction in the average kinetic energy, which is equivalent to a reduction in the temperature.

The first successful observation of Bose–Einstein condensation in an atomic gas was reported for ^{87}Rb in 1995. Similar results were reported for ^{23}Na soon afterward, followed by several other atomic bosons. Figure 10.7 shows some typical data. These pictures are obtained by turning the trapping field off completely and allowing the gas to expand freely. An image of the gas is taken

[3] The value of T_c in a magnetic trap differs from the one given in Eq. (10.21) due to the effect of the trapping potential. This level of detail need not concern us here.

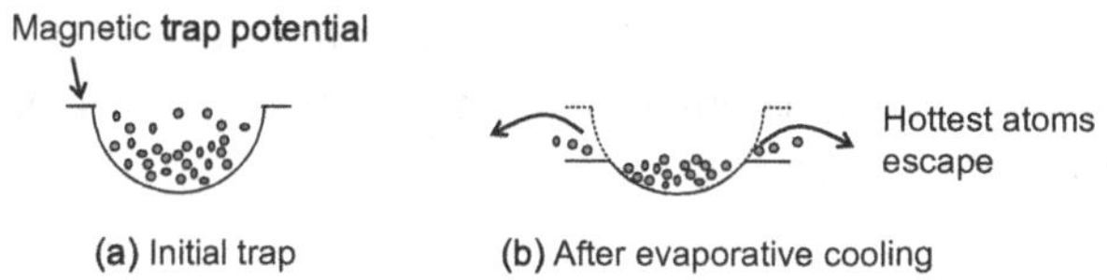

Figure 10.6 Evaporative cooling. (a) A laser-cooled gas of atoms is held in a magnetic trap. (b) The trap potential is reduced by decreasing the magnetic field strength, so that the hottest atoms can escape. This reduces the temperature, in the same way that evaporation cools a liquid.

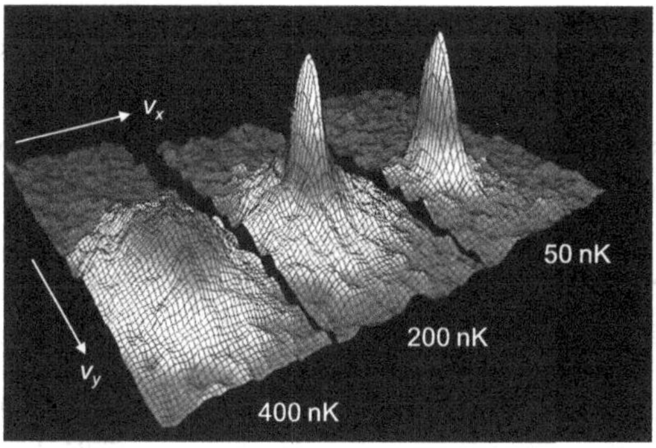

Figure 10.7 Bose–Einstein condensation in rubidium atoms. The three figures show the measured velocity distribution as the gas is cooled through T_c on going from left to right. Above T_c, we have a broad Maxwell–Boltzmann, but as the gas condenses, the fraction of atoms in the zero velocity state at the origin increases dramatically. Image from http://jila.colorado.edu/bec/, with technical details given in Anderson et al. (1995).

at a later time, and the velocity distribution can be inferred from the expansion that has occurred. A broad spread of velocities is observed at 400 nK, which is characteristic of a classical Maxwell–Boltzmann distribution. A peak at zero velocity appears at 200 nK, indicating the onset of BEC. At 50 nK, practically all of the atoms have condensed, following the general trend shown in Figure 10.5(a).

A key implication of BEC is that the atoms in the condensed state should share a common wave function, leading to enhanced coherence. This point has been proven by demonstrating that atomic beams emanating from a condensate can form interference patterns when they overlap. Such coherent atomic beams are sometimes called **atom lasers** in analogy to the difference between the coherence of the light from a laser beam and that from a thermal light source. Such interference cannot be observed for a normal atomic beam, due to the random phases of the atoms.

Exercises

10.1 The Maxwell–Boltzmann distribution $p(v)$ gives the probability that the *speed* of an atom with mass m in a gas at temperature T lies between between v and $v + \mathrm{d}v$. The probability is:

$$p(v)\,\mathrm{d}v = \left(\frac{m}{2\pi k_{\mathrm{B}}T}\right)^{3/2} \exp\left(-\frac{mv^2}{2k_{\mathrm{B}}T}\right) 4\pi v^2\,\mathrm{d}v\,.$$

Show that the most probable and rms speeds are, respectively, $\sqrt{2k_{\mathrm{B}}T/m}$ and $\sqrt{3k_{\mathrm{B}}T/m}$. Relate your answer for the rms speed to Eq. (10.2).

10.2 The velocity distribution within a collimated atomic beam differs from the Maxwell–Boltzmann distribution because the atomic flux is proportional to the velocity of the atoms, so that:

$$p^{\mathrm{beam}}(v)\,\mathrm{d}v \propto vp(v)\,\mathrm{d}v \propto v^3 \exp\left(-\frac{mv^2}{2k_{\mathrm{B}}T}\right) \mathrm{d}v\,.$$

Show that the most probable and rms speeds are, respectively, $\sqrt{3k_{\mathrm{B}}T/m}$ and $\sqrt{4k_{\mathrm{B}}T/m}$.

10.3 Repeat Example 10.1, but using the most probable speed in an atomic beam rather than the rms velocity in the oven.

10.4 In a laser cooling experiment, a gas of ^{87}Rb atoms is cooled by using the 780 nm transition which has an Einstein A coefficient of $3.4 \times 10^7\ \mathrm{s}^{-1}$.

(a) What is the maximum cooling force per atom?
(b) What are the Doppler-limit and recoil-limit temperatures?
(c) What is the average speed of the atoms at the two temperatures?

10.5 A beam of ^{133}Cs atoms is emitted in the $+x$ direction from an oven at 500°C and is cooled by a laser beam directed in the $-x$ direction. The laser is tuned to near resonance with the $6s\,^2S_{1/2} \leftrightarrow 6p\,^2P_{3/2}$ transition at 852 nm, which has a radiative lifetime of 31 ns. Estimate:

(a) the decelerating force applied to the atoms by the laser;
(b) the time taken to cool the atoms to their minimum temperature;
(c) the distance the atoms travel in this time; and
(d) the final temperature and atomic speed.

10.6 The density of liquid ^{4}He is $120\,\mathrm{kg\,m^{-3}}$. What is T_c?

10.7 Potassium ($Z = 19$) has three natural isotopes: ^{39}K, ^{40}K, and ^{41}K. Which of these would be suitable for use in a BEC experiment? Where appropriate, calculate T_c for a gas with 1.0×10^{18} atoms $\mathrm{m^{-3}}$.

10.8 Calculate T_c for a gas of ^{23}Na atoms with a density of $5 \times 10^{18}\,\mathrm{m^{-3}}$. Estimate the de Broglie wavelength at this temperature, and compare it to the mean particle separation.

10.9 BEC-like behavior has been observed for pairs of fermionic atoms, e.g., $(^6\mathrm{Li})_2$. How can this be possible?

11
Atomic Physics Applied to the Solid State

Solids are made up of atoms bound together in crystals, and the understanding of their quantized states is a subject in its own right, namely solid-state physics. In this chapter, we briefly look to see how the general principles developed in atomic physics can be applied to solid-state systems. This will enable us to obtain a basic understanding of light emission in solids.

The focus of the chapter will be restricted to two main examples of optically active solid-state materials:

(i) **Semiconductors**: Semiconductors lie at the heart of modern technology. The silicon chip underpins the electronics industry, while the optoelectronics industry exploits the optical properties of compound semiconductors such as GaAs. Our task here will be to apply simple principles of atomic physics to understand the electronic states of impurities in semiconductors, and the mechanisms of light emission and detection.

(ii) **Ions doped into optical hosts**: Here we consider materials such as ruby, where chromium is lightly doped into Al_2O_3, with the Cr^{3+} ions substituting for the Al^{3+} ions in the crystal. Pure Al_2O_3 is a colorless, transparent crystal, and the characteristic red color of ruby arises from transitions associated with the Cr^{3+} ions. Our task will be to understand how the transitions of the Cr^{3+} ions in the crystal relate to the atomic states of Cr^{3+} ions in isolation.

In both cases, it will not be possible to give a comprehensive treatment; the aim of the chapter is to explain a few basic principles that can lay the foundations for further study. This author has written another book in which these topics are explained in much greater depth. See Fox (2010).

11.1 Solid-State Spectroscopy

Chapter 3 developed the basic principles governing optical transitions in atoms. In this section, we shall see how these principles carry over to solid-state systems.

11.1.1 Selection Rules

The electric-dipole (E1) interaction is the strongest term in the light-matter Hamiltonian, as discussed in Section 3.3. The selection rules that follow from analysis of the E1 perturbation and the wave functions of atomic states were derived in Section 3.4, and are summarized in Table 3.1. These selection rules carry over directly to optical transitions in solid-state systems. However, we must bear in mind that some of the selection rules were derived by assuming that the angular dependence of the wave functions is described by spherical harmonics, which in turn assumes that the central-field approximation holds. (See Section 4.1.) In the solid state, the wave functions can get distorted by the crystal, which means that they are no longer pure atomic-like states. The net result is that some transitions that would be forbidden for isolated atoms become weakly allowed in the solid-state. In other cases where E1 processes are forbidden, weaker, higher-order processes may also occur. (See Section 3.5.)

A case in point is the d $\rightarrow$ d transitions that are important in the spectroscopy of transition-metal ions. (See Section 11.5.1.) Electric-dipole transitions are forbidden by the $\Delta l = \pm 1$ selection rule, and the transition would have to proceed by an M1 or E2 process in the free ion. (See M1 and E2 selection rules in Table 12.1.) However, when the ion is doped into a crystal, the perturbation of the crystal field can mix odd-parity states with the D-states. The E1 matrix element may therefore no longer be zero, although it will always be small, as it relies on the weak admixture of the nondominant states. The end result is that E1 transitions can occur, but at a low rate, which is, nevertheless, larger than the M1 or E2 rate. The low transition rate gives rise to long, excited-state lifetimes (e.g. $\sim$ μs–ms), which can be exploited for storing energy in solid-state lasers.

The strongest E1 selection rule that carries across to the solid state is the **parity** rule. This follows directly from the odd-parity nature of the electric-dipole operator, and implies that the initial and final states must have different parities. If the states have the same parity, then the transition would have to occur by a higher-order process such as M1 or E2.

At the fundamental level, the photon carries one unit of angular momentum. The emission of a photon must therefore change the angular momentum of the system by one unit, which implies $|\Delta J| \leq 1$, with $J = 0 \rightarrow 0$ forbidden. This also applies to the components of the angular momentum, for example, when a magnetic field is applied. If M_J is a good quantum number, then conservation of angular momentum requires that $\Delta M_J = \pm 1$ when observing along the axis of the field (Faraday geometry).

The spin selection rules, namely $\Delta S = \Delta M_S = 0$, follow from the fact that spin does not appear in the electric-dipole interaction. However, as discussed in Section 7.3, spin-orbit coupling creates a perturbation proportional to $\boldsymbol{L} \cdot \boldsymbol{S}$, which can mix two different spin states via a common L state and result in a weak breakdown of the spin selection rules. The spin-orbit coupling increases with Z (see Section 7.3.3), resulting in stronger mixing in heavy atoms. This fact can be exploited, for example, in organic light-emitting diodes, where the doping of a heavy metal into the organic compound facilitates spin-forbidden transitions, and hence improves the efficiency of the device.

11.1.2 Linewidths

The mechanisms that cause line broadening in atoms were discussed in Section 3.7. The three main processes were:

- Lifetime broadening, also called natural broadening;
- Doppler broadening; and
- Collisional broadening, also called pressure broadening.

Of these three, only the first carries over directly to the solid state, since it is a fundamental consequence of the radiative emission process. The other two do not apply directly, since the atoms are locked into a lattice, and therefore cannot move around and collide with each other. On the other hand, there are other processes that replace them.

The equivalent of Doppler broadening in the solid-state is **environmental broadening**. It was pointed out in Section 3.7 that Doppler broadening is an example of an inhomogeneous broadening mechanism. This means that individual atoms emit at slightly different frequencies, causing a spread in the emission wavelengths, and hence broadening of the emission line. In a solid, the environment in which the atoms find themselves may not be entirely uniform, which can cause small shifts in the emission wavelength through the interaction between the atom and the local environment – for example, the local electric field. A good example is the difference between the linewidths of the $^4F_{3/2} \rightarrow {}^4I_{11/2}$ laser transition of the Nd^{3+} ion when doped into a YAG crystal or into phosphate glass. The YAG crystal is far more uniform than the glass, and the linewidth of the transition is significantly smaller.

Another mechanism that can cause inhomogeneous broadening in solids is local fluctuations in composition. Blue and green light-emitting diodes are generally made from the compound semiconductor $Ga_xIn_{1-x}As$, with the emission wavelength depending on the value of x. The growth process of the crystal controls the average value of x very accurately, but there are fluctuations of x on a microscopic scale, resulting in a spread of emission wavelengths and hence broad emission lines.

The equivalent of collisional broadening in the solid-state is nonradiative decay or phonon scattering. In the first case we consider the possibility that the atoms de-excite from the upper level to the lower level by making a **nonradiative transition**. One way this could happen is to drop to the lower level by emitting phonons (i.e., heat) instead of photons, typically via intermediate **trap states**. To allow for this possibility, we must rewrite Eq. (3.24) in the following form:

$$\frac{\mathrm{d}N_2}{\mathrm{d}t} = -AN_2 - \frac{N_2}{\tau_{\mathrm{NR}}} = -\left(A + \frac{1}{\tau_{\mathrm{NR}}}\right)N_2 = -\frac{N_2}{\tau}, \qquad (11.1)$$

where A is the Einstein A coefficient for the transition, and τ_{NR} is the nonradiative decay time. This shows that nonradiative transitions shorten the lifetime of the excited state according to:

$$\frac{1}{\tau} = A + \frac{1}{\tau_{\mathrm{NR}}} = \frac{1}{\tau_{\mathrm{R}}} + \frac{1}{\tau_{\mathrm{NR}}}, \qquad (11.2)$$

where τ_{R} is the radiative lifetime. We thus expect additional lifetime broadening according to Eq. (3.32), when the nonradiative decay rate is comparable to or faster than the radiative decay.

The other mechanism for collisional-type lifetime broadening is phonon scattering within a band of states. The phonon interaction times in solids are often very fast, especially at room temperature, and can cause substantial broadening of the emission lines. Solid-state spectroscopists therefore often work at low temperatures (e.g., liquid He temperature, namely 4.2 K) where the emission and absorption lines are narrower due to the inhibition of thermally activated phonon processes.

Example 11.1 The radiative lifetime of the near infrared fluorescence band in $Co{:}KMgF_3$ is 3.3 ms. The measured lifetime of the excited state is 2.5 ms at 1.6 K and 0.25 ms at 300 K. Calculate the nonradiative lifetime at both temperatures, and account for its change with temperature.

Solution: The radiative lifetime is not expected to change with T, and so the variation of τ is caused by a change in the nonradiative lifetime. On substituting into Eq. (11.2) with $\tau_R = 3.3$ ms at both temperatures, we find:

$$1.6\,\mathrm{K}: \quad \tau_{NR} = \left(\frac{1}{\tau} - \frac{1}{\tau_R}\right)^{-1} = \left(\frac{1}{2.5} - \frac{1}{3.3}\right)^{-1} = 10\,\mathrm{ms}\,,$$

$$300\,\mathrm{K}: \quad \tau_{NR} = \left(\frac{1}{\tau} - \frac{1}{\tau_R}\right)^{-1} = \left(\frac{1}{0.25} - \frac{1}{3.3}\right)^{-1} = 0.27\,\mathrm{ms}\,.$$

The shortening of τ_{NR} is caused by the increase of phonon-assisted processes with increasing T.

11.2 Semiconductors

11.2.1 Electronic States

The atoms in a solid are packed very close to each other, with the interatomic separation approximately equal to the size of the atoms. Hence the outer orbitals of the atoms overlap and interact strongly with each other. This broadens the discrete electronic levels of the free atoms into bands, as illustrated schematically in Figure 11.1(a). The inner core orbitals do not overlap and so remain discrete even in the solid state.

The electronic states of crystals are described by the band theory of solids. This subject is covered extensively in all solid-state physics texts, and we only summarize a few key points here. In any atom, there will be a sequence of energy states with increasing energy. As discussed in Section 1.3, there will be a number of occupied electron shells, followed by the outermost valence shells and excited states. The valence shells may, or may not, be full. In the case of a semiconductor-like silicon, the valence orbitals are the 3s and 3p shells, which together contain 4 valence electrons. In the formation of the solid, these shells evolve into electronic bands, with energy gaps between them. The highest occupied and lowest unoccupied bands are called the **valence band** and **conduction band**, respectively, as shown in Figure 11.1(b). The bonding in semiconductors and insulators works in such a way that the valence band is completely filled with electrons at absolute zero, and the conduction band is empty. (Solids with partially filled bands give rise to metallic behaviour and are not our concern here.) The energy gap between them is called the **band gap**, E_g, with the magnitude of E_g determining whether the crystal shows insulator or semiconductor electrical behavior. In general, any crystal with E_g larger than about 4 eV would be classified as an insulator.

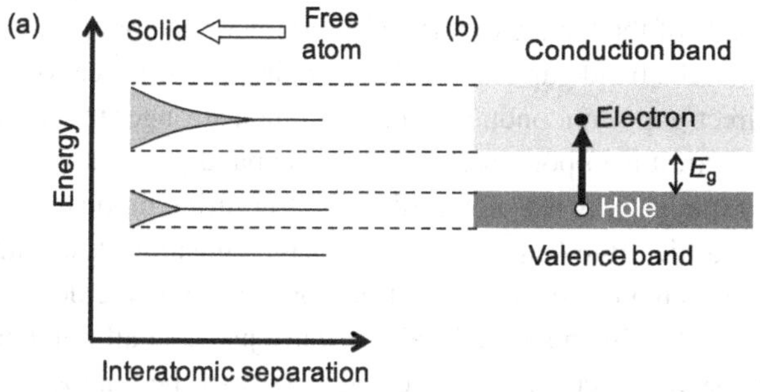

Figure 11.1 (a) Schematic diagram of the formation of electronic bands in a solid from the condensation of free atoms. As the atoms are brought closer together to form the solid, their outer orbitals begin to overlap with each other. These overlapping orbitals interact strongly, and broad bands are formed. (b) Optical transitions between the valence band and the conduction band, separated by the band-gap energy E_g. Free holes and electrons are created in the respective bands.

11.2.2 Interband Transitions

Optical transitions between bands are called **interband transitions**. The minimum amount of energy to promote an electron from the valence to the conduction band is equal to E_g, and this sets the lower limit of the photon energy that can be absorbed. Hence, the absorption spectrum consists of a continuous *band* with a lower energy threshold of E_g, which contrasts with the discrete absorption *lines* of atoms.

The optical transition leaves an empty electron state in the valence band and free electron in the conduction band, as shown in Figure 11.1(b). The empty state in the valence band is called a **hole** and behaves likes a positive particle. The electrons in the conduction band and the holes in the valence band are not completely free particles. Their motion is affected by the crystal lattice, and this results in them behaving as if they have an **effective mass** that is different from the free electron mass. The values of the effective mass for electrons and holes are not the same, and vary from semiconductor to semiconductor.

The strength of the optical transitions between the valence and conduction bands is determined by a number of factors:

(i) The valence and the conduction bands of important opto-electronic semiconductors like GaAs are derived from p-like and s-like atomic orbitals, respectively, which means that electric-dipole transitions between them are allowed.

(ii) The details of the band structure of the semiconductor are important. In particular, the band gap is classified as being either **direct** or **indirect**, with direct-gap semiconductors having much stronger transition rates. In this context, it is important to note that the band gap of silicon is *indirect*, which explains why it is not used in light-emitting diodes.

(iii) Fermi's golden rule (Eq. [3.4]) includes the density of final states $g(h\nu)$. This factors in both the density of photon states and the density of electron states. In an atom, there is usually just a small number of electron states determined by the degeneracy of the final level. In solids, however, the density of electron states can be very large, as each band contains at least as many electronic states as the number of atoms in the crystal.

These three factors lead to extremely large absorption strengths in direct-gap semiconductors at photon energies that exceed the band gap. In some ways, this simply reflects the very large density of absorbing atoms in the solid. The net result is that sizeable optical effects can be obtained in very thin samples, allowing for the production of compact optical devices that form the basis of the modern opto-electronics industry.

The process of spontaneous optical emission in a semiconductor is shown schematically in Figure 11.2. The process begins with the injection of electrons and holes into their respective bands. This can be done by absorbing a photon with energy greater than E_g as shown in Figure 11.1(b). Alternatively, the

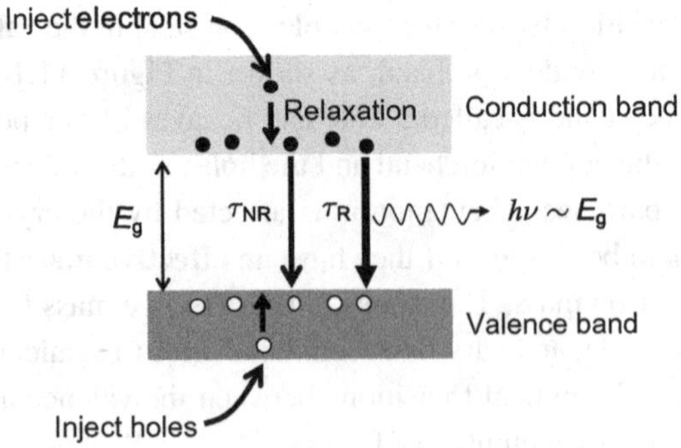

Figure 11.2 Optical emission in a semiconductor. Electrons and holes are injected into the conduction and valence bands respectively, which then relax to the bottom of their respective bands before recombining by emitting of a photon with energy $\sim E_g$. The radiative recombination competes with nonradiative processes, and this determines the quantum efficiency of the process.

charge carriers can be injected by electrical means. (See Section 11.2.3 below.) In the former case, the number of electrons and holes is identical, but this is not necessarily so in the latter, as the electrical injection efficiency may differ between the two types of charge carrier.

The next step in the process is the relaxation of the carriers to the bottom of their bands. In the case of electrons, this means going to the bottom of the conduction band, whereas for holes it means moving to the *top* of the valence band. In both cases, the relaxation proceeds by emission of phonons. The electron–phonon coupling is generally very large, and so this occurs on very rapid ($\sim$ 100 fs) timescales.

The final step is the emission of the photon as the electron drops down to the empty state in the valence band where the hole is. The photon emission therefore destroys a free electron and a hole, and is consequently called electron-hole recombination. On account of the large matrix element and high density of states, the radiative lifetime τ_R in a direct-gap semiconductor like GaAs is in the $\sim$ns range. This is much longer than the phonon emission times, and explains why the electrons and holes are able to relax to the bottom of their bands before emitting. The net result is that the semiconductor always emits photons with energy $h\nu$ very close to E_g, irrespective of how the carriers were initially injected. The band gap thus determines the lower threshold for the interband absorption, but the energy of the emission.

The emission of the photon has to compete with other possible decay channels in which electrons and holes recombine nonradiatively. The **quantum efficiency** η gives the ratio of photons emitted to the number of electron-hole pairs injected, and is defined as:

$$\eta = \frac{\text{radiative decay rate}}{\text{total decay rate}} = \frac{1/\tau_R}{1/\tau} = \frac{1}{1 + \tau_R/\tau_{NR}}, \qquad (11.3)$$

where τ and τ_{NR} are defined in eqns (11.1) and (11.2). The quantum efficiency therefore depends on the ratio of τ_R to τ_{NR}, with high efficiency requiring fast radiative emission and/or slow nonradiative recombination.

The spontaneous emission of photons by a solid is generally called **luminescence**. When the luminescence is triggered by the optical injection of electrons and holes, it is subcategorized as **photoluminescence**. The corresponding name for electrically driven emission is **electroluminescence**.

11.2.3 Light-Emitting Diodes

The principles of electroluminescence discussed above underpin the workings of **light-emitting diodes** (LEDs). These devices are based on p-n junctions,

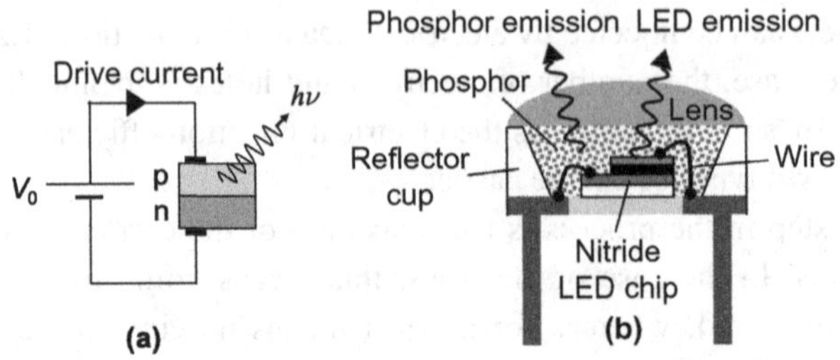

Figure 11.3 (a) A light-emitting diode (LED). The energy of the photon emitted is equal to the band gap, E_g, of the semiconductor from which the diode is made. (b) Schematic diagram of a white-light LED.

which use doping techniques to adapt the electrical properties of the semiconductor. The principles of doping are most easily understood by considering the elemental semiconductors, silicon and germanium. These materials come from group IV (14) in the periodic table, and thus have four valence electrons per atom. The incorporation of impurities from group V (15) of the periodic table adds one extra electron per dopant atom, forming n-type material. Alternatively, the incorporation of impurities from group III (13) takes away one electron for each dopant atom, forming p-type material. In n-type materials, there are free electrons in the conduction band, while in p-type materials there are free holes in the valence band. When p-type and n-type materials are joined together, a p-n diode is formed.

The application of a positive voltage to the p-region with respect to the n-region causes a current to flow, as shown schematically in Figure 11.3(a). Holes flow toward the junction from the p-side and electrons from the n-type region. Note that this conserves the current flow through the device, as the two charge carriers have opposite signs. The electrons and holes meet at the junction, and recombine, emitting photons with energy equal to the band gap E_g. Photons cannot be emitted from other parts of the device, as it is necessary to have both an electron in the conduction band and a hole in the valence band for emission to occur, and this only happens at the junction where holes are injected from the p-side and electrons from the n-side.

The wavelength of an LED is determined by the band gap of the semiconductor at the junction, with $\lambda = hc/E_g$. The most efficient LEDs are made from the direct-gap material GaAs and its variants. GaAs itself has a band gap of 1.42 eV at room temperature, which leads to the emission of infrared photons around 870 nm. The addition of Al and/or P to GaAs brings the wavelength down to the red end of the visible spectral region. Blue and green

LEDs are made with alloys of $Ga_xIn_{1-x}N$ and its variants. Unfortunately, silicon cannot be used in LEDs on account of its indirect band gap, which leads to a long radiative lifetime and a low efficiency due to competition with nonradiative Auger processes.

The lighting industry has been revolutionized in recent years by the advent of white-light LEDs. Figure 11.3(b) shows a schematic diagram of a typical white-light LED. The device contains a blue-emitting LED chip based on nitride semiconductors surrounded by an appropriate **phosphor** material. The purpose of the phosphor is to convert some of the blue photons emitted by the nitride chip into red or green photons to produce a red-green-blue (RGB) balance that appears white.

The phosphor materials that are used in white-light LEDs typically incorporate rare-earth ions (e.g., Eu^{2+}) doped into transparent ceramics. (See Section 11.5.2.) These absorb blue photons, and have emission lines at green and red wavelengths. The red and green photons emitted after absorption of blue photons from the nitride chip combine with unabsorbed blue photons to produce white light. These white-light LEDs are the basis of the solid-state lighting industry that is gradually superseding traditional industries based on incandescent and fluorescent lamps.

Example 11.2 The alloy semiconductor $Al_xGa_{1-x}As$ has a direct band gap for $x \leq 0.43$ that varies with composition according to: $E_g(x) = (1.420 + 1.087x + 0.438x^2)$ eV. What would be the wavelength of an LED made from $Al_{0.2}Ga_{0.8}As$?

Solution: The photons will be emitted at the band gap energy, which for $x = 0.2$ is equal to 1.655 eV. The wavelength will therefore be $hc/(1.655\,\text{eV}) = 749\,\text{nm}$.

11.2.4 Semiconductor Diode Lasers

Semiconductor diodes are by far the most common lasers in everyday use, finding applications, for example, in laser printers, DVD and Blu-ray players, laser pointers, bar-code readers, and optical fiber communication systems. The laser consists of a semiconductor p-n diode cleaved into a small chip, as shown in Figure 11.4. As with the LED discussed above, electrons are injected into the n-region, and holes into the p-region. The drive voltage must be $\gtrsim E_g/e$, where e is the electron charge. At the junction between the n- and p-regions, we have both electrons in the conduction band and holes (i.e., empty states) in the valence band. This creates population inversion between the conduction

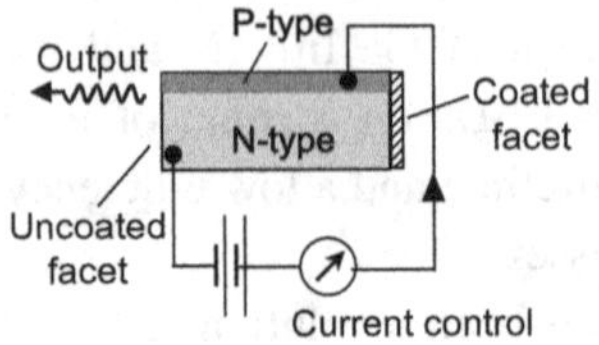

Figure 11.4 Schematic diagram of an edge-emitting semiconductor diode laser.

and valence bands, and gain is produced at the band gap energy E_g of the semiconductor. Diode lasers can be considered three-level systems, since the lower level is fully occupied in the unpumped system: the semiconductor has strong absorption at the laser wavelength until a sufficient number of electrons are pumped out of the valence band to the conduction band.

The easiest way to make a cavity is to use the cleaved facets of the chips, leading to edge emission, as shown in Figure 11.4. The refractive index of a typical semiconductor is in the range 3–4, which gives about 30% reflectivity at each facet. This is enough to support lasing, even in crystals as short as ~ 1 mm, because the gain in the semiconductor crystal is so high. Reflective coatings can also be applied (especially to the rear facet) to prevent unwanted losses and reduce the threshold. Other configurations are also possible in which mirrors are incorporated within the semiconductor wafer above and below the active regions, giving raise to vertical emission from the chip.

As explained in Sections 11.2.2 and 11.2.3, the semiconductor must have a *direct* band gap to be an efficient light emitter. Silicon is therefore not used in laser diodes, on account of its indirect band gap. Instead, the laser diode industry is based mainly on direct-gap compound semiconductors such as GaAs, which has $E_g \sim 1.42$ eV (870 nm). Through the use of alloys of GaAs, the band gap can be shifted into the red spectral region for making laser pointers, or further into the infrared to match the wavelength for lowest losses in optical fibers (1550 nm). Blue laser diodes for use in Blu-ray systems are made from the wide-band gap III–V semiconductor GaN and its alloys.

The power conversion efficiency of electricity into light in a diode laser is very high, with figures of 25% typically achieved. Since the laser chips are so small, it is possible to make high power diode lasers by running many GaAs chips in parallel. Laser power outputs over 20 W can easily be achieved in this way. These high power laser diodes can be used for pumping other solid-state lasers.

11.2.5 Photodiodes

The bias voltage connected to a p-n junction can be connected the other way round, with positive voltage applied to the n-region, as shown in Figure 11.5. In this *reverse* bias configuration, there is no current in the circuit in the absence of incoming photons. Instead, the voltage dropped across the diode generates a strong electric field at the junction. When the diode is illuminated, interband absorption can occur if $h\nu > E_g$, creating electron-hole pairs at the junction. The electrons are swept through the p-type region toward the positive terminal of the power supply, and the holes toward the negative terminal connected to the p-type region. This generates a current in the circuit, and its measurement enables the photon flux to be determined. The device thus acts as a photo-detector. Since the detector is based on a p-n diode, it is frequently called a **photodiode**.

The vast majority of the photo-detectors operating in the world are made from silicon. Its indirect band gap gives it a smaller absorption coefficient than direct gap materials like GaAs, but this deficiency can easily be offset by using thicker absorbing layers. The abundance and convenience of silicon then makes it preferable to manufacturers than more expensive compound semiconductors. The band gap of silicon is 1.1 eV, and so it can serve as an efficient detector for all wavelengths shorter than $\sim$ 1100 nm. This includes the entire visible band from 400–700 nm, and the charge-coupled-device (CCD) chips found in digital cameras are usually made from silicon.

An interesting variant of the photodiode is made by replacing the power supply with an electrical load. The photocurrent generated by absorption of photons then produces electrical power in the load. This is the basis of photovoltaic power generation in solar cells. The power efficiency of the process is limited by conflicting demands on the choice of band gap. A large gap leads to the larger voltages, as the voltage across the load cannot exceed E_g/e, since there would then be negligible field across the diode to generate the current. On the other hand, a small gap leads to a larger current, since only

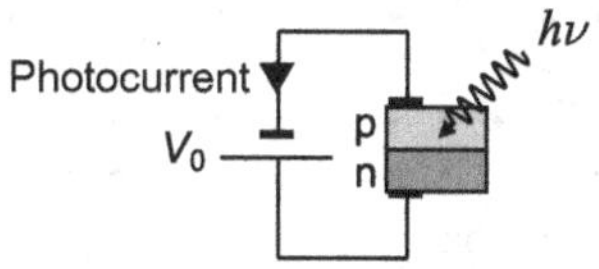

Figure 11.5 A photodiode. The photon is absorbed if its energy exceeds E_g.

a fraction of the solar spectrum with $h\nu > E_g$ can be absorbed. At present, most of the world's solar cells are made from silicon, with power conversion efficiencies of up to $\sim 25\%$ being possible in the best devices.

11.3 Solid-State Hydrogenic Systems

The quantized states of hydrogen atoms were discussed in Chapter 2. In this section, we see how these principles can be applied to two important topics is semiconductor physics, namely impurity states and excitons, which both can be treated as hydrogenic systems. (See Section 2.4.)

11.3.1 Impurity States in Semiconductors

Consider an n-type group IV(14) semiconductor in which a **donor atom** from group V(15) substitutes for one of the silicon or germanium atoms. We assume that the fifth electron of the group V atom is released into the crystal, leaving a positively charged donor ion in the lattice. The electron is attracted back to the positive ion and forms a hydrogenic system, as shown schematically in Figure 11.6(a). The quantized energy levels are given by Eq. (2.8) and the Bohr radius by Eq. (2.17). In applying these formulas, we must remember that the electron behaves as if it has an effective mass of m_e^*, and also include the relative dielectric constant ϵ_r of the host crystal. On the other hand, we do not have to consider the reduced mass of the system, as the positive ion is locked into the crystal and cannot move. We can also take $Z = 1$ as the ion is

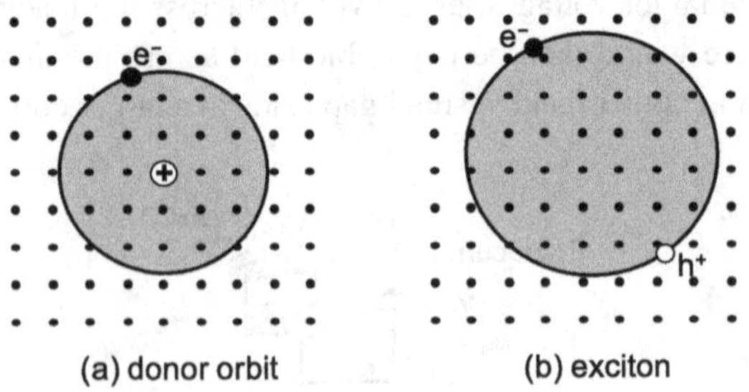

Figure 11.6 (a) Quantized electron states surrounding a positively charged donor ion. (b) An exciton consisting of a free electron bound to a free hole. In both cases, the array of black dots represents the crystal lattice.

singly charged. The net result is that the binding energy and radius are given respectively by:

$$E_n = -\frac{m^* e^4}{8\epsilon_r^2 \epsilon_0^2 h^2 n^2} \approx -\frac{m_e^*}{m_e \epsilon_r^2} \frac{R_H}{n^2}, \tag{11.4}$$

and

$$r_n = \frac{\epsilon_r m_e}{m_e^*} n^2 a_0, \tag{11.5}$$

where $R_H \approx 13.6\,eV$ is the hydrogen Rydberg energy, and a_0 is the hydrogen Bohr radius. With typical values of $m_e^* \sim 0.1 m_e$, and $\epsilon_r \sim 10$, we find binding energies and radii of $\sim$ 0.01 eV and 5 nm, respectively. The radius is much larger than the separation of the atoms, and justifies the use of the dielectric constant to model the crystal lattice.

The energy of the quantized **donor levels** is measured relative to the bottom of the conduction band. With binding energies of $\sim$ 0.01 eV, the electrons are easily excited into the conduction band at room temperature, where $k_B T \sim 0.025$ eV, generating free electrons with a density that is determined by the doping level. The electrons from the donor atoms then control the conductivity of the n-type material. Similar arguments can be applied to **acceptor** impurities in p-type material, where acceptor levels are formed just above the valence band, generating free holes at room temperature when electrons from the valence band are thermally promoted into the vacant acceptor states.

Example 11.3 Silicon has an electron effective mass of $0.85 m_e$, and $\epsilon_r = 16$. Find the binding energy and Bohr radius of the ground-state donor level.

Solution: The binding energy and radius are worked out from Eqs. (11.4) and (11.5), respectively. For the ground state, we put $n = 1$. We then find:

$$E = \left| -\frac{0.85}{16^2} \frac{R_H}{1^2} \right| = 3.3 \times 10^{-3} R_H = 0.045\,eV,$$

$$r = \frac{16}{0.85} 1^2 a_0 = 19\, a_0 = 1.0\,nm.$$

11.3.2 Excitons

An **exciton** consists of a free electron bound to a free hole, similar to positronium, as shown schematically in Figure 11.6(b). Excitons are typically formed in pure (i.e., undoped) semiconductors by optical absorption at the band gap, where a free electron is created in the conduction band and a free hole in the valence band. The electrons and holes have negligible excess energy, and can bind together. The binding energy and radius are again worked out by applying the hydrogenic model:

$$E_n = -\frac{me^4}{8\epsilon_r^2\epsilon_0^2h^2n^2} \approx -\frac{m}{m_e\epsilon_r^2}\frac{R_H}{n^2}, \tag{11.6}$$

and

$$r_n = \frac{\epsilon_r m_e}{m} n^2 a_0 , \tag{11.7}$$

where m is the reduced mass of the electron-hole system, and ϵ_r is the relative dielectric constant of the semiconductor. The reduced mass m is worked out from the effective masses of the electrons and holes according to Eq. (A.5):

$$\frac{1}{m} = \frac{1}{m_e^*} + \frac{1}{m_h^*} . \tag{11.8}$$

The binding energies are even smaller than in donor states on account of the effect of the hole effective mass on m. This means that exciton states are often only observed clearly at low temperatures. They appear as a hydrogenic series of absorption lines just below the fundamental absorption edge at E_g, with the nth exciton level occurring at a photon energy of:

$$h\nu = E_g - \frac{m}{m_e\epsilon_r^2}\frac{R_H}{n^2} . \tag{11.9}$$

Excitons can also be observed in emission. The electrons and holes that have relaxed to the bottom of their bands after injection bind together to form excitons, which then emit at the exciton transition energy given in Eq. (11.9).

Example 11.4 The electron and hole effective masses of GaAs are $0.067m_e$ and $0.2m_e$, respectively, and $\epsilon_r = 12.8$. What is the binding energy and Bohr radius of the ground-state exciton?

Solution: We must first use Eq. (11.8) to work out the reduced electron-hole mass:

$$\frac{1}{m} = \frac{1}{m_e}\left(\frac{1}{0.067} + \frac{1}{0.2}\right) = \frac{20.0}{m_e} ,$$

implying $m = 0.050m_e$. The binding energy and radius are then worked out from Eqs. (11.6) and (11.7) with $n = 1$:

$$E = \left| -\frac{0.050}{12.8^2} R_H \right| = 3.0 \times 10^{-4} R_H = 4.2\,\text{meV} ,$$

$$r = \frac{12.8}{0.05} a_0 = 2.6 \times 10^2 a_0 = 14\,\text{nm} .$$

11.4 Quantum-Confined Semiconductor Structures

Advanced semiconductor growth techniques have enabled the engineering of structures in which the electrons and holes are confined to regions of the crystal that are smaller than their de Broglie wavelengths, i.e., ∼nm length scales. (See Exercise 11.7.) This leads to behavior in which the electrons and holes are free in some directions, but quantum confined in the others. A general classification is given in Figure 11.7. Starting from the bulk crystal, the electrons are free to move in all three directions, and we therefore have normal, 3-dimensional (3-D) physics. If the electrons are trapped in a very thin layer, we have a quantum well where the electrons are free to move in only two dimensions, and we therefore have 2-D physics. The progression continues through quantum wires (1-D physics) to quantum dots (0-D physics). The effect of the quantum confinement on the electronic properties is a huge subject, and here we just briefly consider two aspects that are interesting from the perspective of atomic physics.

11.4.1 The Quantum-Confined Stark Effect

The Stark effect for atomic states was discussed in Section 8.4. Here we consider the **quantum-confined Stark effect**, which describes the effect of a strong electric field on the exciton states in a semiconductor quantum well. The electric field is applied by using a reverse-biassed p-n diode, as shown in Figure 11.5, and the quantum well is located at the junction. The voltage from the power supply is dropped across the narrow junction region, thereby generating a large electric field that is controlled by the reverse bias.

The normal behavior for the ground-state of an atom is a small, quadratic red shift with increasing field, as discussed in Section 8.4.1. This effect is hard to observe in bulk semiconductors, as the excitons are very unstable to applied electric fields due to their low binding energy. (See Exercise 11.8.) The electrons and holes are pushed in opposite directions, and the exciton then

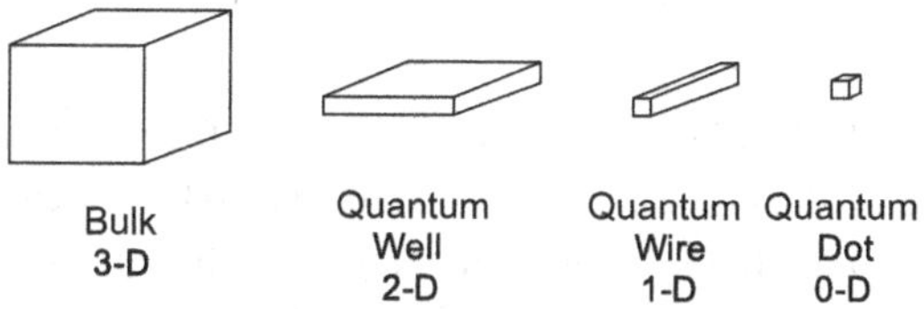

Figure 11.7 Progression of quantum confinement, starting from the bulk and progressing to quantum dots.

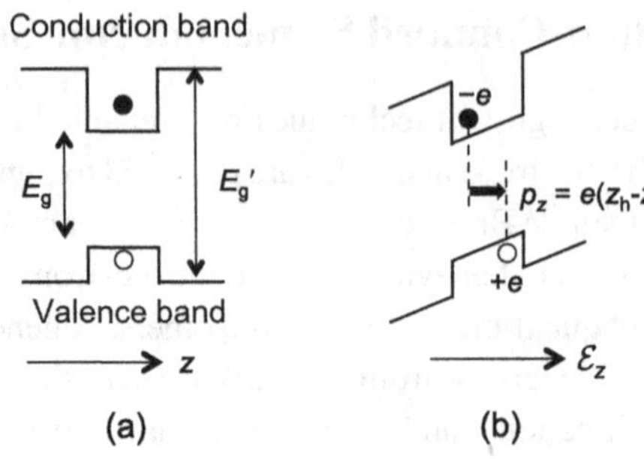

Figure 11.8 The quantum confined Stark effect. (a) A quantum well is formed when a thin layer of a semiconductor with a band gap E_g is sandwiched between layers of another semiconductor with a larger band gap E_g'. (b) Effect of a strong electric field applied along the z direction (i.e., perpendicular to the layers). The electrons and holes are pushed in opposite directions, creating a dipole p_z parallel to the field. In both (a) and (b), the filled and open circles represent the expectation values of z for the electron and hole wave functions, respectively.

easily gets ripped apart by the field. This effect is called **field ionization**. It can also be observed in atoms, but only at extremely high field strengths.

The situation in a semiconductor quantum well is very different. Consider the case of the quantum well shown in Figure 11.8(a). The quantum well is formed by sandwiching a thin layer of a semiconductor with a band gap of E_g between layers of another semiconductor with a larger band gap E_g'. This then gives rise to spatial discontinuities in the conduction and valence band energies as shown in the figure. The excitons that are formed by optical transitions across the smaller band gap are then trapped in the z-direction by the finite potential well created by the band discontinuities. The excitons remain free to move in the perpendicular 2-D x–y plane.

When an electric field $\mathcal{E}_z$ is applied along the z-direction, the energy of the electrons is shifted by $qV = -eV$, where V is the electro-static potential associated with the field via:

$$\mathcal{E}_z = -\frac{\mathrm{d}V}{\mathrm{d}z}. \tag{11.10}$$

For a uniform field, V varies linearly with z, causing the potential well to tilt as shown in Figure 11.8(b). The excitons that are created by optical transitions are relatively stable to the field, because the barriers of the quantum well prevent them from being ripped apart easily. The electrons are pushed to one side, and the holes to the other, creating a dipole of magnitude $p_z = e(z_\mathrm{h} - z_\mathrm{e})$ where z_h and z_e represent the expectation values of z for the electron and hole wave functions, respectively. For a quantum well of width d, the dipole will have a magnitude of eCd, where C is a dimensionless parameter < 1 that increases with the field. Since p_z is roughly proportional to $\mathcal{E}_z$, the energy shift is quadratic and negative, as in Eq. (8.32). The magnitude of the quadratic red-shift is much larger than in atoms, on account of the larger dipole: C can approach ~ 0.1 at large fields, and with $d \sim 10$ nm, the electron–hole separation can be much larger than the size of an atom. The large, voltage-controllable red-shift of the exciton absorption line is widely used for making electro-optical modulators.

11.4.2 Quantum Dots

A quantum dot is formed when the electrons and holes are trapped in all three dimensions, as shown schematically on the right-hand side of Figure 11.7. This is typically achieved when a nano-crystal of one semiconductor (e.g., InAs, $E_g = 0.42$ eV at 4 K) is formed within another semiconductor with a larger band gap (e.g., GaAs, $E_g = 1.52$ eV at 4 K). Such structures can form spontaneously during the epitaxial (i.e., layered) crystal-growth process when the right conditions are achieved. Note that the shape of the nano-structure is not necessarily cubic as suggested by Figure 11.7. In fact the InAs/GaAs quantum dots formed during epitaxy typically have approximately hemispherical shapes. The important point is that the dimensions are small in all three directions, but still substantially larger than atomic sizes.

With confinement in all three dimensions, the quantized states of the electrons and holes have discrete energies instead of the continuous energy bands that usually characterize the solid state. In this sense, they can be considered as solid-state atoms. The advantage compared to real atoms is that the energy levels and wave-functions can be engineered by the size, composition, and shape of the quantum dot. In particular, the wave functions spread over $\sim$ nm length scales, which compares to the $\sim$Å length scales of atoms. This results in larger optical matrix elements, and correspondingly stronger light–matter coupling. For example, the radiative lifetime of an InAs/GaAs quantum dot is around 1 ns, which is more than an order of magnitude faster than a typical atomic transition (e.g., 16 ns for the sodium D-lines at 589 nm.)

The quasi-atomic nature of the states in quantum dots makes them attractive for cutting-edge applications in quantum technologies. For example, if the emission spectra of a single quantum dot can be isolated, then it can be used as a single-photon source. The operating principle of such sources is the same as for atoms, where the excitation of a *single* atom leads to the emission of just one photon of a particular color for each excitation cycle. The quantum dot with its faster radiative lifetime can produce more single photons per second than the atom. Furthermore, the dot can be integrated into advanced solid-state devices to produce, for example, a single-photon LED in which exactly one photon is emitted in response to each drive pulse. Such devices are required for applications such as quantum cryptography, in which the security of data transmission is guaranteed by the laws of quantum mechanics.

11.5 Ions Doped in Crystals

A number of solid-state lasers, and also phosphors that are used in solid-state lighting, are based on optically active ions doped into crystalline or glass hosts. The host is usually transparent at the emission wavelength, and the optical transitions that are used in the technological applications derive from the ions. In understanding how the emission spectra compare to those of free ions, there are two main effects that have to be considered:

(i) **The crystal field**: The active ion is surrounded by the ions that form the host material, and this generates local electric fields that perturb the energy levels.
(ii) **Phonon coupling**: The active ion is coupled to the phonons (i.e., vibrational modes) of the crystal through the time dependence of the local electric fields as the crystal ions vibrate about their means positions.

There are two main classes of material that we need to consider: transition-metal and rare-earth-metal dopants. The way in which the crystal field and phonon coupling affects the energy levels is very different in the two cases, and so we consider them separately. Supplementary notes on three solid-state lasers — ruby, Nd^{3+}, and Ti:sapphire — are available online.

11.5.1 Transition Metals

The elements that lie in the middle of the periodic table are called transition metals. The key aspect of their atomic physics is the filling of d-shells. For the sake of simplicity, we focus our attention on the fourth row of the periodic table, namely elements 21–30.

The usual sequence for filling electronic shells is shown in Figure 4.1, and this leads to ground-state electronic configurations of [Ar] $3d^n4s^2$, where

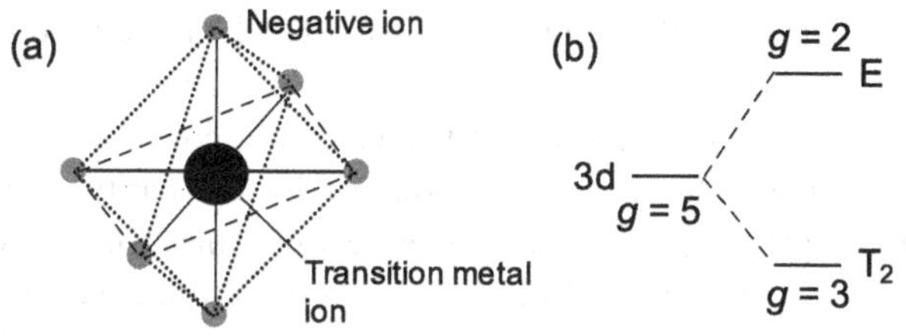

Figure 11.9 (a) A transition metal ion (large back dot) surrounded by negative ions (grey dots) in an octahedral lattice. (b) Splitting of the d-states in an octahedral crystal field. The value of g gives the degeneracy of the orbital angular momentum states.

$n = (Z - 20)$.[1] For example, the electronic configuration of Co ($Z = 27$) is [Ar] $3d^7 4s^2$. When the Co^{2+} ion is formed, the outermost 4s electrons are lost giving a configuration of [Ar] $3d^7$. We thus have an ion with seven electrons in an unfilled 3d shell outside the filled shells of the argon configuration. These outermost-shell d-electrons are very sensitive to the electric fields of their crystal environment, and it is this interaction that determines the dominant features of the optical spectra.

Consider, as an example, the Ti:sapphire crystal in which Ti^{3+} ions ($Z = 22$, configuration [Ar] $3d^1$) are doped into Al_2O_3. Hund's rules would give a ground state of ${}^2D_{3/2}$ for a free ion. However, the Ti^{3+} ions in the crystal occupy the sites of the Al^{3+} ions, and find themselves surrounded by six negatively charged O^{2-} ions in an octahedral arrangement, as shown in Figure 11.9(a).[2] The five m_l states of the d shell are split by the crystal field into a triplet and doublet, as shown in Figure 11.9(b). Since the splitting is caused by the crystal field, rather than the residual-electrostatic and spin-orbit interactions that determine the angular momentum states of the free ion, the LS-coupling regime no longer applies, and Hund's rules are no longer applicable for determining the ground state. Instead, a notation derived from the symmetry of the crystal has to be used, and the states are labeled T_2 and E. Furthermore, strong phonon coupling broadens the split d-states into **vibronic bands**, which give rise to absorption and emission bands rather than sharp lines. The broad emission bands of Ti:sapphire are ideally suited for making widely tuneable CW lasers. (See Table 9.1.) Alternatively, the large gain

[1] Chromium ($Z = 24$) and copper ($Z = 29$) are exceptions, with configurations of [Ar] $3d^5 4s^1$ and [Ar] $3d^{10} 4s^1$, respectively. See Section 4.2.

[2] The crystal structure of Al_2O_3 is actually trigonal, and so the environment is not exactly octahedral. However, the distortion from octahedral symmetry is relatively small, and the main gist of the argument is valid.

bandwidth associated with the broad emission band can be used to generate ultrashort laser pulses. (See Eq. [9.35].)

It was pointed out in Section 11.1.1 that electric-dipole transitions between d-states are normally forbidden, for example by the parity selection rule. However, the crystal field distorts the wave functions, which then gives a probability for E1 transitions to occur in proportion to the admixture of odd parity states. Since the admixture is generally small, the probability is low, and the radiative lifetimes are correspondingly long (e.g., 3 ms for the laser transition in ruby).

The way in which the d-states split and the magnitude of the splittings depends on the symmetry and nature of the host, and can therefore vary significantly from crystal to crystal. An interesting example is the difference between Cr^{3+} doped into beryl ($Be_3Al_2(SiO_3)_6$) and sapphire (Al_2O_3). The former is emerald, which has absorption bands in the blue and red, giving it a green color. The latter is ruby, and the absorption bands are in the green/blue spectral region, giving a red coloration. (The Latin word *ruber* means "red.") One of the emission lines of ruby is at 694.3nm, and is used in lasers based on Cr^{3+}:Al_2O_3. (See § 9.7.)

11.5.2 Rare Earths

The rare-earth metals are usually found at the bottom of periodic tables. Specifically, we are dealing with elements 57–71. Since the first of these is lanthanum (La), they are also called **lanthanides**. These elements are important in solid-state lasers, phosphors, and magnets. The key point of their atomic physics is the filling of the 4f shell.

Let us focus on one technologically important lanthanide element, namely neodymium ($Z = 60$). By applying the rules of Figure 4.1, we can work out that the ground-state configuration of the neutral atom is [Xe] $4f^4\,6s^2$. The corresponding configuration of the Nd^{3+} ion that is used in solid-state lasers is [Xe] $4f^3$. There is an important difference here with the transition metals, in that the 4f electrons are strongly shielded from the crystal field. This happens because there is a high probability that the 4f electrons lie inside other shells. For example, the 5s and 5p shells have lower energy than the 4f shell, but have significant probability density outside it. (See, e.g., Eq. [2.59], which shows that the average radius increases with n and decreases with l.) The end result is that the states are labeled by the angular momentum nomenclature of atomic physics, and Hund's rule can be used to determine that the ground state is $^4I_{9/2}$. Moreover, the transitions tend to be lines rather than bands.

A particularly important transition of Nd^{3+} occurs between the $^4F_{3/2}$ and $^4I_{11/2}$ excited states. The bottom level lies $2111\,cm^{-1}$ above the ground-state, and is one of the other spin-orbit-split J states of the ^4I term. The $^4F_{3/2} \rightarrow {}^4I_{11/2}$ transition violates both the $|\Delta L| \leq 1$ and $|\Delta J| \leq 1$ selection rules, and is thus E1-forbidden for free ions. However, the perturbation of the crystal field distorts the wave functions, and this relaxes the selection rules. For example, for Nd^{3+} in yttrium aluminium garnet (YAG), the Einstein A coefficient is $4.3 \times 10^3\,s^{-1}$, which gives a radiative lifetime of 0.23 ms. The transition occurs at 1064 nm, and is the basis of the 4-level Nd:YAG laser. The long upper state lifetime is beneficial for achieving population inversion, and also for storing energy. As a consequence, Nd:YAG lasers can generate very high output powers.

The Nd^{3+} ion can be doped into many other crystalline or glass hosts. However, in contrast to transition-metal ions, this does not strongly affect the wavelength, due to the shielding of the 4f electrons from the crystal field. The transition is, of course, not completely immune to perturbation by the crystal. For example, the wavelength of the $^4F_{3/2} \rightarrow {}^4I_{11/2}$ line shifts to 1054 nm when Nd^{3+} is doped into phosphate glass. As mentioned in Section 11.1.2, the linewidth in the glass host is significantly larger than in YAG on account of the inhomogeneity of the noncrystalline environment. The larger linewidth of the Nd:glass transition is exploited in ultrafast pulsed lasers. (See Eq. [9.35].)

The white-light LED illustrated in Figure 11.3(b) includes a **phosphor** material to generate red and green light after absorption of blue photons. These phosphors frequently contain lanthanide elements doped in ceramic hosts. Europium ions, in both their divalent and trivalent forms, are frequently used, along with cerium. These rare-earth phosphors also find widespread applicant in fluorescent lighting, where they absorb blue and ultraviolet light (e.g., from a mercury discharge lamp) and reemit green and red photons to produce a red-green-blue white-light balance.

Exercises

11.1 Calculate the quantum efficiency of the $Co:KMgF_3$ crystal considered in Example 11.1 at 1.6 K and 300 K.

11.2 The radiative lifetime of the laser transition in titanium-doped sapphire is 3.9 μs. The lifetime of the excited state is measured to be 3.1 μs at 300 K and 2.2 μs at 350 K. Find the nonradiative lifetimes and quantum efficiencies at the two temperatures. Explain why it is necessary to use water cooling of the laser crystal in a Ti:sapphire laser.

11.3 The band gap of the alloy semiconductor $GaAs_{1-x}P_x$ is given approximately by $E_g(x) = (1.42 + 1.36x)$ eV, and is direct for $x \leq 0.45$ and indirect for $x > 0.45$.

(a) Estimate the shortest wavelength that can be produced efficiently by a $GaAs_{1-x}P_x$ light-emitting diode.
(b) Estimate the composition of the alloy in an LED emitting at 670 nm.

11.4 A laser beam with power 1 mW and wavelength 632.8 nm is incident on a photodiode. What is the maximum photocurrent that can be generated?

11.5 The values of the electron effective mass and relative dielectric constant of GaAs are $0.067m_e$ and 12.8, respectively.

(a) Calculate the binding energy and Bohr radius of the $n = 1$ donor level in n-type GaAs.
(b) Find the wavelength of the $n = 1 \rightarrow 2$ donor level transition. In what spectral region does this transition lie?

11.6 CdTe is a direct-gap semiconductor with $E_g = 1.605$ eV. The electron and hole effective masses are $0.099\,m_e$ and $0.3\,m_e$ respectively. The relative dielectric constant is 9.0.

(a) Calculate the binding energy and Bohr radius of the $n = 1$ exciton.
(b) Calculate the wavelength of the $n = 1$ exciton transition.

11.7 The size, d, of a nanostructure where quantum confinement effects are significant can be estimated by finding the value of d equivalent to the de Broglie wavelength associated with free thermal motion of the particle. Estimate d for an electron with effective mass of $0.1m_e$ at room temperature and 4 K. Repeat the calculation for a hole with $m^* = 0.5m_e$.

11.8 Use the Bohr model to show that the magnitude of the electric field between the electron and hole in the ground state of an exciton is equal to $2E/er$, where E is its binding energy, and r its Bohr radius. Estimate this field for the GaAs exciton considered in Example 11.4. Assuming that the voltage in a diode is dropped over a region of $\sim 1\,\mu$m, what voltage does this correspond to?

11.9 The exciton in a quantum well shifts from 850 to 860 nm in an electric field of 1.0×10^7 V/m. What is the average separation of the electron and hole at this field strength?

11.10 Optical amplifiers for 1550 nm telecommunication wavelength systems can be made by doping Er^{3+} ions ($Z = 68$) into optical fibers.

(a) What is the electronic configuration of Er^{3+}?
(b) What is the ground-state level?
(c) The amplifier operates by stimulated emission between vibronic bands associated with the $^4I_{13/2} \rightarrow {}^4I_{15/2}$ transition. What causes the splitting of the $^4I_{13/2}$ and $^4I_{15/2}$ levels? What type of transition is it?

11.11 A white-light LED contains a phosphor emitting at 650 nm. Calculate the maximum possible energy conversion efficiency for the phosphor when it is excited by a blue LED operating at 450 nm.

12
Atomic Physics in Astronomy

The subjects of atomic physics and astronomy have developed together over centuries. The understanding of stars, galaxies, nebulae, planets, and so on relies on detailed spectroscopic analysis of the atoms they contain. There is healthy feedback between the two disciplines, with astronomical observations prompting new research in atomic physics, and developments in atomic physics and spectroscopy leading to new understanding of astrophysical processes.

It is not possible to do justice to such a broad subject in a single chapter such as this. The purpose here is to highlight some of the ways the principles developed in the book apply in the astrophysical context, and to point out where interesting differences are observed compared to lab-based experiments. The reader is referred to specialist books for a more comprehensive treatment of the subject.[1]

12.1 Astrophysical Environments

The atoms in astrophysical sources are frequently found in extreme environments that are very different to those in Earth-based laboratories. The conditions inside an atomic discharge tube are usually benign compared to those found in stars. At the same time, the atom densities can be orders of magnitudes higher than those found in nebulae. All of this has an effect on the spectra that are observed. The underlying principles of atomic physics are

[1] See Tennyson (2011) for an excellent introductory text, or Pradham and Nahar (2011) for a more advanced treatment.

the same, but the spectra can appear very different due to the change of the environment of the atoms. The two main differences that have to be considered relate to the *temperature* and the *density*.

Astrophysical Temperatures

The gas in an atomic discharge tube or an oven might reach a temperature of a few hundred degrees celsius. By contrast, the surface temperature of the sun is 5800 K, with the corona reaching 10^6 K. Other stars can be hotter. At such high temperatures, the thermal energy $k_B T$ is more than sufficient to dissociate molecules. A hydrogen cylinder on Earth will contain molecular hydrogen H_2, but the temperatures in stars are sufficient to break the molecular bond and dissociate H_2 into atomic hydrogen. Hence the spectra of stars are dominated by atomic hydrogen, whereas a standard hydrogen lamp will emit the spectrum of molecular hydrogen.

Another consequence of the high temperatures in stars is the abundance of highly ionized atoms. These multiply charged ions, which might be quite hard to produce in the laboratory, are formed by repeatedly stripping off the electrons as explained in Section 1.2. It might be quite feasible to observe all the ionization states of an atom in different astrophysical environments. Take, for example, the case of iron, which has $Z = 26$. At low temperatures, the spectra of neutral iron (Fe I) would dominate, but as the temperature is raised, all the ionization states up to the bare nucleus (Fe XXVII) will be observed. Analysis of the ionization states that prevail can therefore give useful information about the temperature of the star.

Not all astrophysical objects are very hot. The regions of space in between stars (e.g., the gas clouds in the interstellar medium) are expected to be cold, as there is no nearby source of heat. This means that some of the atoms will form molecules, giving rise to molecular rather than atomic spectra. Another consequence is that the atom will be in its ground state, so that absorption will dominate over emission. (See discussion in Section 1.4.)

Astrophysical Densities

The atom densities found in some astrophysical environments (e.g., interstellar gas clouds, outer regions of an atmosphere) can be extremely small by comparison with those in normal laboratory conditions on Earth. This has two main consequences:

(i) The gas might be cold enough for molecules to be stable, but the density is so low that the probability of atoms colliding and associating to form a molecule is small. The atoms and molecules are therefore not in proper thermal equilibrium, and their relative populations will not be governed by the normal rules of thermal physics, e.g., Boltzmann's law.
(ii) The low density means that the time between collisions is long. In normal laboratory conditions, an atom in a metastable state would probably de-excite by colliding with another atom or with the walls of the discharge tube. This will not occur in the astrophysical environment, leading to the observation of electric-dipole forbidden lines that are not usually observed in the laboratory. See Section 12.2.2.

12.2 Astrophysical Spectra

The vast majority of the information that has been accumulated about objects outside the solar system is gained from analysis of the electromagnetic radiation that they emit.[2] For objects within the solar system, we can also gain information from space missions, but observation by telescope still provides a great wealth of information. As explained above, the underlying principles that determine the spectra are the same as those developed in Chapters 1–8, but the extreme environments that apply in astrophysics can lead to some interesting differences.

12.2.1 General Features

The spectrum emitted by an astrophysical source may contain either absorption or emission lines, or both. Absorption lines are observed when a continuous spectrum passes through an atomic gas and is absorbed at the characteristic frequencies of the transitions. The continuous spectrum might originate from many different sources, the most obvious being black-body radiation from hot matter. Emission lines are observed when the atoms have been promoted to excited states, and then emit as the electrons drop to lower states. One important way in which this might happen is when an ion recaptures an electron, which then cascades through the excited states down to the ground state, as illustrated in Figure 12.1. In astronomy, these transitions are called **recombination lines**.

[2] Neutrino and gravitational wave astronomy are two examples of nonelectromagnetic observational techniques. Both of these branches of astronomy are in their infancy.

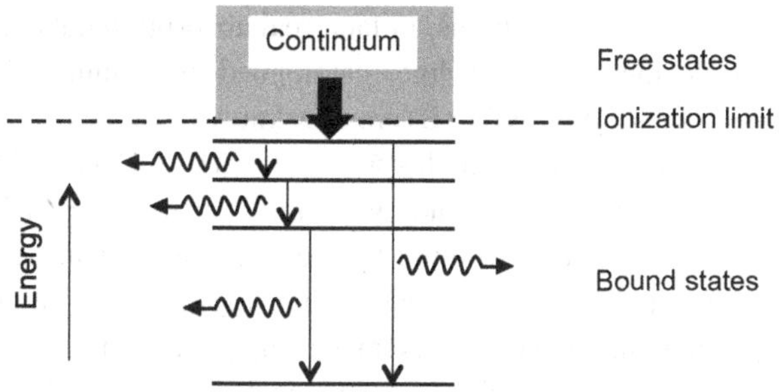

Figure 12.1 Recombination lines. An ionized atom captures an electron into an excited state from the continuum above the ionization limit. The electron then cascades to lower levels and emits radiation in each transition.

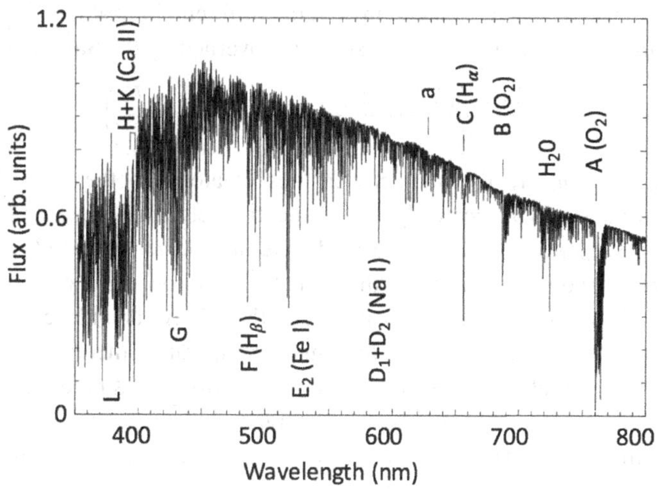

Figure 12.2 Spectrum of the sun observed from a mountaintop, with Fraunhofer absorption lines labeled. Data from Kurucz et. al. (1984), adapted by P.A. Crowther, University of Sheffield.

The classic example of astrophysical absorption lines is the spectrum of the sun. Dark lines in the solar spectrum were first observed by William Hyde Wollaston in 1802, and were then studied in detail by Joseph von Fraunhofer in 1814. Figure 12.2 shows a high-resolution spectrum of the sun. The envelope of the spectrum approximates to a black-body source at 5,800 K, as appropriate for the surface temperature. The black-body continuum is modulated by a myriad of absorption lines from different elements and molecules; this gives the impression of being noise, but is, in fact, real absorption data. The labeled dips correspond to the absorption lines catalogued by Fraunhofer, using his notation. Some of these (e.g., A and B) are now known to arise from absorption in the Earth's atmosphere by molecules such as O_2, H_2O, CO_2, and OH. These are called **telluric lines**, from the Latin word *tellus* meaning "earth." The other labeled lines correspond to absorption by elements in the sun's atmosphere. Prominent among these are lines C and F that originate from hydrogen, the most common element in the sun. Also worthy of note is the dip labeled D_1+D_2, which corresponds to the 3s→3p fine-structure doublet of sodium. (See Section 7.5.) In fact, the notation of calling the $ns \rightarrow np$ transition of an alkali a D-line, when ns is the ground state, originates from Fraunhofer's catalogue.

The spectrum from an astronomical source might contain absorption or emission lines from an element in several different ionization states. The relative abundance of the ionization states is governed by the **Saha equation**:

$$\frac{n_{i+1} n_{\mathrm{e}}}{n_i} = \frac{(2\pi m_{\mathrm{e}} k_{\mathrm{B}} T)^{3/2}}{h^3} \frac{2 g_{i+1}}{g_i} \exp\left(-\frac{E_{i+1} - E_i}{k_{\mathrm{B}} T}\right). \tag{12.1}$$

Here, n_{i+1} and n_i are the number densities of atoms in the $(i+1)^{\mathrm{th}}$ and i^{th} ionization states respectively, n_{e} is the free electron density, T is the temperature, g_{i+1} and g_i are the ground-state degeneracies, and $(E_{i+1} - E_i)$ is the energy required to remove the $(i+1)^{\mathrm{th}}$ electron. As the temperature increases, the atom loses more electrons and the relative abundance of higher ionization states increases. In fact, for a given temperature there is a particular ionization stage that is dominant, with low ionization states predominating at lower temperatures, and vice versa for the higher ionization states.

The ability to resolve fine-structure features in the spectra depends on the spectral linewidths. The Doppler linewidth given in Eq. (3.43) is proportional to $\sqrt{T}$, and so increases with increasing temperature. Hence the amount of detail that can be resolved depends on the temperature of the atoms. Other factors that affect the linewidth include the pressure (see Section 3.9) and whether the source (e.g., a star or galaxy) is rotating or not. In the latter case, the lines get broadened through the variation of the velocities (and hence Doppler shifts; see Section 12.2.4) of the atoms in the spinning object relative to an observer on Earth.

12.2.2 Forbidden Transitions

It was pointed out in Section 12.1 that the atom density in some astronomical objects is extremely low. For example, the atom density in nebulae is so low that the time between collisions might typically be in the range 10–10^4 s. This contrasts with typical laboratory conditions, where the collision rate (either between atoms, or with the containing vessel) is much faster. The consequence is that some lines that are forbidden by E1 selection rules are commonly observed in astrophysics, whereas they are hard to observe in the laboratory.

The observation of forbidden lines is a consequence of the presence of **metastable states**. These are excited states with no allowed E1 transitions to lower-lying states. Two examples were given in Section 6.5 in the discussion of helium, namely the $^1S_0$ and $^3S_1$ levels of the 1s2s configuration. E1 transitions to the 1s1s $^1S_0$ ground state are forbidden by a variety of rules. Both involve an s→s transition and hence contravene the $\Delta l = \pm 1$ and parity change rules. The former is a $J = 0 \rightarrow 0$ transition and thus is forbidden by the ΔJ rule, while the latter contravenes the spin selection rule. The result is that these states have long lifetimes. In the laboratory, they de-excite in a collision, as used to good effect in the He:Ne laser. (See Section 9.6.) In a low-density nebula, by contrast, collisions would not be an option, and the atom would have to relax to the ground state by other processes, such as forbidden transitions. (See Section 12.5.)

The electric-dipole selection rules for hydrogen were discussed in Section 3.3, and then extended to multi-electron atoms in Section 5.8. They are summarized in Table 12.1, along with the rules for the higher-order magnetic dipole (M1) and electric quadrupole (E2) transitions mentioned in Section 3.5. M1 and E2 transitions have low probabilities, and are only observed when E1 transitions are forbidden and the time between collisions is longer than $1/A$, where A is the Einstein A coefficient for the transition.

The rules on parity and ΔJ in Table 12.1 are rigorous, in the absence of nuclear spin effects. The parity rules follow from symmetry arguments,

Table 12.1 *Selection rules for one-photon E1, M1, and E2 transitions. After Corney (1977).*

	Electric dipole E1	Magnetic dipole M1	Electric quadrupole E2
Parity	Changes	Unchanged	Unchanged
ΔJ	$0, \pm 1$	$0, \pm 1$	$0, \pm 1, \pm 2$
	Not $0 \leftrightarrow 0$	Not $0 \leftrightarrow 0$	Not $0 \leftrightarrow 0, 0 \leftrightarrow 1, \frac{1}{2} \leftrightarrow \frac{1}{2}$
ΔS	0	0	0
Electron jumps	One	None	One or none
Δn	Any	0	Any
Δl	± 1	0	$0, \pm 2$

while the rules on ΔJ follow from conservation of angular momentum: E1 and M1 processes involve the emission of a photon with one unit of angular momentum, while for E2 processes the angular momentum change is two units. The remaining four rules are based on approximations. For example, we have already seen in Section 11.1.1 that spin-orbit coupling or other mechanisms can cause mixing of spin-states, thereby leading to a partial breakdown of the spin selection rules. A transition that satisfies the E1 selection rules apart from spin is called an **intercombination line**.

Well-known examples of forbidden transitions occur in O III (i.e., O^{2+}). The ground-state configuration of O III is the same as carbon, namely $1s^22s^22p^2$. As discussed in Section 5.9, this has five angular momentum states: two singlets ($^1S_0$ and $^1D_2$) and three triplets ($^3P_0$, $^3P_1$, and $^3P_2$). Hund's rules give $^3P_0$ as the ground state, with the others being excited states, as shown in Figure 12.3(a). The singlet terms are metastable with no allowed E1 transitions to the ground state, as no electron jumps shell and parity is unchanged. The green color of the Orion nebula originates from two forbidden transitions starting from the $^1D_2$ state, namely $^1D_2 \rightarrow {}^3P_1$ at 495.9 nm and $^1D_2 \rightarrow {}^3P_2$ at 500.7 nm.[3] These satisfy both the M1 and E2 rules, apart from spin, with the M1 process having the higher probability. The transitions therefore proceed mainly by spin-forbidden M1 transitions, and have Einstein A coefficients of 6.2×10^{-3} and 1.8×10^{-2} s^{-1} respectively. These low A coefficient values should be contrasted with the values in the range 10^6–10^9 s^{-1} that are typical for allowed E1 transitions. Despite the low A coefficients, the lines are extremely strong in the spectrum of the Orion nebula, as shown in Figure 12.3(b). The 500.7 nm line is about three time stronger on account of its larger A coefficient. The $^1D_2 \rightarrow {}^3P_0$ transition at 493.1 nm is not observed, as it can only proceed via the weaker spin-forbidden E2 process, and has a much lower A coefficient of $2.4 \times 10^{-6}\,s^{-1}$. The spin-allowed $^1S_0 \rightarrow {}^1D_2$ transition at 436.3 nm is considered in Example 12.1. Analysis of its intensity ratio compared to the 500.7 nm line gives information about the relative populations of the $^1S_0$ and $^1D_2$ states, and hence of the temperature of the nebula. Note that these forbidden transitions can only be observed in low-density conditions such as those found in the nebula, where the collision time τ_c satisfies $\tau_c \gg A^{-1}$.

Other interesting examples of forbidden transitions occur in the Auroræ (i.e., the Northern and Southern lights). The O_2 molecules in the upper atmosphere

[3] The green color of the Orion nebula is not always apparent in the images that are readily available, as astronomers frequently add *false color*. The green color to the eye can be verified relatively easily with a simple optical telescope of the type used by amateur astronomers. The reader might also wonder why the oxygen atoms are ionized. This is because the nebula is a star-forming region, where large amounts of ionizing ultraviolet radiation are being generated.

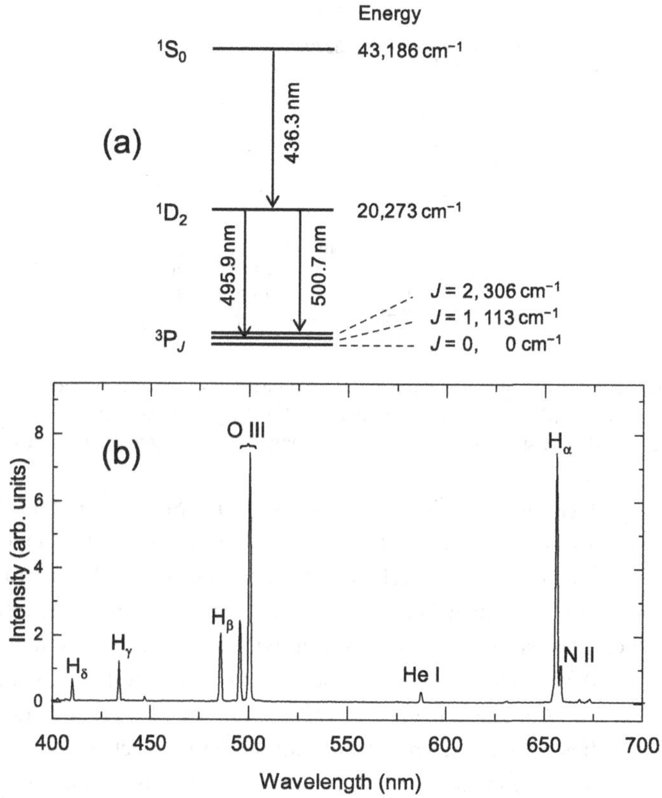

Figure 12.3 (a) Level diagram for the ground-state configuration of O III, namely $1s^22s^22p^2$. (b) Visible emission from the Orion nebula. The $^1D_2 \rightarrow {}^3P_1$ and $^1D_2 \rightarrow {}^3P_2$ forbidden transitions of O III at 495.9 nm and 500.7 nm, respectively, are both strong, along with several hydrogen Balmer lines and lines from other elements. Data from Osterbrock and Ferland (2006). The notation for the hydrogen lines is explained in Section 12.4.1.

are dissociated into oxygen atoms by ultraviolet radiation from the sun, which are then promoted to excited states by collisions with charged particles ejected by the sun. Collisions between oxygen atoms are unlikely in the rarefied conditions in the upper atmosphere, and the excited states therefore only decay by radiative transitions. The ground-state configuration of oxygen is $1s^22s^22p^4$, and has the same five levels as O III, but with the order of the triplet levels reversed, as shown in Figure 12.4. (See Exercise 12.7.) Many atoms decaying from upper levels end up in the $^1S_0$ and $^1D_2$ metastable states, and transitions involving these states cause the characteristic colors

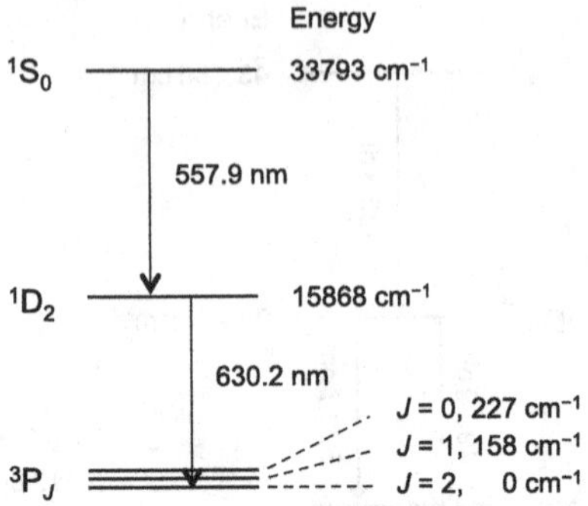

Figure 12.4 Levels of the ground-state configuration of neutral oxygen (O I), namely $1s^22s^22p^4$. The forbidden transitions for the green and red auroral lines are indicated. The splitting of the lower ^3P_J manifold is exaggerated for clarity.

of the Auroræ. The green light originates from the $^1S_0 \to {}^1D_2$ transition at 558 nm. In this transition, no electron jumps shell, parity is unchanged, and $\Delta J = 2$, which contravenes the E1 rules but is allowed for E2. The transition is therefore of electric-quadrupole nature, with an Einstein A coefficient of $1.26\,s^{-1}$. Such a transition can only be observed in a rarefied atmosphere where $\tau_c \gg A^{-1} = 0.79\,s$. The auroral red line seen at higher altitudes is the spin-forbidden 630 nm transition from the $^1D_2$ level to the $^3P_2$ ground state. This transition satisfies both the E2 and M1 selection rules on electron jumps, parity, and ΔJ, but not the spin rule. The A coefficient is therefore smaller than for the green line, being only $5.6 \times 10^{-3}\,s^{-1}$ for the stronger M1 process.

Example 12.1 The $^1S_0 \to {}^1D_2$ transition within the $1s^22s^22p^2$ configuration of O III (O^{2+}) occurs at 436.3 nm and has an Einstein A coefficient of $1.71\,s^{-1}$. (See Figure 12.3[a].) (a) What type of transition is this? (b) Account qualitatively for the value of the Einstein coefficient.
Solution:

(a) No electron jumps shell in this transition and the parity is unchanged, so it cannot be an E1 process. The total angular momentum J changes by two, which excludes M1 transitions and points to an electric quadrupole process. Spin is preserved, and we thus conclude that the transition is an E2 process.
(b) The transition has E2 character and will therefore have a much smaller A coefficient than allowed E1 transitions (typically 10^6–$10^9\,s^{-1}$ at optical frequencies). On the other hand, it is a spin-allowed process and is therefore faster than the spin-forbidden $^1D_2 \to {}^3P_J$ transitions discussed above, which have A values in the range 10^{-6}–$10^{-2}\,s^{-1}$. The value of $1.71\,s^{-1}$ lies between these two limits.

12.2.3 Spectral Regions

Astronomers make observations across the full electromagnetic spectrum, from radio and microwave frequencies, through the infrared, visible, and ultraviolet spectral regions, and ultimately to X-ray and γ-ray frequencies. Astronomy began, of course, by measurements at visible frequencies, where the color of the star relates to its surface temperature. The development of detectors that are sensitive to radiation outside the visible spectrum, together with the deployment of telescopes above the atmosphere (which is opaque to many wavelengths), opened new windows on the cosmos. A general point can be made that radiation at low frequencies comes from cold regions, whereas high frequencies come from hot regions. We would therefore expect radio and microwave astronomy to be most useful for looking at cold, interstellar regions, while X-ray astronomy will give useful information at very hot regions, for example in a stellar corona.

We have seen throughout this book that different types of transitions have frequencies in different spectral bands. The transitions of the valence electrons of atoms in neutral or low ionization states generally occur in the infrared, visible, or ultraviolet spectral regions, while the transitions of inner-shell electrons in heavy atoms occur at X-ray frequencies. However, the presence of very high temperatures in astrophysics (e.g., $\sim 10^6$ K in the solar corona) can produce highly ionized atoms, where the reduced screening of the nuclear charge increases the frequency of valence shell transitions in proportion to Z_{eff}^2. For example, the wavelength of the $n = 2 \rightarrow 1$ transition of the hydrogenic atom $\mathrm{A}^{(Z-1)+}$ is less than 1Å for $Z > 35$. At the lower end of the spectrum, the hydrogen hyperfine line at 21 cm is much studied, as well as microwave- and radio-frequency waves emitted as electrons cascade through highly excited states. See Sections 12.4.2 and 12.4.3.

12.2.4 Doppler Shifts

The wavelength of the light emitted by a moving atom is shifted by the Doppler effect. If the velocity of the atom is in the nonrelativistic range (i.e., $v \ll c$), the shifted wavelength is given by:

$$\lambda' = \lambda \left(1 - \frac{v}{c}\right), \tag{12.2}$$

where v is the velocity component towards the observer. When the velocity is higher, this is modified by relativistic corrections, to:

$$\lambda' = \lambda \left(\frac{c - v}{c + v}\right)^{1/2}. \tag{12.3}$$

It is easy to show that Eq. (12.3) reduces to 12.2 when $v \ll c$, and in both cases a blue or redshift is observed depending, respectively, on whether the source is moving toward or away from the observer. A measurement of the shifted wavelength therefore enables the velocity component of the source relative to the Earth to be determined.

In astronomy it is standard practice to define the Doppler shift in terms of the redshift z defined as:

$$z = \frac{\lambda'}{\lambda} - 1 . \tag{12.4}$$

Sources moving away from the observer therefore have positive z values.

Example 12.2 The wavelength of the hydrogen Balmer line at 656.3 nm emitted by a galaxy is measured to be 660 nm. What is the redshift parameter and the velocity of the star relative to Earth?

Solution: The redshift is found by substituting into Eq. (12.4):

$$z = \frac{\lambda'}{\lambda} - 1 = \frac{660}{656.3} - 1 = +0.00564 .$$

The velocity is found by substituting into Eq. (12.2):

$$\frac{\lambda'}{\lambda} = 1 + z = 1.00564 = \left(1 - \frac{v}{c}\right) .$$

Hence $v = -0.00564c = -1.69 \times 10^6$ m/s. The galaxy is moving *away* from the Earth.

Example 12.3 The wavelength of the hydrogen $n = 1 \rightarrow 2$ line at 121.6 nm emitted by a galaxy is measured to be 404 nm. What is the velocity of the galaxy relative to the Earth?

Solution: In this case, the Doppler shift is very large, which means that the galaxy is moving at a relativistic velocity. We therefore have to use Eq. (12.3) instead of Eq. (12.2). We can square Eq. (12.3) and rearrange to obtain:

$$\frac{v}{c} = -\frac{(\lambda'/\lambda)^2 - 1}{(\lambda'/\lambda)^2 + 1} = -\frac{(404/121.6)^2 - 1}{(404/121.6)^2 + 1} = -0.834 .$$

This implies that the galaxy is receding at a velocity $0.834c$. The redshift is $z = (400/121.6) - 1 = +2.29$.

12.3 Information Gained from Analysis of Astrophysical Spectra

The work of astronomers involves careful analysis of the spectra emitted by sources throughout the universe. It is not possible here to give a comprehensive list of the wealth of information that can be obtained from the spectra, but a few useful general remarks can be made.

Composition and Abundance

Every element has a unique spectrum, both in its neutral form and in its various ionized states. This fact provides a method for determining the composition of astronomical sources such as stars, galaxies, and nebulae. In the case of stars, dark absorption lines are observed on top of the continuum caused by black-body emission, with the spectrum of the sun with its Fraunhofer lines being the classic example. (See Figure 12.2.) Following pioneering work by Cecilia Payne-Gaposchkin in the 1920s, it was realized that analysis of the lines provides information about the elements that are present in the star, their ionization state, and their relative abundance.

An interesting historical example relates to the discovery of helium. Most of the absorption lines observed in the solar spectrum could be matched up with known spectra, with the strongest ones being attributed to hydrogen, which is the most abundant element in the sun. However, the line at 587.49 nm could not be explained, and so was attributed to a new, unknown element by Norman Lockyer in 1868. The element was named helium, after *helios* in Greek, meaning the sun. It was only several years later that helium was isolated on Earth, and the mystery line at 587.49 nm confirmed as originating from the second element of the periodic table. We now know that helium is present in large quantities in the sun as the product of hydrogen fusion.

Temperature

Information about the temperature T of an astronomical source can be obtained by a number of methods. The relative strengths of different absorption and emission lines is determined by the occupation of different levels, which in turn depends on the temperature. In thermal equilibrium, the probability that a level is occupied is given by Boltzmann's law:

$$p_i = \frac{1}{\mathcal{Z}} g_i \exp\left(-\frac{\Delta E_i}{k_{\mathrm{B}} T}\right), \tag{12.5}$$

where g_i is the degeneracy, ΔE_i is its energy relative to the ground state, and $\mathcal{Z}$ is the partition function given by:

$$\mathcal{Z} = \sum_i g_i \exp\left(-\frac{\Delta E_i}{k_B T}\right) . \tag{12.6}$$

The establishment of thermal equilibrium requires energy exchange between the atoms, and this cannot automatically be assumed, as it usually can be in normal laboratory conditions. For example, in low-density media such as nebulae, the atoms might interact so infrequently that equilibrium is not established. However, if the atoms are in equilibrium, then the occupancies of the levels will obey Eq. (12.5).

As an example, consider the hydrogen Balmer lines that have the $n = 2$ state as the lower level. The observation of the line in absorption requires that there should be a significant population in the $n = 2$ level. The occupancies of the $n = 2$ and $n = 1$ levels are equal at temperatures approaching 10^5 K (see Exercise 12.3), but at this high temperature, there is also a very high probability that the atom will be ionized. The Balmer lines are therefore strongest at a lower temperature of $\sim$10,000 K, where the population ratio is $\sim 10^{-5}$. Such temperatures occur in A-type stars such as Sirius and Vega. The large abundance of hydrogen makes the Balmer absorption lines detectable, despite the small $n = 2 : 1$ population ratio.

The point about there being an optimal temperature for the observation of the Balmer lines is an example of the way spectra change as their ionization state changes. As T increases, the excited states get occupied, leading to an increase in the intensity of the appropriate spectral lines. However, the probability of ionization also increases, as determined by the Saha equation given in Eq. (12.1), and this ultimately leads to a decrease in the intensity of the line. The process then repeats itself for the lines of the next ionization state as T increases further. The intensity of a particular line of each ionization state therefore peaks at a characteristic temperature, and the observation of specific lines and analysis of their relative intensity ratio enables T to be determined.

Motion

The motion of an astronomical source can be detected through the Doppler shift of spectral lines, as discussed in Section 12.2.4. This technique was used, famously, by Edwin Hubble in 1929 to measure the velocity of a large number of galaxies by analysis of Doppler-shifted hydrogen lines. He concluded that all the galaxies are receding relative to our own, and hence that the universe is expanding.

Many galaxies rotate, and this leads to a spread of Doppler shifts being observed. Analysis of the Doppler shifts of the hydrogen 21 cm line (see Section 12.4.2) from rotating galaxies has given strong evidence for the existence of dark matter in the universe since the 1960s.

In more recent years, an interesting application of the method has been in the discovery of exoplanets. The center of mass of a planetary system lies close to the star, which comprises the bulk of the mass. However, it does not coincide exactly, and the presence of orbiting planets causes the star itself to orbit about the combined center of mass. This motion can be detected by careful analysis of the spectral lines, enabling the presence of planets to be deduced from the observation of periodic Doppler shifts.

Magnetic Fields

The magnetic field that an atom experiences can be deduced by observing the splitting of spectral lines. In the sun, the typical field strengths are quite low ($B \sim 10^{-3}$ T), and so the weak-field Zeeman limit is appropriate. At such low fields, the Zeeman splitting is smaller than the Doppler linewidth, and so all that is observed is a slight additional broadening.

At the other extreme, the magnetic field in some astronomical sources can be so large that the strong-field Paschen–Back limit is reached. (See Section 8.1.3.) For example, the magnetic field strength of white dwarf stars can be ~ 100 T, which is well into the strong-field limit for hydrogen, where the spin-orbit coupling is small. (See Figure 7.3.) Fields of this magnitude are larger than those found in superconducting magnets, and can only be reached in a few specialized laboratories around the world that develop pulsed magnets.

Tests of Fundamental Theories

It is taken for granted that the laws of physics are the same throughout the whole universe at all times, but astrophysics provides a means to test this hypothesis. Hubble's law of the expanding universe (i.e., velocity proportional to distance) implies that the spectral lines observed from distant objects will be shifted compared to those on Earth. Since all the lines are shifted in the same way by the Doppler effect, the rest-frame frequencies of the transitions can be deduced. It is found that the hydrogen transitions in very distant astronomical sources are identical to those on Earth to within experimental error. This implies that all the laws that went in to the derivation of the quantized energies of hydrogen (e.g., Coulomb forces, quantum mechanics), and also the values of the fundamental constants (m_e, ϵ_0, e, h), are the same throughout the universe. Moreover, since the radiation from such distant objects takes billions of years to reach the Earth, it also implies that the fundamental laws and constants are independent of time.

One particular question that these experiments seek to answer is whether the dimensionless fine-structure constant $\alpha = e^2/2\epsilon_0 hc \approx 1/137$ has changed during the lifetime of the universe. This is a very active research area, and current best estimates set an upper limit of the fractional change at $\lesssim 10^{-5}$ over the last 10–12 billion years.

12.4 Hydrogen Spectra

Hydrogen comprises about 90% of the atomic matter in the universe by number (75% by mass), making it central to astronomical spectroscopy. On Earth, hydrogen would usually be found in the molecular form (H_2), but this is frequently not the case in astrophysics. In hot regions inside stars, the molecules are dissociated, while in the rarefied conditions in nebulae, the probability of finding another atom to form a molecule is low. Hence the spectra of *atomic* hydrogen is prevalent in astrophysics.

12.4.1 Optical Frequency Transitions

Hydrogen can be identified both from absorption or emission lines. The wavelengths of the transitions are given by the standard hydrogenic formula (see Eq. [2.23]):

$$\frac{1}{\lambda} = \frac{m}{m_e} R_\infty \left(\frac{1}{n_1^2} - \frac{1}{n_2^2} \right), \tag{12.7}$$

where m is the reduced mass, and R_∞ is the Rydberg constant (109,737 cm^{-1}). This formula is accurate to about four significant figures in the absence of fine-structure corrections. The lines are named according to the historical nomenclature given in Table 12.2. Hence the Lyα line is $1 \leftrightarrow 2$, Lyβ is $1 \leftrightarrow 3$, Hα is $2 \leftrightarrow 3$, Hγ is $2 \leftrightarrow 5$, etc.

Hydrogen absorption spectra are observed when light generated in the hot core of stars passes through cooler atmospheres containing large amounts of hydrogen. The observation of higher absorption series (e.g., Pfund lines) implies that a significant number of atoms are in excited states, and hence that the atoms are hot. In fact, as noted in Section 12.3, the analysis of the intensity

Table 12.2 *Nomenclature for hydrogen lines. Lines in which* n *changes by 1, 2, 3, ... are labeled* α, β, γ, *... respectively.*

Series	Abbreviation	n_1	n_2	Spectral region
Lyman	Ly	1	≥ 2	Ultraviolet
Balmer	H	2	≥ 3	Visible
Paschen	P	3	≥ 4	Infrared
Brackett	Br	4	≥ 5	Infrared
Pfund	Pf	5	≥ 6	Infrared
Humphreys	Hu	6	≥ 7	Infrared

ratios of the different lines gives important information about the temperature of the atmosphere.

Emission lines can be produced by a number of different mechanisms, one of which being the recombination process in H II regions, where many of the hydrogen atoms are ionized. (The name H II refers to the H^+ ion; see Section 1.2.) Recombination radiation is generated when the H^+ ions (i.e., bare protons) recapture electrons, with photons being emitted as the electrons relax to the ground state. (See Figure 12.1.) The first Balmer line Hα occurs at 656.3 nm and is responsible for the red color of the solar chromosphere that is visible during an eclipse. The observation of the Balmer emission lines requires occupancy of states above the $n = 2$ shell, which is perfectly feasible at the high temperatures present in the chromsphere (~6,000–20,000 K). The red Hα line can also be seen in photographs of H II regions (e.g., the Orion nebula).

The formula in Eq. (12.7) describes discrete spectral lines. There is, however, one important decay mechanism that leads to the emission of a continuum of radiation. This is the decay of the 2s excited state. E1 transitions to the 1s ground state are forbidden by the parity and Δl selection rules, and the most efficient decay channel is by two-photon emission. In this process, two photons are emitted at the same time, subject to energy conservation, as shown in Figure 12.5. The energies of the two photons must therefore satisfy:

$$\hbar\omega_1 + \hbar\omega_2 = E_{2s} - E_{1s} = 0.75R_{\mathrm{H}} = 10.2\,\mathrm{eV}\,. \tag{12.8}$$

Any combination of frequencies $\{\omega_1, \omega_2\}$ compatible with Eq. (12.8) is allowed, and this produces a continuous emission spectrum. The probability of two-photon decay is low, as it proceeds via a *virtual* intermediate state (i.e., not a real state). This is shown by the fact that the 2s decay rate of $8.2\,\mathrm{s}^{-1}$ (lifetime = 0.12 s) is about eight orders of magnitude slower than for the 2p levels, which

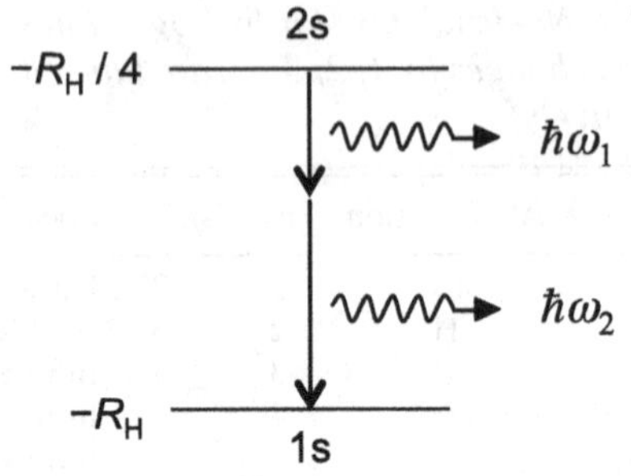

Figure 12.5 Decay of the 2s excited state of hydrogen to the 1s ground state by two-photon emission. The energies of the two photons must add up to the energy difference of the 2s and 1s states, namely $3R_H/4$.

have allowed E1 transitions to the 1s ground state. The low emission rate makes the two-photon continuum hard to detect in normal laboratory conditions, as it is very easy for the atom to scatter to the nearly degenerate 2p states in a collision, and then decay radiatively to the ground state by the allowed E1 channel. However, the 2s $\rightarrow$ 1s two-photon continuum can be observed from rarefied astrophysical environments such as nebulae, where collisions are improbable.

12.4.2 Radio-Frequency Transitions

A whole branch of astronomy focuses on radio-frequency emitters. Among these, the hydrogen 21 cm line ($\nu = 1.42\,\text{GHz}$) is by far the most important. This wavelength corresponds to the hyperfine transition between the $F = 1$ and 0 states of the 1s $^2S_{1/2}$ ground state. (See Section 7.8.2.) The emission of 21 cm radiation from interstellar hydrogen was predicted by Hendrik van de Hulst in 1944, and then observed by Harold Ewen and Edward Purcell in 1951. The transition has an extremely low Einstein A coefficient of $2.9 \times 10^{-15}\,\text{s}^{-1}$ (lifetime $>$ 10 million years) due to its M1 nature and the ν^3 scaling of spontaneous emission (see Eq. [9.11]). This means that, on average, only one out of $\sim 3 \times 10^{14}$ atoms will emit per second. Nevertheless, the transition is easily observed due to the negligible collision probability in the interstellar regions, and the fact that the total number of atoms being observed is immense. As mentioned in Section 12.3, the analysis of the 21 cm line from rotating galaxies was important for the discovery of dark matter.

Another important source of radio-frequency radiation is transitions between high n states, as illustrated in Figure 12.6. These can be part of

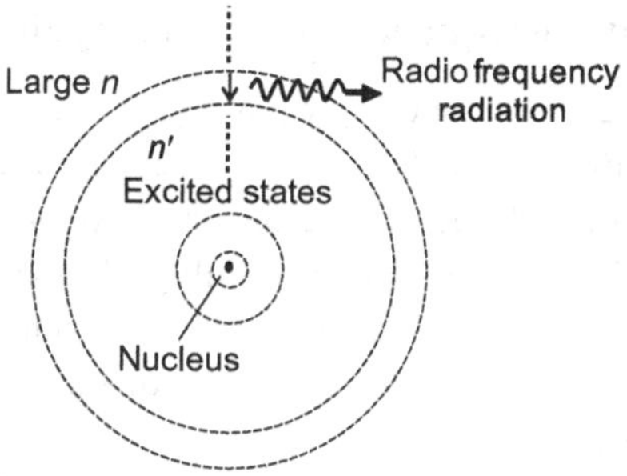

Figure 12.6 Radio-frequency transitions in hydrogen between highly excited states with large values of n.

the recombination radiation emitted as the electrons cascade through the very highly excited states of the hydrogen atoms. On inserting $n_1 = n$ and $n_2 = n + \Delta n$ into Eq. (2.22), with $n \gg \Delta n$, we find:

$$h\nu = R_{\mathrm{H}} \left| \frac{1}{n^2} - \frac{1}{(n+\Delta n)^2} \right| \approx R_{\mathrm{H}} \frac{2|\Delta n|}{n^3} \,. \tag{12.9}$$

The value of $R_{\mathrm{H}}/h \equiv (m/m_{\mathrm{e}})cR_{\infty}$ is 3.288×10^{15} Hz, where m is the reduced mass of hydrogen. Hence we find $\nu = 6.479$ GHz for $n = 100$ and $\Delta n = 1$. The Einstein A coefficient for an E1 transition scales as ν^3 (see Eq. [9.11]), and so it is to be expected that the transition rates for these radio-frequency transitions will be slow compared to E1 transitions at optical frequencies, with typical values lying in the range $\sim 10^{-3}$–10^{1} s^{-1}. The transitions are generally observed as recombination radiation from H II regions, i.e., regions where the hydrogen atoms are predominantly ionized.

A simple criterion based on the atom density can be given as to whether it might be possible to observe a particular radio-frequency recombination line or not. We see from Eq. (2.17) that the atomic Bohr radius is equal to $n^2 a_{\mathrm{H}}$, where $a_{\mathrm{H}} = (m_{\mathrm{e}}/m)a_0 \approx a_0$ is the Bohr radius for the hydrogen ground state. The volume per atom is thus approximately $V_n \sim 4\pi r_n^3/3$, and the nth shell will then be stable against collisions when the atom density $N < 1/V_n$. For $n = 100$, we have $r_n = 5.29 \times 10^{-7}$ m, $V_n \sim 6.2 \times 10^{-19}$ m^3, and hence $N < 1.6 \times 10^{18}$ m^{-3}. This can be compared, for example, to the $\sim 10^{25}$ m^{-3} particle density in the earth's atmosphere at sealevel, giving an idea of how rarefied the medium (e.g., nebula, interstellar space) must be to allow the observation of these transitions.

12.4.3 Radio-Frequency Spectra of Rydberg Atoms

We noted in Section 2.4 that the energies of highly excited states of all atoms converge to the hydrogen limit. This happens because the valence electron is in a large radius orbit, with the other electrons in tightly bound states close to the nucleus. (See Figure 2.6.) The probability of inner shell penetration is then small, so that the system reduces to the hydrogenic case with a single electron orbiting a positive charge at the core of the atom. Such atoms are called Rydberg atoms.

The frequency of the radio-frequency transitions can be worked out from the standard hydrogen formulae (see Eqs. [2.9] and [2.10]) in the same limit as Eq. (12.9), giving

$$\nu = \frac{m}{m_e} Z_{\text{eff}}^2 c R_\infty \left| \frac{1}{n^2} - \frac{1}{(n+\Delta n)^2} \right| \approx \frac{m}{m_e} Z_{\text{eff}}^2 c R_\infty \frac{2|\Delta n|}{n^3} . \tag{12.10}$$

For a neutral Rydberg atom, we have a nucleus with charge $+Ze$ surrounded by $(Z-1)$ inner shell electrons, giving $Z_{\text{eff}} = 1$. The frequency of the radio frequency lines is then the same as hydrogen, apart from the reduced mass factor. This can gives shifts of up to 0.05%, which are easily detectable by accurate radio-frequency spectroscopy. The size of the shift enables the mass of the atom to be determined, giving important clues to help identify the element responsible for the line.

12.5 Helium Spectra

Helium with $Z = 2$ comprise about 25% of the universe by mass, and is therefore very important in astrophysics. As noted in Section 12.3, its discovery was related to analysis of the solar spectrum. Helium atoms have two ionization states: He I (neutral helium) and He II (the He^+ ion). He III is the He^{2+} ion, which is a bare α particle. The spectra of He II is hydrogenic, scaled by a factor of $Z^2 = 4$ and the appropriate reduced mass factor m/m_e. (See Eqs. [2.9] and [2.10].) The latter factor enables hydrogen transitions from $n \leftrightarrow n'$ to be distinguished from helium transitions from $2n \leftrightarrow 2n'$. This comparison is shown for the hydrogen Balmer lines in Table 12.3. The $4 \leftrightarrow$ (odd n) lines for He II have no counterpart in the hydrogen spectrum, and are called the Pickering series. When Edward Pickering first observed them in the spectrum of the very hot star Zeta Puppis in 1896, he mistakenly attributed them to hydrogen, using half-integer values of n. It was not until work by Alfred Fowler in 1912 that their true origin from He^+ was established.

Table 12.3 *Comparison of the wavelength, λ, of the visible hydrogen Balmer lines and the He II Pickering series lines.*

Hydrogen		He II	
Line	λ (nm)	Line	λ (nm)
2 – 5 (Hγ)	434.13	4 – 10	433.89
		4 – 9	454.16
2 – 4 (Hβ)	486.13	4 – 8	485.95
		4 – 7	541.15
2 – 3 (Hα)	656.28	4 – 6	656.04
		4 – 5	1012.4

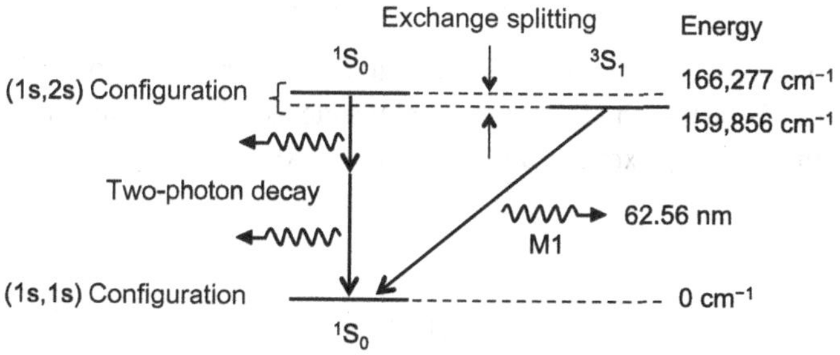

Figure 12.7 Decay of the 1s2s singlet and triplet metastable states of helium by two-photon and M1 processes, respectively. The M1 transition is spin-forbidden, and therefore has a very low probability.

The division of the spectrum of neutral helium atom (He I) into singlet and triplet transitions was described in Chapter 6. One interesting additional feature that occurs in astrophysics compared to lab-based experiments is the radiative decay of the metastable 1s2s configuration. As discussed in Section 6.5, the 1s2s configuration has two energy states, namely $^3S_1$ and $^1S_0$, split by exchange effects. The ground state $1s^2$ configuration just has a singlet $^1S_0$ level. E1 transitions from both states to the ground state are forbidden, and so the decay proceeds by higher-order process, as shown in Figure 12.7.

- The $1s2s\,^3S_1 \rightarrow 1s^2\,^1S_0$ transition at 62.56 nm is parity-forbidden for E1 transitions, and also contravenes the spin selection rule. The transition proceeds by a spin-forbidden M1 transition (see Section 12.2.2), with a small Einstein A coefficient of $1.27 \times 10^{-4}\,s^{-1}$. This gives the triplet state a lifetime of about 2.2 hours. The transition is in the extreme ultraviolet spectral region, and is not much studied. However, the transition probability increases with Z, and the equivalent transitions in helium-like ions (e.g., the 6.74 Å soft X-ray transition in Si XIII) have been observed from hot sources, e.g., supernovæ remnants.

- The $1s2s\,^1S_0 \rightarrow 1s^2\,^1S_0$ transition has $J = 0 \rightarrow 0$, and therefore cannot occur by the emission of a single photon, since that would contravene conservation of angular momentum. Therefore, the $1s2s\,^1S_0$ state decays by two-photon emission, in an analogous process to the $2s \rightarrow 1s$ transition of hydrogen discussed in Section 12.4.1. (See Figure 12.5.) Angular momentum can be conserved if the two photons have opposite spins. The photon wave-numbers are constrained by energy conservation to add up to the energy of the excited state relative to the ground state:

$$\overline{\nu}_1 + \overline{\nu}_2 = 166,277\,\text{cm}^{-1}, \quad (12.11)$$

which produces a continuum of radiation, just as for the hydrogen 2s decay. The decay is faster than that of the triplet state as it is not spin-forbidden, giving the $1s2s\,^1S_0$ excited state a lifetime of 20 ms.

Exercises

12.1 Work out the temperature at which the Doppler linewidth of the D-lines in neutral sodium (Na I) would be equal to their splitting ($17.2\,\text{cm}^{-1}$). The atomic mass of sodium is 23.0, and the center wavelength of the D-lines is 589.3 nm. Given that the first ionization energy of sodium is 5.14 eV, is this an issue?

12.2 The 4d shell of Ca^+ ($Z = 20$, ground state configuration $[\text{Ar}]\,4s^1$) lies below the 4p shell, and is therefore metastable. How does an electron in the 4d shell relax to the ground state?

12.3 Calculate the temperature at which the population of the $n = 2$ shell of hydrogen is (a) 10% of that of the ground state, and (b) equal to the population of the ground state.

12.4 What would be the wavelength of the hydrogen $n = 4 \rightarrow 2$ transition emitted by a source moving away from the Earth with speed $0.700c$?

12.5 The Lyα line ($n = 1 \leftrightarrow 2$) from a quasar is observed at 826 nm. What is the redshift of the quasar relative to the Earth?

12.6 Jupiter is the largest planet in the solar system, with a mass 1047 times smaller than the sun. Its average distance from the sun is 7.78×10^{11} m, and it has a period of 11.9 years. Imagine that you are an alien astronomer observing the solar system from a planet orbiting a neighboring star. Estimate the maximum Doppler shift of the sun's emission lines that you could observe due to the orbit of Jupiter.

12.7 Work out the allowed angular momentum states of the $1s^2 2s^2 2p^4$ ground-state configuration of oxygen. (Hint: think about holes in a filled shell.) Use Hund's rules to find out which one has the lowest energy.

12.8 The 569.4 nm line observed from the solar corona originates from the Ca XV $1s^2 2s^2 2p^2\ {}^3P_0 \leftrightarrow {}^3P_1$ transition. What type of transition is it?

12.9 A magnetic white dwarf star emits two lines at 492.4nm and 483.4 nm. These have been identified as the $\Delta M = 0$ and -1 transitions of the Hβ line, respectively. Estimate the magnetic field.

12.10 One of the photons emitted in the decay of the 2s level of hydrogen has a wavelength of 250 nm. What is the wavelength of the other photon?

12.11 Consider the $n = 51 \to 50$ recombination line of He I. Calculate the shift of this line relative to hydrogen, expressing your answer as both an absolute frequency and as a fractional shift.

12.12 The hydrogen $116 \to 114$ radio-frequency transition from a nebula is observed at 8649.1 MHz. A subsidiary peak is observed at 8652.7 MHz. Find the reduced mass of the atoms responsible for the subsidiary peak, and use this information to identify the most likely element responsible for the subsidiary peak.

12.13 The rest-frame frequencies of some hydrogen radio-frequency lines are (a) 9.173 GHz; (b) 99.22 GHz; (c) 5.675 GHz. Suggest possible origins for the transitions.

12.14 As mentioned in Section 12.4.2, the $101 \to 100$ rest-frame transition of hydrogen occurs at 6.479 GHz. What level of precision would be required to distinguish this transition from another one with $\Delta n = 2$?

Appendix A The Reduced Mass

The reduced mass is a very useful concept for dealing with the relative motion of two particles, such as the nucleus and the electron in a hydrogen atom. It allows us to separate the motion of the center of mass of the whole atom from the internal motion associated with the quantized orbits of the electron around the nucleus. It is the latter that is our concern when using the Bohr model or solving the Schrödinger equation.

Let $\boldsymbol{r}_1$ and $\boldsymbol{r}_2$ be the position vectors of the two particles, which have masses of m_1 and m_2 respectively. The center of mass coordinate $\boldsymbol{R}$ and the relative co-ordinate $\boldsymbol{r}$ are defined by:

$$\begin{aligned} M\boldsymbol{R} &= m_1\boldsymbol{r}_1 + m_2\boldsymbol{r}_2\,, \\ \boldsymbol{r} &= \boldsymbol{r}_1 - \boldsymbol{r}_2\,, \end{aligned} \tag{A.1}$$

where $M = (m_1 + m_2)$ is the total mass. As the names suggest, these give the position of the center of mass and the relative separation of the two particles respectively. The reverse relationships are:

$$\begin{aligned} \boldsymbol{r}_1 &= \boldsymbol{R} + \frac{m_2}{M}\boldsymbol{r}\,, \\ \boldsymbol{r}_2 &= \boldsymbol{R} - \frac{m_1}{M}\boldsymbol{r}\,. \end{aligned} \tag{A.2}$$

We assume that the only force acting on the particles comes from their mutual interaction, so that the potential energy only depends on their separation. For example, in the case of a hydrogen atom, the two charged particles interact through the Coulomb interaction, with $V(r) = -e^2/4\pi\epsilon_0 r$. In classical mechanics, we can write the total energy (i.e., the Hamiltonian) as the sum of the kinetic energies of the particles and the potential energy due to their mutual interaction:

$$H = \frac{1}{2}m_1 v_1^2 + \frac{1}{2}m_2 v_2^2 + V(r)\,. \tag{A.3}$$

It is easily verified from Eq. (A.2) that:

$$\begin{aligned} (\dot{\boldsymbol{r}}_1)^2 &= \dot{\boldsymbol{R}}^2 + \frac{2m_2}{M}\dot{\boldsymbol{R}}\dot{\boldsymbol{r}} + \left(\frac{m_2}{M}\right)^2 \dot{\boldsymbol{r}}^2 \\ (\dot{\boldsymbol{r}}_2)^2 &= \dot{\boldsymbol{R}}^2 - \frac{2m_1}{M}\dot{\boldsymbol{R}}\dot{\boldsymbol{r}} + \left(\frac{m_1}{M}\right)^2 \dot{\boldsymbol{r}}^2\,. \end{aligned}$$

Hence:

$$H = \frac{1}{2}m_1(\dot{r}_1)^2 + \frac{1}{2}m_2(\dot{r}_2)^2 + V(r)$$
$$= \frac{1}{2}M\dot{R}^2 + \frac{1}{2}m\dot{r}^2 + V(r)\,, \tag{A.4}$$

where the **reduced mass** $m = m_1 m_2/M$ is defined by:

$$\frac{1}{m} = \frac{1}{m_1} + \frac{1}{m_2}\,. \tag{A.5}$$

Equation A.4 shows that the energy is equal to the kinetic energy of the center of mass, plus the energy (i.e., kinetic energy plus potential energy) of the relative motion of a particle of mass m, namely the reduced mass. In other words, we can separate the motion into the free motion of the whole system, plus the internal energy in terms of the relative coordinates and the reduced mass.

In quantum mechanics, the Hamiltonian is given by:

$$\hat{H} = -\frac{\hbar^2}{2m_1}\nabla_1^2 - \frac{\hbar^2}{2m_2}\nabla_2^2 + V(r)\,, \tag{A.6}$$

where:

$$\nabla_i^2 = \frac{\partial^2}{\partial x_i^2} + \frac{\partial^2}{\partial y_i^2} + \frac{\partial^2}{\partial z_i^2}\,. \tag{A.7}$$

To transform this to the center of mass and relative coordinates, we need to work with the Cartesian co-ordinates:

$$x = x_1 - x_2\,,$$
$$X = \frac{m_1}{M}x_1 + \frac{m_2}{M}x_2\,.$$

We start by finding the first derivatives:

$$\frac{\partial}{\partial x_1} = \frac{\partial X}{\partial x_1}\frac{\partial}{\partial X} + \frac{\partial x}{\partial x_1}\frac{\partial}{\partial x} = \frac{m_1}{M}\frac{\partial}{\partial X} + \frac{\partial}{\partial x}\,,$$
$$\frac{\partial}{\partial x_2} = \frac{\partial X}{\partial x_2}\frac{\partial}{\partial X} + \frac{\partial x}{\partial x_2}\frac{\partial}{\partial x} = \frac{m_2}{M}\frac{\partial}{\partial X} - \frac{\partial}{\partial x}\,.$$

This implies that the second derivative with respect to x_1 is:

$$\frac{\partial^2}{\partial x_1^2} = \left(\frac{m_1}{M}\frac{\partial}{\partial X} + \frac{\partial}{\partial x}\right)\left(\frac{m_1}{M}\frac{\partial}{\partial X} + \frac{\partial}{\partial x}\right),$$
$$= \frac{m_1^2}{M^2}\frac{\partial^2}{\partial X^2} + 2\frac{m_1}{M}\frac{\partial^2}{\partial X \partial x} + \frac{\partial^2}{\partial x^2}\,.$$

Similarly:

$$\frac{\partial^2}{\partial x_2^2} = \frac{m_2^2}{M^2}\frac{\partial^2}{\partial X^2} - 2\frac{m_2}{M}\frac{\partial^2}{\partial X \partial x} + \frac{\partial^2}{\partial x^2}\,.$$

Therefore:

$$-\frac{\hbar^2}{2m_1}\frac{\partial^2}{\partial x_1^2} - \frac{\hbar^2}{2m_2}\frac{\partial^2}{\partial x_2^2} = -\frac{\hbar^2}{2M}\frac{\partial^2}{\partial X^2} - \frac{\hbar^2}{2m}\frac{\partial^2}{\partial x^2},$$

where m is the reduced mass defined in Eq. (A.5). Similar results can be derived for the y- and z-components, leading to:

$$\hat{H} = \hat{H}_R + \hat{H}_r, \tag{A.8}$$

where:

$$\begin{aligned}\hat{H}_R &= -\frac{\hbar^2}{2M}\nabla_R^2,\\ \hat{H}_r &= -\frac{\hbar^2}{2m}\nabla_r^2 + V(r).\end{aligned} \tag{A.9}$$

This shows that the Hamiltonian is the sum of:

- The Hamiltonian $\hat{H}_R$ of a free particle of mass M with position coordinates of the center of mass; and
- The Hamiltonian $\hat{H}_r$ that describes the relative motion of the two particles, behaving as if they had mass m, namely the reduced mass.

This is our final result. It shows that we can separate the motion of hydrogenic atoms into the motion of the center of mass, that moves freely throughout space, and the internal motion that is governed by the potential energy $V(r)$, and acts like a particle of mass m. Hence the mass m that appears in the Bohr model in Section 2.1 and in the hydrogen Schrödinger equation in Section 2.2 is the reduced mass. This separation works for any two-particle system with a central potential that depends only on the particle separation r.

Appendix B Mathematical Solutions for the Hydrogen Schrödinger Equation

This appendix deals with the more mathematical aspects of the Schrödinger equation for hydrogen that were omitted from the main discussion in Chapter 2.

B.1 The Angular Equation

The eigenfunctions of the angular momentum operator are found by solving equation 2.36, namely:

$$\hat{L}^2 F(\theta,\phi) \equiv -\hbar^2 \left[\frac{1}{\sin\theta}\frac{\partial}{\partial\theta}\left(\sin\theta\frac{\partial}{\partial\theta}\right) + \frac{1}{\sin^2\theta}\frac{\partial^2}{\partial\phi^2} \right] F(\theta,\phi) = CF(\theta,\phi)\,. \quad \text{(B.1)}$$

For reasons that will become clearer later, the constant C is usually written in the form:

$$C = l(l+1)\hbar^2\,. \quad \text{(B.2)}$$

At this stage, l can take any value, real or complex. We can separate the variables by writing:

$$F(\theta,\phi) = f(\theta)g(\phi)\,. \quad \text{(B.3)}$$

On substitution into Eq. (B.1) and canceling the common factor of $\hbar^2$, we find:

$$-\frac{1}{\sin\theta}\frac{\mathrm{d}}{\mathrm{d}\theta}\left(\sin\theta\frac{\mathrm{d}f}{\mathrm{d}\theta}\right) g - \frac{1}{\sin^2\theta} f \frac{\mathrm{d}^2 g}{\mathrm{d}\phi^2} = l(l+1)fg\,. \quad \text{(B.4)}$$

Multiply by $-\sin^2\theta/fg$ and rearrange to obtain:

$$\frac{\sin\theta}{f}\frac{\mathrm{d}}{\mathrm{d}\theta}\left(\sin\theta\frac{\mathrm{d}f}{\mathrm{d}\theta}\right) + \sin^2\theta\, l(l+1) = -\frac{1}{g}\frac{\mathrm{d}^2 g}{\mathrm{d}\phi^2}\,. \quad \text{(B.5)}$$

The left-hand side is a function of θ only, while the right-hand side is a function of ϕ only. The equation must hold for all values of θ and ϕ and hence both sides must be equal to a constant. On writing this arbitrary separation constant as m^2, we then find:

$$\sin\theta\frac{\mathrm{d}}{\mathrm{d}\theta}\left(\sin\theta\frac{\mathrm{d}f}{\mathrm{d}\theta}\right)+l(l+1)\sin^2\theta\, f=m^2f\,, \tag{B.6}$$

and:

$$\frac{\mathrm{d}^2g}{\mathrm{d}\phi^2}=-m^2g\,. \tag{B.7}$$

The equation in ϕ is easily solved to obtain:

$$g(\phi)=A\mathrm{e}^{im\phi}\,, \tag{B.8}$$

where A is a constant. The wave function must have a single value for each value of ϕ, and hence we require:

$$g(\phi+2\pi)=g(\phi)\,, \tag{B.9}$$

which implies that the separation constant m must be an integer. Using this fact in Eq. (B.6), we then have to solve:

$$\sin\theta\frac{\mathrm{d}}{\mathrm{d}\theta}\left(\sin\theta\frac{\mathrm{d}f}{\mathrm{d}\theta}\right)+[l(l+1)\sin^2\theta-m^2]f=0\,, \tag{B.10}$$

with the constraint that m must be an integer. On making the substitution $u=\cos\theta$ and writing $f(\theta)=P(u)$, Eq. (B.10) becomes:

$$\frac{\mathrm{d}}{\mathrm{d}u}\left((1-u^2)\frac{\mathrm{d}P}{\mathrm{d}u}\right)+\left[l(l+1)-\frac{m^2}{1-u^2}\right]P=0\,. \tag{B.11}$$

Equation (B.11) is known as either the Legendre equation or the associated Legendre equation, depending on whether m is zero or not. Solutions only exist if l is an integer $\geq |m|$ and $P(u)$ is a polynomial function of u. This means that the solutions to Eq. (B.10) are of the form:

$$f(\theta)=P_l^m(\cos\theta)\,, \tag{B.12}$$

where $P_l^m(\cos\theta)$ is a polynomial function in $\cos\theta$ called the (associated) **Legendre polynomial** function.

Putting this all together, we then find:

$$F(\theta,\phi)=\text{normalization constant}\times P_l^m(\cos\theta)\,\mathrm{e}^{im\phi}\,, \tag{B.13}$$

where m and l are integers, and m can have values from $-l$ to $+l$. The correctly normalized functions are called the **spherical harmonic** functions $Y_{lm}(\theta,\phi)$.

It is apparent from Eqs. (B.1) and (B.2) that the spherical harmonics satisfy:

$$\hat{\boldsymbol{L}}^2Y_{lm}(\theta,\phi)=l(l+1)\hbar^2\,Y_{lm}(\theta,\phi)\,. \tag{B.14}$$

Furthermore, on substituting from Eq. (2.40), it is also apparent that:

$$\hat{L}_zY_{lm}(\theta,\phi)=m\hbar\,Y_{lm}(\theta,\phi)\,. \tag{B.15}$$

The integers l and m that appear here are called the orbital and magnetic quantum numbers respectively. Some of the spherical harmonic functions are listed in Table 2.2. Equations B.14–B.15 show that the square of the magnitude of the angular momentum and its z-component are equal to $l(l+1)\hbar^2$ and $m\hbar$ respectively, as consistent with Figure 2.3.

B.2 The Radial Equation

The radial equation for hydrogen is given in Eq. (2.34) and reads as follows:

$$-\frac{\hbar^2}{2m}\frac{1}{r^2}\frac{\mathrm{d}}{\mathrm{d}r}\left(r^2\frac{\mathrm{d}R(r)}{\mathrm{d}r}\right)+\frac{\hbar^2 l(l+1)}{2mr^2}R(r)-\frac{Ze^2}{4\pi\epsilon_0 r}R(r)=ER(r)\,, \tag{B.16}$$

where l is an integer ≥ 0. We first put this in a more user-friendly form by introducing the dimensionless radius ρ according to:

$$\rho=\left(\frac{8m|E|}{\hbar^2}\right)^{1/2} r\,. \tag{B.17}$$

The modulus sign around E is important here because we are seeking bound solutions where E is negative. The radial equation now becomes:

$$\frac{\mathrm{d}^2R}{\mathrm{d}\rho^2}+\frac{2}{\rho}\frac{\mathrm{d}R}{\mathrm{d}\rho}+\left(\frac{\lambda}{\rho}-\frac{1}{4}-\frac{l(l+1)}{\rho^2}\right)R=0\,, \tag{B.18}$$

where:

$$\lambda=\frac{1}{4\pi\epsilon_0}\frac{Ze^2}{\hbar}\left(\frac{m}{2|E|}\right)^{1/2}\,. \tag{B.19}$$

We first consider the behavior at $\rho\to\infty$, where Eq. (B.18) reduces to:

$$\frac{\mathrm{d}^2R}{\mathrm{d}\rho^2}-\frac{1}{4}R=0\,. \tag{B.20}$$

This has solutions of $\mathrm{e}^{\pm\rho/2}$. The $\mathrm{e}^{+\rho/2}$ solution cannot be normalized and is thus excluded, which implies that $R(\rho)\sim\mathrm{e}^{-\rho/2}$.

Now consider the behavior for $\rho\to 0$, where the dominant terms in Eq. (B.18) are:

$$\frac{\mathrm{d}^2R}{\mathrm{d}\rho^2}+\frac{2}{\rho}\frac{\mathrm{d}R}{\mathrm{d}\rho}-\frac{l(l+1)}{\rho^2}R=0\,, \tag{B.21}$$

with solutions $R(\rho)=\rho^l$ or $R(\rho)=\rho^{-(l+1)}$. The latter diverges at the origin and is thus unacceptable.

The consideration of the asymptotic behaviors suggests that we should look for general solutions of the radial equation with $R(\rho)$ in the form:

$$R(\rho)=L(\rho)\,\rho^l\,\mathrm{e}^{-\rho/2}\,. \tag{B.22}$$

On substituting into Eq. (B.18) we find:

$$\frac{d^2L}{d\rho^2} + \left(\frac{2l+2}{\rho} - 1\right)\frac{dL}{d\rho} + \frac{\lambda - l - 1}{\rho}L = 0\,. \tag{B.23}$$

We now look for a series solution of the form:

$$L(\rho) = \sum_{k=0}^{\infty} a_k \rho^k\,. \tag{B.24}$$

Substitution into Eq. (B.23) yields:

$$\sum_{k=0}^{\infty}\left[k(k-1)a_k\rho^{k-2} + \left(\frac{2l+2}{\rho} - 1\right)ka_k\rho^{k-1} + \frac{\lambda - l - 1}{\rho}a_k\rho^k\right] = 0\,, \tag{B.25}$$

which can be rewritten:

$$\sum_{k=0}^{\infty}\left[(k(k-1) + 2k(l+1))a_k\rho^{k-2} + (\lambda - l - 1 - k)a_k\rho^{k-1}\right] = 0\,, \tag{B.26}$$

or alternatively:

$$\sum_{k=0}^{\infty}\left[((k+1)k + 2(k+1)(l+1))a_{k+1}\rho^{k-1} + (\lambda - l - 1 - k)a_k\rho^{k-1}\right] = 0\,. \tag{B.27}$$

This will be satisfied if:

$$((k+1)k + 2(k+1)(l+1))a_{k+1} + (\lambda - l - 1 - k)a_k = 0\,, \tag{B.28}$$

which implies:

$$\frac{a_{k+1}}{a_k} = \frac{-\lambda + l + 1 + k}{(k+1)(k+2l+2)}\,. \tag{B.29}$$

At large k we have:

$$\frac{a_{k+1}}{a_k} \sim \frac{1}{k}\,. \tag{B.30}$$

Now the series expansion of e^{ρ} is:

$$e^{\rho} = 1 + \rho + \frac{\rho^2}{2!} + \cdots \frac{\rho^k}{k!} + \cdots\,, \tag{B.31}$$

which has the same limit for a_{k+1}/a_k. With $R(\rho)$ given by Eq. (B.22), we would then have a dependence of $e^{+\rho} \cdot e^{-\rho/2} = e^{+\rho/2}$, which is unacceptable. We therefore conclude that the series expansion must terminate for some value of k. Let n_r be the value of k for which the series terminates. It then follows that $a_{n_r+1} = 0$, which implies:

$$-\lambda + l + 1 + n_r = 0\,, \qquad n_r \geq 0\,, \tag{B.32}$$

or:

$$\lambda = l + 1 + n_r\,. \tag{B.33}$$

We now introduce the **principal quantum number** n according to:

$$n = n_r + l + 1\,. \qquad \text{(B.34)}$$

It follows that:

(i) n is an integer,
(ii) $n \geq l + 1$, and
(iii) $\lambda = n$.

The first two points establish the general rules for the quantum numbers n and l. The third one fixes the energy. On inserting $\lambda = n$ into Eq. (B.19) and remembering that E is negative, we find:

$$E_n = -\frac{me^4}{(4\pi\epsilon_0)^2 2\hbar^2}\frac{Z^2}{n^2}\,. \qquad \text{(B.35)}$$

This is the usual Bohr result. The wave functions are of the form given in Eq. (B.22):

$$R(\rho) = \rho^l L(\rho)\,\mathrm{e}^{-\rho/2}\,. \qquad \text{(B.36)}$$

The polynomial series $L(\rho)$ that satisfies Eq. (B.23) is known as an **associated Laguerre function**. On substituting for ρ from Eq. (B.17) with $|E|$ given by Eq. (B.35), we then obtain:

$$R(r) = \text{normalization constant} \times \text{Laguerre polynomial in } r \times r^l \mathrm{e}^{-r/r_0}\,, \qquad \text{(B.37)}$$

with:

$$r_0 = \left(\frac{\hbar^2}{2m|E|}\right)^{1/2} = \frac{4\pi\epsilon_0\hbar^2}{me^2}\frac{n}{Z} \equiv \frac{n}{Z}a_{\mathrm{H}}\,, \qquad \text{(B.38)}$$

where a_{H} is the Bohr radius of hydrogen. Equation (B.37) is the justification for Eq. (2.50) in the main text.

Appendix C Helium Energy Integrals

The concept of exchange integrals was introduced in Section 6.4 in the discussion of the energy levels of helium. Our task here is to evaluate the three terms that appear in the gross structure energy E:

$$E = E_1 + E_2 + E_{12}\,, \tag{C.1}$$

where the energies are defined in Eqs. (6.14) and (6.15).

We restrict ourselves to configurations of the type (1s, *nl*), since these are the ones that give rise to the excited states that are observed in the optical spectra of neutral helium. From Eq. (6.5) we see that the spatial part of the wave function is given by:

$$\Psi(\boldsymbol{r}_1,\boldsymbol{r}_2) = \frac{1}{\sqrt{2}}\Big(u_{1s}(\boldsymbol{r}_1)u_{nl}(\boldsymbol{r}_2) \pm u_{nl}(\boldsymbol{r}_1)u_{1s}(\boldsymbol{r}_2)\Big),$$

where we take the plus sign for singlets with $S = 0$ and the minus sign for triplets with $S = 1$.

We first tackle E_1, with $\hat{H}_1$ defined in Eq. (6.11):

$$\begin{aligned}
E_1 &= \iint \Psi^* \,\hat{H}_1\, \Psi \,\mathrm{d}^3\boldsymbol{r}_1\mathrm{d}^3\boldsymbol{r}_2 \\
&= \frac{1}{2}\iint \Big(u_{1s}^*(\boldsymbol{r}_1)\, u_{nl}^*(\boldsymbol{r}_2) \pm u_{nl}^*(\boldsymbol{r}_1)\, u_{1s}^*(\boldsymbol{r}_2)\Big) \\
&\qquad\qquad \hat{H}_1\Big(u_{1s}^*(\boldsymbol{r}_1)\, u_{nl}^*(\boldsymbol{r}_2) \pm u_{nl}^*(\boldsymbol{r}_1)\, u_{1s}^*(\boldsymbol{r}_2)\Big)\,\mathrm{d}^3\boldsymbol{r}_1\,\mathrm{d}^3\boldsymbol{r}_2\,,
\end{aligned}$$

where the plus sign applies for singlet states and the minus sign for triplets. This splits into four integrals:

$$\begin{aligned}
E_1 &= \frac{1}{2}\iint u_{1s}^*(\boldsymbol{r}_1)u_{nl}^*(\boldsymbol{r}_2)\hat{H}_1 u_{1s}(\boldsymbol{r}_1)u_{nl}(\boldsymbol{r}_2)\mathrm{d}^3\boldsymbol{r}_1\mathrm{d}^3\boldsymbol{r}_2 \\
&\quad + \frac{1}{2}\iint u_{nl}^*(\boldsymbol{r}_1)u_{1s}^*(\boldsymbol{r}_2)\hat{H}_1 u_{nl}(\boldsymbol{r}_1)u_{1s}(\boldsymbol{r}_2)\mathrm{d}^3\boldsymbol{r}_1\mathrm{d}^3\boldsymbol{r}_2 \\
&\quad \pm \frac{1}{2}\iint u_{1s}^*(\boldsymbol{r}_1)u_{nl}^*(\boldsymbol{r}_2)\hat{H}_1 u_{nl}(\boldsymbol{r}_1)u_{1s}(\boldsymbol{r}_2)\,\mathrm{d}^3\boldsymbol{r}_1\mathrm{d}^3\boldsymbol{r}_2 \\
&\quad \pm \frac{1}{2}\iint u_{nl}^*(\boldsymbol{r}_1)u_{1s}^*(\boldsymbol{r}_2)\hat{H}_1 u_{1s}(\boldsymbol{r}_1)u_{nl}(\boldsymbol{r}_2)\mathrm{d}^3\boldsymbol{r}_1\mathrm{d}^3\boldsymbol{r}_2\,.
\end{aligned}$$

We now use the fact that $u_{nl}(\boldsymbol{r}_1)$ is an eigenstate of $\hat{H}_1$:

$$\hat{H}_1 \, u_{nl}(\boldsymbol{r}_1) = E_{nl} \, u_{nl}(\boldsymbol{r}_1) ,$$

and that $\hat{H}_1$ has no effect on $\boldsymbol{r}_2$, to obtain:

$$\begin{aligned}
E_1 &= \frac{1}{2}E_{1s} \int u_{1s}^*(\boldsymbol{r}_1)u_{1s}(\boldsymbol{r}_1)\mathrm{d}^3\boldsymbol{r}_1 \int u_{nl}^*(\boldsymbol{r}_2)u_{nl}(\boldsymbol{r}_2)\mathrm{d}^3\boldsymbol{r}_2 \\
&\quad + \frac{1}{2}E_{nl} \int u_{nl}^*(\boldsymbol{r}_1)u_{nl}(\boldsymbol{r}_1)\mathrm{d}^3\boldsymbol{r}_1 \int u_{1s}^*(\boldsymbol{r}_2)u_{1s}(\boldsymbol{r}_2)\mathrm{d}^3\boldsymbol{r}_2 \\
&\quad \pm \frac{1}{2}E_{nl} \int u_{1s}^*(\boldsymbol{r}_1)u_{nl}(\boldsymbol{r}_1)\,\mathrm{d}^3\boldsymbol{r}_1 \int u_{nl}^*(\boldsymbol{r}_2)u_{1s}(\boldsymbol{r}_2)\mathrm{d}^3\boldsymbol{r}_2 \\
&\quad \pm \frac{1}{2}E_{1s} \int u_{nl}^*(\boldsymbol{r}_1)u_{1s}(\boldsymbol{r}_1)\mathrm{d}^3\boldsymbol{r}_1 \int u_{1s}^*(\boldsymbol{r}_2)u_{nl}(\boldsymbol{r}_2)\mathrm{d}^3\boldsymbol{r}_2 \\
&= \frac{1}{2}E_{1s} + \frac{1}{2}E_{nl} + 0 + 0 .
\end{aligned}$$

The integrals in the first two terms are unity because the u_{nl} wave functions are normalized, while the last two terms are zero by orthogonality.

The evaluation of E_2 follows a similar procedure:

$$\begin{aligned}
E_2 &= \iint \Psi^* \hat{H}_2 \Psi \mathrm{d}^3\boldsymbol{r}_1 \mathrm{d}^3\boldsymbol{r}_2 , \\
&= +\frac{1}{2} \iint u_{1s}^*(\boldsymbol{r}_1)u_{nl}^*(\boldsymbol{r}_2)\hat{H}_2 u_{1s}(\boldsymbol{r}_1)u_{nl}(\boldsymbol{r}_2)\mathrm{d}^3\boldsymbol{r}_1\mathrm{d}^3\boldsymbol{r}_2 \\
&\quad +\frac{1}{2} \iint u_{nl}^*(\boldsymbol{r}_1)u_{1s}^*(\boldsymbol{r}_2)\hat{H}_2 u_{nl}(\boldsymbol{r}_1)u_{1s}(\boldsymbol{r}_2)\mathrm{d}^3\boldsymbol{r}_1\mathrm{d}^3\boldsymbol{r}_2 \\
&\quad \pm\frac{1}{2} \iint u_{1s}^*(\boldsymbol{r}_1)u_{nl}^*(\boldsymbol{r}_2)\hat{H}_2 u_{nl}(\boldsymbol{r}_1)u_{1s}(\boldsymbol{r}_2)\mathrm{d}^3\boldsymbol{r}_1\mathrm{d}^3\boldsymbol{r}_2 \\
&\quad \pm\frac{1}{2} \iint u_{nl}^*(\boldsymbol{r}_1)u_{1s}^*(\boldsymbol{r}_2)\hat{H}_2 u_{nl}(\boldsymbol{r}_1)u_{1s}(\boldsymbol{r}_2)\mathrm{d}^3\boldsymbol{r}_1\mathrm{d}^3\boldsymbol{r}_2 \\
&= +\frac{1}{2}E_{nl} + \frac{1}{2}E_{1s} + 0 + 0 .
\end{aligned}$$

Finally, we have to evaluate the Coulomb repulsion term, with $\hat{H}_{12}$ defined in Eq. (6.12):

$$\begin{aligned}
E_{12} &= \iint \Psi^* \, \hat{H}_{12} \, \Psi \, \mathrm{d}^3\boldsymbol{r}_1\mathrm{d}^3\boldsymbol{r}_2 \\
&= \iint \Psi^* \, \frac{e^2}{4\pi\epsilon_0 r_{12}} \, \Psi \, \mathrm{d}^3\boldsymbol{r}_1\mathrm{d}^3\boldsymbol{r}_2 \\
&= \frac{1}{2} \iint \left(u_{1s}^*(\boldsymbol{r}_1)\, u_{nl}^*(\boldsymbol{r}_2) \pm u_{nl}^*(\boldsymbol{r}_1)\, u_{1s}^*(\boldsymbol{r}_2) \right) \frac{e^2}{4\pi\epsilon_0 r_{12}} \\
&\qquad\qquad \left(u_{1s}^*(\boldsymbol{r}_1)\, u_{nl}^*(\boldsymbol{r}_2) \pm u_{nl}^*(\boldsymbol{r}_1)\, u_{1s}^*(\boldsymbol{r}_2) \right) \mathrm{d}^3\boldsymbol{r}_1 \, \mathrm{d}^3\boldsymbol{r}_2 ,
\end{aligned}$$

where again the plus sign applies for singlet states and the minus sign for triplets. The four terms are:

$$
\begin{aligned}
E_{12} = &+\frac{1}{2}\frac{e^2}{4\pi\epsilon_0}\iint u_{1s}^*(\boldsymbol{r}_1)u_{nl}^*(\boldsymbol{r}_2)\frac{1}{r_{12}}u_{1s}(\boldsymbol{r}_1)u_{nl}(\boldsymbol{r}_2)\mathrm{d}^3\boldsymbol{r}_1\mathrm{d}^3\boldsymbol{r}_2 \\
&+\frac{1}{2}\frac{e^2}{4\pi\epsilon_0}\iint u_{nl}^*(\boldsymbol{r}_1)u_{1s}^*(\boldsymbol{r}_2)\frac{1}{r_{12}}u_{nl}(\boldsymbol{r}_1)u_{1s}(\boldsymbol{r}_2)\mathrm{d}^3\boldsymbol{r}_1\mathrm{d}^3\boldsymbol{r}_2 \\
&\pm\frac{1}{2}\frac{e^2}{4\pi\epsilon_0}\iint u_{1s}^*(\boldsymbol{r}_1)u_{nl}^*(\boldsymbol{r}_2)\frac{1}{r_{12}}u_{nl}(\boldsymbol{r}_1)u_{1s}(\boldsymbol{r}_2)\mathrm{d}^3\boldsymbol{r}_1\mathrm{d}^3\boldsymbol{r}_2 \\
&\pm\frac{1}{2}\frac{e^2}{4\pi\epsilon_0}\iint u_{nl}^*(\boldsymbol{r}_1)u_{1s}^*(\boldsymbol{r}_2)\frac{1}{r_{12}}u_{1s}(\boldsymbol{r}_1)u_{nl}(\boldsymbol{r}_2)\mathrm{d}^3\boldsymbol{r}_1\mathrm{d}^3\boldsymbol{r}_2 \\
= &+\frac{D}{2}+\frac{D}{2}\pm\frac{J}{2}\pm\frac{J}{2},
\end{aligned}
$$

where D and J are given by Eqs. (6.19) and (6.20) respectively.

The total energy is thus given by:

$$E = E_{1s} + E_{nl} + D \pm J,$$

where the plus sign applies to singlets and the minus sign to triplets. If we assume hydrogenic wave functions, this becomes:

$$E = -4R_{\mathrm{H}} - 4R_{\mathrm{H}}/n^2 + D \pm J.$$

However, this is only an approximation, as the actual potential experienced by the electrons will depart from the strict $1/r$ dependence of hydrogen.

Appendix D Perturbation Theory of the Stark Effect

This appendix gives an explanation of the quadratic and linear Stark shifts by perturbation theory. The basic phenomena were described in Section 8.4 of Chapter 8. We focus specifically on the quadratic shift in an alkali atom, and the linear shift in hydrogen.

D.1 Quadratic Stark Shifts

The energy shift caused by the quadratic Stark effect can be evaluated by applying perturbation theory. The perturbation to the energy of the electrons by a field $\boldsymbol{\mathcal{E}}$ is of the form:

$$\begin{aligned} H' &= -\sum_i (-e\boldsymbol{r}_i) \cdot \boldsymbol{\mathcal{E}}\,, \\ &= e\mathcal{E} \sum_i z_i\,, \end{aligned} \tag{D.1}$$

where the field is assumed to point in the $+z$ direction. This is just the sum of the interaction energies of the electron dipoles with the electric field. In principle, the sum is over all the electrons; but in practice, we need only consider the valence electrons, because the electrons in closed shells are very strongly bound to the nucleus and are therefore very hard to perturb. In writing Eq. (D.1), we take, as always, $\boldsymbol{r}_i$ to be the relative displacement of the electron with respect to the nucleus.

For simplicity, we shall just consider the case of alkali atoms which possess only one valence electron. In this case, the perturbation to the valence electron caused by the field reduces to:

$$H' = e\mathcal{E}z\,. \tag{D.2}$$

The *first-order* energy shift is given by:

$$\Delta E = \langle \psi | H' | \psi \rangle = e\mathcal{E} \langle \psi | z | \psi \rangle\,, \tag{D.3}$$

where:

$$\langle\psi|z|\psi\rangle = \iiint_{\text{all space}} \psi^* z\, \psi\, \mathrm{d}^3\boldsymbol{r}\,. \tag{D.4}$$

Now unperturbed atomic states have definite parities. (See discussion in Section 3.4.) The product $\psi^*\psi = |\psi^2|$ is therefore an even function, while z is an odd function. It is then apparent that:

$$\langle\psi|z|\psi\rangle = \iiint_{\text{all space}} (\text{even function}) \times (\text{odd function})\, \mathrm{d}^3\boldsymbol{r} = 0\,.$$

The first-order energy shift is therefore zero, which explains why the energy shift is quadratic in the field, rather than linear.

The quadratic energy shift can be calculated by *second-order* perturbation theory. In general, the energy shift of the ith state predicted by second-order perturbation theory is given by:

$$\Delta E_i = \sum_{j \neq i} \frac{|\langle\psi_i|H'|\psi_j\rangle|^2}{E_i - E_j}\,, \tag{D.5}$$

where the summation runs over all the other states of the system, and E_i and E_j are the unperturbed energies of the states. The condition of validity is that the magnitude of the perturbation, namely $|\langle\psi_i|H'|\psi_j\rangle|$, should be small compared to the unperturbed energy splittings. For the Stark shift of the valence electron of an alkali atom, this becomes:

$$\Delta E_i = e^2 \mathcal{E}^2 \sum_{j \neq i} \frac{|\langle\psi_i|z|\psi_j\rangle|^2}{E_i - E_j}\,. \tag{D.6}$$

We see immediately that the shift is expected to be quadratic in the field, which is indeed the case for most atoms.

As a specific example, we consider sodium, which has a single valence electron in the 3s shell. We first consider the ground state 3s $^2\mathrm{S}_{1/2}$ term. The summation in Eq. (D.6) runs over all the excited states of sodium, namely the 3p, 3d, 4s, 4p, ... states. Now in order that the matrix element $\langle\psi_i|z|\psi_j\rangle$ should be non-zero, it is apparent that the states i and j must *opposite* parities. In this case, we would have:

$$\langle\psi_i|z|\psi_j\rangle = \iiint_{\text{all space}} (\text{even parity}) \times (\text{odd parity}) \times (\text{odd parity})\, \mathrm{d}^3\boldsymbol{r} \neq 0\,,$$

or

$$\langle\psi_i|z|\psi_j\rangle = \iiint_{\text{all space}} (\text{odd parity}) \times (\text{odd parity}) \times (\text{even parity})\, \mathrm{d}^3\boldsymbol{r} \neq 0\,,$$

since the integrand is an even function in both cases. On the other hand, if the states have the same parities, we have:

$$\langle\psi_i|z|\psi_j\rangle = \iiint_{\text{all space}} (\text{even parity}) \times (\text{odd parity}) \times (\text{even parity})\, \mathrm{d}^3\boldsymbol{r} = 0\,,$$

or

$$\langle\psi_i|z|\psi_j\rangle = \iiint_{\text{all space}} (\text{odd parity}) \times (\text{odd parity}) \times (\text{odd parity})\, \mathrm{d}^3\boldsymbol{r} = 0\,,$$

since both integrands are odd functions. Since the parity varies as $(-1)^l$, the s and d states do not contribute to the Stark shift of the 3s state, and the summation in Eq. (D.6) is only over the p and f excited states. Owing to the energy difference factor in the denominator, the largest perturbation to the 3s state will arise from the first excited state, namely the 3p state. Since this lies above the 3s state, the energy difference in the denominator is negative, and the energy shift is therefore negative. Indeed, it is apparent that the quadratic Stark shift of the ground state of an atom will always be negative, since the denominator will be negative for all the available states of the system. This implies that the Stark effect will always correspond to a *redshift* for the ground-state level.

There is no easy way to calculate the size of the energy shift, but we can give a rough order of magnitude estimate. If we neglect the contributions of the odd parity excited states above the 3p state, the energy shift will be given by:

$$\Delta E_{3s} \approx -e^2\mathcal{E}^2 \frac{|\langle\psi_{3s}|z|\psi_{3p}\rangle|^2}{E_{3p} - E_{3s}}\,.$$

The expectation value of z over the atom must be smaller than a, where a is the atomic radius of sodium, namely 0.18 nm. Hence with $E_{3p} - E_{3s} = 2.1$ eV, we then have:

$$\Delta E_{3s} \lesssim -\frac{e^2 a^2}{E_{3p} - E_{3s}}\mathcal{E}^2\,.$$

On introducing the atomic polarizability defined in Eq. (8.32), we then find that $\alpha_{3s} \lesssim 3.2 \times 10^{-20}\,\mathrm{eV\,m^2\,V^{-2}}$. This predicts a shift of $\lesssim -1 \times 10^{-5}$ eV $(-0.08\,\mathrm{cm}^{-1})$ in a field of 250 kV/cm, which compares reasonably well with the experimental value of -0.6×10^{-5} eV $(-0.05\,\mathrm{cm}^{-1})$.

The order of magnitude calculation given above can also provide a useful estimation of the field strength at which the second-order perturbation approximation breaks down. This will occur when the magnitude of the perturbation become comparable to the unperturbed energy splitting, that is when:

$$e\mathcal{E}|\langle\psi_{3s}|z|\psi_{3p}\rangle| \sim (E_{3p} - E_{3s})\,.$$

On setting $|\langle\psi_{3s}|z|\psi_{3p}\rangle| = a$ as before, we find $\mathcal{E} \sim 10^{10}$ V/m, which is an extremely large field. The second-order perturbation approach will therefore be a good approximation in most practical situations.

Now consider the Stark shift of the 3p state. The 3p state has odd parity, and so the non-zero contributions in Eq. (D.6) will now arise from the even parity ns and nd states:

$$\Delta E_{3p} = e^2\mathcal{E}^2 \left(\frac{|\langle\psi_{3p}|z|\psi_{3s}\rangle|^2}{E_{3p} - E_{3s}} + \frac{|\langle\psi_{3p}|z|\psi_{3d}\rangle|^2}{E_{3p} - E_{3d}} + \frac{|\langle\psi_{3p}|z|\psi_{4s}\rangle|^2}{E_{3p} - E_{4s}} + \cdots \right).$$

The first term gives a positive shift, while all subsequent terms are negative. Therefore, it is not immediately obvious that the Stark shift of excited states like the 3p state will be negative. However, since the energy difference of the excited states tends to get smaller as we go up the ladder of levels, it will generally be the case that the negative terms dominate, and we have a redshift as for the ground state. Moreover, the redshift is generally expected to be larger than that of the ground state for the same reason (i.e., the smaller denominator). In the case of the 3p state of sodium, the largest contribution comes from the 3d state which lies 1.51 eV above the 3p state, even though the 4s state is closer (relative energy +1.09 eV). This is because of the smaller value of the matrix element for the s states.

D.2 Linear Stark Effect

The second-order energy shift given by Eq. (D.6) diverges if an atom possesses degenerate states with opposite parities. This is the case for the l states of hydrogen with the same n. A new approach to calculate the Stark shift must then be taken based on *degenerate* perturbation theory.

Consider first the 1s ground state of hydrogen. This level is unique, and hence the second-order perturbation approach is valid. A small quadratic redshift therefore occurs, as discussed in the previous sub-section.

Now consider the $n = 2$ shell, which has four levels, namely the $m = 0$ level from the 2s term and the $m = -1$, 0, and +1 levels of the 2p term. In the absence of an applied field, these four levels are degenerate.[1] If the atom is in the $n = 2$ shell, it is equally likely to be in any of the four degenerate levels. We must therefore write its wave function as:

$$\psi_{n=2} = \sum_{i=1}^{4} c_i \psi_i \,, \tag{D.7}$$

where the subscript i identifies the quantum numbers $\{n, l, m\}$, that is:

$$\psi_1 \equiv \psi_{2,0,0}\,; \qquad \psi_2 \equiv \psi_{2,1,-1}\,; \qquad \psi_3 \equiv \psi_{2,1,0}\,; \qquad \psi_4 \equiv \psi_{2,1,+1}\,.$$

The first-order energy shift from Eq. (D.3) becomes:

$$\Delta E = e\mathcal{E} \sum_{i,j} c_i c_j \langle \psi_i | z | \psi_j \rangle \,. \tag{D.8}$$

Unlike the case of the ground state, we can see from parity arguments that some of the matrix elements are non-zero. For example, ψ_1 has even parity, but ψ_3 has odd parity. We therefore have:

$$\begin{aligned} \langle \psi_1 | z | \psi_3 \rangle &= \iiint_{\text{all space}} \psi_1^* \, z \, \psi_3 \, \mathrm{d}^3\boldsymbol{r} \,, \\ &= \iiint_{\text{all space}} (\text{even parity}) \times (\text{odd parity}) \times (\text{odd parity}) \, \mathrm{d}^3\boldsymbol{r} \,, \\ &\neq 0 \,. \end{aligned}$$

[1] We are ignoring the small fine-structure splitting of the $n = 2$ shell here because, at sufficiently strong fields, the Stark shift is larger than the fine-structure splitting.

This implies that we can observe a *linear* shift of the levels with the field. It turns out that $\langle\psi_1|z|\psi_3\rangle$ is the only non-zero matrix element. This is because the perturbation $H' = e\mathcal{E}z$ commutes with $\hat{L}_z$, and so the only non-zero matrix elements are those between states with the same m value but opposite parity – that is, between the two $m = 0$ levels derived from the 2s and 2p states.

It can easily be evaluated from the hydrogenic wave functions of the $n = 2$ levels given in Tables 2.2 and 2.3 that:

$$\langle\psi_1|z|\psi_3\rangle = -3a_0\,,$$

where a_0 is the Bohr radius of hydrogen. We then deduce that the field splits the $n = 2$ shell into a triplet, with energies of $-3ea_0\mathcal{E}$, 0, and $+3ea_0\mathcal{E}$ with respect to the unperturbed level. As expected, the splitting is linear in the field.

Appendix E Laser Dynamics

E.1 Interaction with Narrow-Band Radiation

The Einstein B coefficients were introduced in Chapter 9 to consider the interaction of atoms with broad-band radiation, such as black-body radiation: see Figure E.1(a). In this situation, the spectral energy density $u(\nu)$ varies much more slowly with frequency ν than the atomic lineshape function $g(\nu)$, and may effectively be taken as constant over the linewidth of the transition. In a laser, by contrast, the spectral width of the radiation inside the cavity is frequently much narrower than the width of the atomic transition, as illustrated in Figure E.1(b).

The absorption and stimulated emission transition rates for the case of narrow-band radiation, as shown in Figure E.1(b), can be calculated as follows. The spectral lineshape function $g(\nu)\,\mathrm{d}\nu$ gives the probability that a particular atom will absorb or emit in the spectral range $\nu \to \nu + \mathrm{d}\nu$. Hence the number of atoms in the lower level per unit volume that can absorb radiation in this frequency range is $N_1 g(\nu)\,\mathrm{d}\nu$. From the definition of the Einstein B_{12} coefficient given in Eq. (9.4), the absorption rate in this frequency range is therefore:

$$\mathrm{d}W_{12} = B_{12} N_1 g(\nu)\mathrm{d}\nu\, u(\nu)\,. \tag{E.1}$$

The total absorption rate is thus:

$$W_{12} = \int_0^\infty B_{12} N_1 g(\nu) u(\nu)\,\mathrm{d}\nu\,. \tag{E.2}$$

Since the spectral energy density of the radiation inside the laser cavity is much narrower than the width of the atomic transition, we can write it as:

$$u(\nu) = u_\nu\, \delta(\nu - \nu_{\text{laser}})\,, \tag{E.3}$$

where u_ν is the total energy density of the beam (cf. Eq. [9.19]) and $\delta(\nu)$ is the Dirac delta function. The Dirac delta function $\delta(x - x_0)$ takes the value of 0 at all values of x apart from x_0, and is normalized such that $\int_0^\infty \delta(x - x_0)\,\mathrm{d}x = 1$. It can be thought of as the limit of a top-hat function of width Δ and height $1/\Delta$ centred at x_0 in the limit where $\Delta \to 0$. It is easy to show that:

$$\int_0^\infty f(x)\,\delta(x-x_0)\,\mathrm{d}x = f(x_0)\,.$$

On inserting Eq. (E.3) into (E.2), we obtain:

$$\begin{aligned} W_{12} &= \int_0^\infty B_{12}N_1 g(\nu)u_\nu\,\delta(\nu-\nu_{\text{laser}})\,\mathrm{d}\nu\,. \\ &= B_{12}N_1 g(\nu_{\text{laser}})u_\nu\,. \end{aligned} \tag{E.4}$$

The argument for the stimulated emission rate follows similarly, and leads to:

$$W_{21}^{\text{stim}} = B_{21}N_2 g(\nu_{\text{laser}})u_\nu\,. \tag{E.5}$$

E.2 Population Inversion in a Four-Level Laser

We consider an ideal four-level laser system as discussed in Section 9.5, with a level scheme as shown in Figure 9.5. We assume that the atoms are inside a cavity and are being pumped to level 3. The atoms then decay rapidly to the upper laser level (level 2). We can write down the following rate equations for N_1 and N_2, the populations per unit volume of levels 1 and 2:

$$\begin{aligned} \frac{\mathrm{d}N_2}{\mathrm{d}t} &= -\frac{N_2}{\tau_2} - W_{21}^{\text{net}} + \mathbb{R}_2\,, \\ \frac{\mathrm{d}N_1}{\mathrm{d}t} &= +\frac{N_2}{\tau_2} + W_{21}^{\text{net}} - \frac{N_1}{\tau_1}\,. \end{aligned} \tag{E.6}$$

where $\mathbb{R}_2$ is the pumping rate per unit volume into level 2. The various terms account for:

- spontaneous emission from level 2 to level 1 at rate N_2/τ_2;
- net stimulated emission from level 2 to level 1 at rate W_{21}^{net};
- pumping into level 2 at rate $\mathbb{R}_2$; and
- decay from level 1 to the ground state by radiative transitions and/or collisions at rate N_1/τ_1.

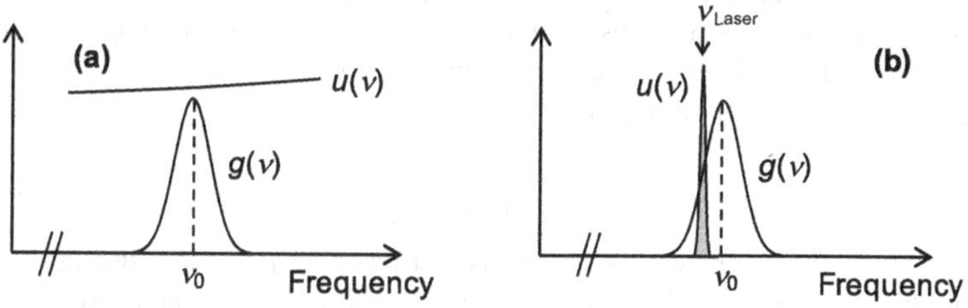

Figure E.1 Interaction of an atomic transition with: (a) broad-band radiation, and (b) narrow-band radiation. Note that the spectral energy densities and the atomic line-shape functions are not drawn on the same vertical scales.

Note that all the rates are per unit volume, and that W_{21}^{net} is the net stimulated transition rate from level 2 to level 1, as given in Eq. (9.20). This is equal to the rate of stimulated emission transitions downward, minus the rate of stimulated absorption transitions upward.

There are two important assumptions implicitly contained in Eq. (E.6):

(i) There is no pumping into level 1.
(ii) The only decay route from level 2 is by radiative transitions to level 1 (i.e., there are no nonradiative transitions between level 2 and level 1, and transitions to other levels are not possible).

It may not always be possible to satisfy these assumptions, but it helps if we can. That is why we described the above scenario as an "ideal" four-level laser.

We can rewrite the net stimulated emission rate given in Eq. (9.20) in the following form:

$$W_{21}^{\text{net}} = B_{21} g(\nu) \frac{n}{c} I \Delta N \equiv W \Delta N \,, \tag{E.7}$$

where $W = B_{21} g(\nu) n I / c$, I is the intensity inside the laser cavity, and ΔN is the population inversion density given by Eq. (9.21). Note that W is just a rate, with units s^{-1}, in contrast to W_{21}^{net}, which has units $\text{s}^{-1}\,\text{m}^{-3}$. In steady-state conditions, the time derivatives in Eq. (E.6) must be zero. We can thus solve Eq. (E.6) for N_1 and N_2 using Eq. (E.7) to obtain:

$$\begin{aligned} N_1 &= \mathbb{R}_2 \tau_1 \,, \\ N_2 &= \left(\mathbb{R}_2 - W_{21}^{\text{net}}\right) \tau_2 = (\mathbb{R}_2 - W \Delta N)\, \tau_2 \,. \end{aligned} \tag{E.8}$$

On subtracting $(g_2/g_1)N_1$ from N_2, we get the population inversion density as:

$$\Delta N \equiv N_2 - \frac{g_2}{g_1} N_1 = (\mathbb{R}_2 - W \Delta N)\, \tau_2 - \frac{g_2}{g_1} \mathbb{R}_2 \tau_1 \,. \tag{E.9}$$

On solving for ΔN, we then find:

$$\Delta N = \frac{\mathbb{R}_2}{W + 1/\tau_2} \left(1 - \frac{g_2 \tau_1}{g_1 \tau_2}\right). \tag{E.10}$$

This shows that the population inversion is directly proportional to the pumping rate into the upper level. Note, however, that it is not possible to achieve population inversion (i.e., $\Delta N > 0$) unless $\tau_2 > (g_2/g_1)\tau_1$. The reason behind this condition is most clearly seen if $g_2 = g_1$, when it becomes $\tau_2 > \tau_1$, i.e., the lower level must empty faster than the decay of the upper level. If this condition is not met, atoms will pile up in the lower laser level, and this will destroy the population inversion.

Equation (E.10) can be rewritten as:

$$\Delta N = \frac{\mathbb{R}}{W + 1/\tau_2} \,, \tag{E.11}$$

where $\mathbb{R} = \mathbb{R}_2 (1 - g_2 \tau_1 / g_1 \tau_2)$. This is the net pumping rate after allowing for the unavoidable accumulation of atoms in the lower level because τ_1 is non-zero.

References

Anderson, M. H., Ensher, J. R., Matthews, M. R., Wieman, C. E., and Cornell, E. A. 1995. Observation of Bose–Einstein Condensation in a Dilute Atomic Vapor, *Science* **269**, 198–201.

Cornell, E. 1996. Very Cold Indeed: The Nanokelvin Physics of Bose–Einstein Condensation, *J. Res. Natl. Inst. Stand. Technol.* **101**, 419–436.

Corney, A. 1977. *Atomic and Laser Spectroscopy*, Oxford University Press, Oxford.

Deslattes, R. D., Kessler Jr., E. G., Indelicato, P., et al. X-Ray Transition Energies (version 1.2). [Online] Available: http://physics.nist.gov/XrayTrans. National Institute of Standards and Technology, Gaithersburg, MD.

Feynman, R. P., Leighton, R. B., and Sands, M. 1963. *The Feynman Lectures on Physics*, Vol. I, 1-2, Addison-Wesley, Reading, MA.

Foot, C. J. 2004. *Atomic Physics*, Oxford University Press, Oxford.

Fox, A. M. 2010. *Optical Properties of Solids*, 2nd edn, Oxford University Press, Oxford.

Hooker, S. and Webb, C. 2010. *Laser Physics*, Oxford University Press, Oxford.

Hubbell, J. H. and Seltzer, S. M. 2004. Tables of X-Ray Mass Attenuation Coefficients and Mass Energy-Absorption Coefficients (version 1.4). [Online] Available: http://physics.nist.gov/xaamdi. National Institute of Standards and Technology, Gaithersburg, MD.

Jackson, J. D. 1998. *Classical Electrodynamics*, (3rd edition), Wiley, New York, NY.

Kramida, A., Ralchenko, Y., Reader, J., and NIST ASD Team. 2016. NIST Atomic Spectra Database (version 5.4). [Online] Available: http://physics.nist.gov/asd. National Institute of Standards and Technology, Gaithersburg, MD.

Kurucz, R. L., Furenlid, I., Brault, J., and Testerman, L. 1984. *Solar Flux Atlas from 296 to 1300 nm*, National Solar Observatory, Sunspot, NM.

Mandl, F. 1988. *Statistical Physics*, 2nd edn, Wiley, Chichester, UK.

Metcalf, H. J. and van der Straten, P. 1999. *Laser Cooling and Trapping*, Springer-Verlag, New York, NY.

Osterbrock, D. E. and Ferland, G. J. 2006. *Astrophysics of Gaseous Nebulae and Active Galactic Nuclei*, 2nd edn, University Science Books, Sausalito, CA.

Phillips, W. D. 1998. Laser Cooling and Trapping of Neutral Atoms, *Rev. Mod. Phys.* **70**, 721–741.

Pradhan, A. K. and Nahar, S. N. 2011. *Atomic Astrophysics and Spectroscopy*, Cambridge University Press, Cambridge, UK.

Sansonetti, J. E., Martin, W. C., and Young, S. L. 2005. Handbook of Basic Atomic Spectroscopic Data (version 1.1.2). [Online] Available: http://physics.nist.gov/Handbook. National Institute of Standards and Technology, Gaithersburg, MD.

Schawlow, A. L. and Townes, C. H. 1958. Infrared and Optical Masers. *Phys. Rev.* **112**, 1940–1949.

Silfvast, W. T. 2004. *Laser Fundamentals*, 2nd edn, Cambridge University Press, Cambridge, UK.

Slater, J. C. 1964. Atomic Radii in Crystals. *J. Chem. Phys.* **41**, 3199–3204.

Tennyson, J. 2011. *Astronomical Spectroscopy*, 2nd edn, World Scientific, Singapore.

Thompson, A. C. (editor). 2009. *X-Ray Data Booklet*, 3rd edn [online]. Available: http://xdb.lbl.gov. Lawrence Berkeley National Laboratory, Berkeley, CA.

Woodgate, G. K. 1980. *Elementary Atomic Structure*, 2nd edn, Oxford University Press, Oxford, UK.

Index

Fundamental constants

Bohr radius ($\epsilon_0 h^2/\pi e^2 m_e$)	a_0	5.29177×10^{-11} m
speed of light in vacuum	c	299 792 458 m s^{-1}
elementary charge	e	1.602177×10^{-19} C
acceleration due to gravity	g	9.80665 m s^{-2}
electron g factor ($-g_s$)	g_e	−2.002319
proton g factor	g_p	5.58569
Planck constant	h	6.62607×10^{-34} J s
Dirac constant ($h/2\pi$)	$\hbar$	1.05457×10^{-34} J s
Boltzmann constant	k_B	1.38065×10^{-23} J K^{-1}
electron mass	m_e	9.10938×10^{-31} kg
neutron mass	m_n	1.67493×10^{-27} kg
proton mass	m_p	1.67262×10^{-27} kg
Avogadro constant	N_A	6.02214×10^{23} (mol)$^{-1}$
molar gas constant	R	8.31446 J (mol)$^{-1}$ K^{-1}
Rydberg constant ($m_e e^4/8\epsilon_0^2 h^3 c$)	R_∞	1.097373×10^7 m^{-1}
Rydberg energy	$R_\infty hc$	13.60569 eV
Rydberg energy of hydrogen	R_H	13.598 eV
molar volume of ideal gas	V_m	22.4×10^{-3} m^3 (mol)$^{-1}$
fine-structure constant ($e^2/2\epsilon_0 hc$)	α	$7.29735 \times 10^{-3} \approx (137.036)^{-1}$
electron gyromagnetic ratio ($eg_s/2m_e$)	γ_e	1.76086×10^{11} s^{-1} T^{-1}
permittivity of free space	ϵ_0	8.854188×10^{-12} F m^{-1}
permeability of free space	μ_0	$4\pi \times 10^{-7}$ H m^{-1}
Bohr magneton ($e\hbar/2m_e$)	μ_B	9.27401×10^{-24} A m^2 *or* J T^{-1}
nuclear magneton ($e\hbar/2m_p$)	μ_N	5.05078×10^{-27} A m^2 *or* J T^{-1}

Conversion factors

1 u (unified atomic mass unit)	$= 1.660539 \times 10^{-27}$ kg
1 Å (angstrom)	$= 10^{-10}$ m
1 eV (electron Volt)	$= 1.602176 \times 10^{-19}$ J = 8065.54 cm^{-1}
1 atmosphere	$= 1.01325 \times 10^5$ Pa

Printed in the United States
By Bookmasters